International Thermodynamic Tables
of the Fluid State—7
Propylene (Propene)

INTERNATIONAL UNION OF PURE AND APPLIED CHEMISTRY

IUPAC Secretariat: Bank Court Chambers, 2-3 Pound Way
Cowley Centre, Oxford OX4 3YF, UK

International Thermodynamic Tables of the Fluid State

Volume 1. Argon

Volume 2. Ethylene

Volume 3. Carbon Dioxide

Volume 4. Helium

Volume 5. Methane

Volume 6. Nitrogen

Volume 7. Propylene (Propene)

A full list of titles in the Chemical Data
Series is to be found on page 401.

NOTICE TO READERS

Dear Reader

If your library is not already a standing-order customer or
subscriber to the Chemical Data Series, may we recommend that
you place a standing order or subscription order to receive
immediately upon publication all new volumes published in this
valuable series. Should you find that these volumes no longer
serve your needs your order can be cancelled at any time
without notice.

Robert Maxwell
Publisher at Pergamon Press

INTERNATIONAL UNION OF
PURE AND APPLIED CHEMISTRY

CHEMICAL DATA SERIES NO. 25

DIVISION OF PHYSICAL CHEMISTRY
COMMISSION ON THERMODYNAMICS
THERMODYNAMIC TABLES PROJECT

International Thermodynamic Tables of the Fluid State—7 Propylene (Propene)

Edited and compiled at the Project Centre, London, UK, by

S. ANGUS
B. ARMSTRONG
K. M. de REUCK

PERGAMON PRESS

OXFORD · NEW YORK · TORONTO · SYDNEY · PARIS · FRANKFURT

U.K.	Pergamon Press Ltd., Headington Hill Hall, Oxford OX3 0BW, England
U.S.A.	Pergamon Press Inc., Maxwell House, Fairview Park, Elmsford, New York 10523, U.S.A.
CANADA	Pergamon of Canada, Suite 104, 150 Consumers Road, Willowdale, Ontario M2J 1P9, Canada
AUSTRALIA	Pergamon Press (Aust.) Pty. Ltd., P.O. Box 544, Potts Point, NSW 2011, Australia
FRANCE	Pergamon Press SARL, 24 rue des Ecoles, 75240 Paris, Cedex 05, France
FEDERAL REPUBLIC OF GERMANY	Pergamon Press GmbH, 6242 Kronberg/Taunus, Hammerweg 6, Federal Republic of Germany

First edition 1980

British Library Cataloguing in Publication Data
Thermodynamic Tables Project
International thermodynamic tables of the fluid
state.
7 : Propylene (propene).—(International Union of
Pure and Applied Chemistry. Chemical data series;
no. 25).
1. Fluids—Thermal properties—Tables
2. Thermodynamics—Tables
I. Title II. Angus, Selby III. Armstrong, Barrie
IV. De Reuck, K. M. V. Series
536'.7   QC145.4.T5   79–41457

ISBN 0–08–022373–7

C

Printed and bound in Great Britain by William Clowes (Beccles) Limited,
Beccles and London

Contents

List of Figures

Tables in Text

Preface

The Thermodynamic Tables Project was inaugurated in 1963 by the Commission on Thermodynamics and Thermochemistry of the Division of Physical Chemistry of the International Union of Pure and Applied Chemistry with the object of compiling internationally agreed values of the equilibrium thermodynamic properties of liquids and gases of interest to both scientists and technologists. The range to be covered for each fluid is that for which there exist reliable experimental data, and the agreed values will be issued as tables upon the basis of which users may produce equations suited to their own special requirements. Tables on argon, ethylene, carbon dioxide, helium, methane and nitrogen have already appeared.

The general policy of the Project is formulated by a Sub-Committee appointed by IUPAC and currently consisting of:

Dr. J. D. Cox (Chairman)	Professor F. Kohler
Dr. S. Angus	Dr. A. Kozlov
Dr. H. V. Kehiaian	Dr. S. Malanowski
Professor J. Kestin	Dr. H. J. White, Jr.

The work of the Project is co-ordinated and assisted by a small permanent Project Centre under the direction of Dr. S. Angus, located at Imperial College of Science and Technology, London.

London
October 1979

J. D. COX
on behalf of the IUPAC Sub-Committee on
Thermodynamic Tables

Préface

Le Projet relatif aux Tables de Valeurs Thermodynamiques a été mis en place en 1963 par la Commission de Thermodynamique et de Thermochimie, de la Division de Chimie Physique de l'Union Internationale de Chimie Pure et Appliquée, dans le but de rassembler des valeurs acceptées sur le plan international, de propriétés thermodynamiques d'équilibre de liquide et de gaz présentant un intérêt à la fois pour les scientifiques et les techniciens. Le domaine couvert pour chaque fluide est celui pour lequel il existe des valeurs expérimentales valables; les valeurs acceptées seront publiées sous forme de tables, à l'aide desquelles les utilisateurs pourront formuler des équations adaptées à leurs propres besoins.

Des tables pour l'argon, l'éthylène, le dioxyde de carbone, l'hélium, le méthane et l'azote ont déjà été publiées.

L'orientation générale du Projet est définie par une sous-commission désignée par l'IUPAC et composée actuellement de

M. le Docteur J. D. Cox (Président)	M. le Professeur F. Kohler
M. le Docteur S. Angus	M. le Docteur A. D. Kozlov
M. le Docteur H. B. Kehiaian	M. le Docteur S. Malanovski
M. le Professeur J. Kestin	M. le Docteur H. J. White, Jr.

Le développement du Projet est assumé et coordonné par un petit centre permanent d'étude sous la direction du Docteur S. Angus, situé à l'Imperial College of Science and Technology, à Londres.

J. D. COX
au nom de la Sous-commission
des Tables Thermodynamiques
de l'IUPAC

Londres
Octobre 1979

Vorwort

Das Projekt "Thermodynamische Tafeln" wurde 1963 von der Kommission für Thermodynamik und Thermochemie der Abteilung für Physikalische Chemie der International Union of Pure and Applied Chemistry begonnen mit dem Ziel, international anerkannte Werte für die thermodynamischen Gleichgewichtseigenschaften von Flüssigkeiten und Gasen zusammenzustellen, die für Wissenschaft und Technik von Interesse sind. Der Zustandsbereich, der für jedes Fluid erfaßt werden soll, wird dabei durch den Existenzbereich zuverlässiger experimenteller Werte bestimmt. Die anerkannten Werte sollen in Tafeln veröffentlicht werden; darauf aufbauend können die Benutzer Gleichungen aufstellen, die für ihre besonderen Bedürfnisse geeignet sind. Tafeln fur Argon, Äthylen, Kohlendioxyd, Helium, Methan und Stickstoff sind schon herausgegeben worden.

Die allgemeine Zielsetzung des Projekts wird von einer durch die IUPAC eingesetzten Unterkommission bestimmt, die zur Zeit aus folgenden Mitgliedern besteht:

Dr. J. D. Cox (Vorsitzender)   Professor F. Kohler
Dr. S. Angus   Dr. A. D. Kozlov
Dr. H. V. Kehiaian   Dr. S. Malanowski
Professor J. Kestin   Dr. H. J. White, Jr.

Die Arbeit am Projekt wird von einem kleinen, permanent besetzten Projektzentrum koordiniert und unterstützt, welches unter der Leitung von Dr. S. Angus am Imperial College of Science and Technology in London besteht.

London
October 1979

J. D. COX
im Namen des IUPAC-Subkomitees
für Thermodynamische Tafeln

Предисловие

Проект термодинамических таблиц был создан в 1963 г. Комиссией по термодинамике и термохимии отдела физической химии Международной унии теоретической и прикладной химии с цельюсоставить во всем мире утвержденные величины равновесия термодинамических свойств жидкостей и газов, которые представляют интерес как для ученых, так и для технических работников. Диапазон для каждой жидкости определен наличием надежных экспериментальных данных; утвержденные величины будут изданы в виде таблиц, по которым могут потребители составлять уравнения, удовлетворяющие их специальные требования.

Таблицы по аргону, этилену, окиси углеряда, гелию, у мегани азоту были уже опубликованы.

Характеристика этого проекта формулируется субкомиссией, наэхаченной IUPAC, состав которой в настоящее время следующий:

Д-р Й. Д. Кокс (председатель)   Проф. Ф. Колер
Д-р С. Ангус   Д-р А. Д. Козлов
Д-р Х. В. Кехиайан   Д-р С. Малановски
Проф. Й. Кестин   Д-р Х. Й. Уайт, мл

Работа над этим проектом координируется с помощью небольшого постоянного проектного центра под руководством д-ра С. Ангуса; центр находится в Империал Колидж в Лондоне.

Лондон
октябрь, 1979

Й. Д. КОКС
от имени субкоммиссии IUPAC
по термодинамическим таблицам

Introduction

The Aliphatic Hydrocarbons Working Panel comprises:

Dr. D. R. Douslin	Bartlesville Energy Research Center, USA
Professor P. T. Eubank	Texas A & M University, USA
Mr. R. D. Goodwin	National Bureau of Standards (Cryogenics Division), Boulder, USA
Dr. F. Lazarre	Société Nationale des Pétroles d'Aquitaine, France
Professor J. E. Powers	University of Michigan, USA
Professor D. S. Viswanath	Indian Institute of Science, Bangalore, India
Professor V. A. Zagoruchenko	Odessa Institute of Marine Engineers (OIIMF), USSR

In 1972 the IUPAC Thermodynamic Tables Project published tables of the thermodynamic properties of ethylene[51] which showed that the experimental base was smaller than had been realized. The need for accurate tables is such that a research project was put into effect in the USA, backed by several firms who are major users and makers of ethylene in order to provide accurate data over a wider region. The extent of knowledge of the thermodynamic properties of propylene is not dissimilar to that for ethylene, and the use of propylene in industry is increasing steadily, so it seemed timely to add propylene to the list of fluids studied.

Three studies on propylene have been published recently. One is by a group at the Kiev Institute of Technology,[1] who critically examined all data published before 1968 and used a combination of analytical and graphical techniques to produce tables of the thermodynamic and transport properties which accurately represent the available experimental data.

The second, by Bender[46] of Kaiserlautern University, showed that accurate tables could be constructed from a single equation of state, containing 20 coefficients, which is valid over a fairly limited range of parameters.

The third by Juza, Sifner and Hoffer[47] of the Institute of Thermomechanics, Prague, also produces accurate tables over a limited range from a single equation of state containing 33 coefficients.

On the basis of these publications, the IUPAC Project Centre has produced tables from an equation of state which takes in the widest range of accurate experimental data. The equation was constructed using an adaptation by de Reuck and Armstrong[50] of the regression technique developed by Wagner.[49, 49a, 49b, 49c] This regression method has made it possible to represent the data over a considerably extended range of parameters with an equation of state containing 22 coefficients.

Thermodynamic consistency between the various properties tabulated is assured if the available data are used to prepare an equation representing the Helmholtz free energy as a function of density and temperature, from which the other properties can be evaluated by mathematical manipulation. The Helmholtz free energy cannot be measured directly, but properties derived from it can be written as

$$X(\rho_2, T_2) = X_0 + \int_{T_1}^{T_2} X_1(T)C(T)\,dT + \int_{\rho_1}^{\rho_2} X_2(\rho, T)\,d\rho,$$

where $C(T)$ is a function representing the variation of a heat capacity with temperature; $X_1(T)$ is a simple multiplier depending on the property X; $X_2(\rho, T)$ is derived from an equation of state, $P = P(\rho, T)$, and X_0 is a constant; and all these terms are accessible to experiment.

It is convenient to take $C(T)$ as the variation of the isobaric heat capacity in the ideal gas state. When this is done, the lower limit of the density integral, ρ_1, becomes zero. The ideal gas heat capacity function, $C(T)$, the temperature integrals resulting from it and the choice of X_0 are all dealt with in Section 2.1.

The major portion of Section 1 of the text of this book is concerned with the discussion of the available data, and the way in which equations of state, $P(\rho, T)$, were constructed from them is dealt with in Section 2. Section 3 explains how the tabulated properties were calculated from the chosen equation. In addition, the available data and equations correlating them are discussed for the saturation curve and the melting curve.

The tables presented here give the volume, entropy, enthalpy, isobaric heat capacity, compression factor, fugacity/pressure ratio, Joule–Thomson coefficient, ratio of the heat capacities and speed of sound as a function of pressure and temperature; and the pressure, entropy, internal energy and isochoric heat capacity as functions of density and temperature for the gas and liquid states from 0.025 MPa to 1000 MPa at temperatures from 90 K to 475 K, and from 0.025 MPa to 10 MPa, 475 K to 575 K. Zero pressure tables are given, as are tables of the properties of the fluid phases along the saturation curve and the melting curve.

Equations which reproduce these properties within these limits are given, as well as estimates of uncertainty of some of the principal properties listed.

Introduction

Le Groupe de travail "Hydrocarbures aliphatiques" est composé de:

M. le Docteur D. R. Douslin	Bartlesville Energy Research Center, E–U
M. le Professeur P. T. Eubank	Texas A & M University, E–U
M. le Docteur R. D. Goodwin	National Bureau of Standards (Cryogenics Division), Boulder, E–U
M. le Docteur F. Lazarre	Société Nationale des Pétroles d'Aquitaine, France
M. le Professeur J. E. Powers	University of Michigan, E–U
M. le Professeur D. S. Viswanath	Indian Institute of Science, Bangalore, Inde
M. le Professeur V. A. Zagoruchenko	Odessa Institute of Marine Engineers (OIIMF), URSS

Les tables des propriétés thermodynamiques de l'éthylène[51] ont été publiées en 1972, dans le cadre du Projet relatif aux Tables de Grandeurs Thermodynamiques de l'IUPAC.

Il a été mise en évidence, à cette occasion, que les données de bases expérimentales étaient moins importantes qu'on aurait pu l'espérer.

Le besoin de données précises était tel, qu'un projet de recherches, appuyé par plusieurs Sociétés, qui produisent et utilisent l'éthylène, fut lancé aux E-U, dans le but de fournir des données dans un domaine plus large. Le domaine de connaissance des propriétés thermodynamiques du propylène n'est pas très différent de celui de l'éthylène; comme l'utilisation du propylène dans l'industrie croît régulièrement, il a semblé opportun d'ajouter le propylène à la liste des fluides étudiés.

Trois études sur le propyléne ont été publiées récemment:

La première, provenant de l'Institut de Technologie de Kiev,[1] fait un examen critique de toutes les données publiées avant 1968, et en utilisant une combinaison de techniques analytique et graphique, propose des tables des propriétés thermodynamiques et de transport, qui représentent avec précision les données expérimentales disponibles.

La seconde, dont l'auteur est Bender[46] de l'Université de Kaiserslautern, montre que des tables précises pourraient être construites à partir d'une seule équation d'état, comportant 20 coefficients, et valable dans un domaine assez limité.

La troisième, par Juza, Sifner et Hoffer[47] de l'Institut de Thermomécanique, Prague, produit aussi de tables exactes dans un domaine limité à partir d'une seule équation d'état comportant 33 coefficients.

Sur la base de ces publications, le Centre d'Etude IUPAC publie des tables obtenues à partir d'une équation d'état qui tient compte d'un maximum de données expérimentales précises. L'équation a été construite en utilisant une adaptation par de Reuck et Armstrong[50] de la technique de régression développée par

Wagner.[49, 49a, 49b, 49c] Cette méthode de régression a fourni une équation d'état contenant 22 coefficients, et qui représente les données dans un domaine très étendu.

Un moyen d'assurer la cohérence thermodynamique des diverses grandeurs tabulées est de construire, à partir des points expérimentaux, une équation donnant l'enthalpie libre de Helmholtz en fonction de la densité molaire et de la température. Les autres grandeurs sont ensuites dérivées d'un traitement mathématique de cette équation.

L'enthalpie libre ne peut pas être mesurée directement mais toute grandeur qui en dérive peut s'écrire:

$$X(\rho_2, T_2) = X_0 + \int_{T_1}^{T_2} X_1(T)C(T)\,\mathrm{d}T + \int_{\rho_1}^{\rho_2} X_2(\rho, T)\,\mathrm{d}\rho,$$

où $C(T)$ est une représentation mathématique de la variation d'une chaleur spécifique en fonction de la température, $X_1(T)$ est un simple facteur multiplicatif de la fonction de la nature de la grandeur X, $X_2(\rho, T)$ est obtenu à partir d'une équation d'état $P = P(\rho, T)$, et X_0 est une constante; tous les paramètres précités sont mesurables.

Il est commode de choisir $C(T)$ comme la variation de la chaleur spécifique à pression constante dans l'état gazeux parfait. Dans ce cas la limite inférieure ρ_1 de l'intégrale de densité devient égale à zéro.

On trouvera dans le paragraphe 2.1 l'équation donnant la chaleur spécifique dans l'état gazeux idéal, les intégrales de température qui en résultent, et le choix retenu pour X_0.

La majeure partie de la Section 1 du texte de ce livre est relative à la discussion des données disponibles, la Section 2 traitant de la manière dont les équations d'état ont été construites à partir de ces données. La Section 3 précise comment ont été calculées les propriétés à partir de l'équation choisie. De plus, on y présente une discussion des données disponibles relatives à la courbe de saturation et à la courbe de fusion, ainsi que des équations les corrélant.

Les tables présentées ici donnent le volume, l'entropie, l'enthalpie, la chaleur spécifique à pression constante, le facteur de compressibilité, le rapport de la fugacité à la pression, le coefficient de Joule–Thomson, le rapport des chaleurs spécifiques et la vitesse du son en fonction de la pression et de la température; sont données également la pression, l'entropie, l'énergie intense et la chaleur spécifique à volume constant en fonction de la densité et de la température pour les volumes phases gazeuse et liquide de 0,025 MPa à 1000 MPa à des températures de 90 K à 475 K, et de 0,025 MPa à 10 MPa, pour des températures de 475 K à 575 K.

On donne également des tables de valeurs à pression nulle ainsi que des tables de propriétés des deux phases le long de la courbe de saturation et de la courbe de fusion.

On donne des représentations analytiques de ces propriétés dans les limites citées ainsi qu'une estimation des incertitudes relatives aux principales grandeurs tabulées.

Einleitung

Der Arbeitsausschuß für aliphatische Kohlenwasserstoffe besteht aus:

Dr. D. R. Douslin	Bartlesville Energy Research Center, USA
Professor P. T. Eubank	Texas A & M University, USA
Dr. R. D. Goodwin	National Bureau of Standards (Cryogenic Division), Boulder, USA
Dr. F. Lazarre	Société Nationale des Pétroles d'Aquitaine, Frankreich
Professor J. E. Powers	University of Michigan, USA
Professor D. S. Viswanath	Indian Institute of Science, Bangalore, Indien
Professor V. A. Zagoruchenko	Odessa Institute of Marine Engineers (OIIMF), UdSSR

Im Jahre 1972 veröffentlichte das IUPAC-Projekt "Thermodynamische Tafeln" Tabellen der thermodynamischen Eigenschaften von Äthylen[51], welche zeigten, daß die experimentelle Basis dafür schwächer war als ursprünglich angenommen. Das Bedürfnis nach genauen Tabellen ist so stark, daß ein Forschungsprojekt in den USA ins Leben gerufen Wurde—mit Unterstützung einiger Firmen, welche Äthylen herstellen oder benutzen—, um genaue Daten über ein größeres Gebiet zur Verfügung zu haben. Das Ausmaß unserer Kenntnis der thermodynamischen Eigenschaften von Propylen ist nicht sehr verschieden von dem für Äthylen, und der Einsatz von Propylen in der Industrie wächst ständig, daher schien es an der Zeit, Propylen auf die Liste der untersuchten Fluide zu setzen.

Drei Arbeiten über Propylen sind kürzlich erschienen. Die eine[1] des Instituts für Technologie, Kiew, prüfte kritisch alle vor 1968 publizierten Daten und benutzte eine Kombination von analytischen und graphischen Techniken, um Tabellen der thermodynamischen Eigenschaften und der Transporteigenschaften herzustellen, welche die verfügbaren experimentellen Daten genau wiedergeben.

Die zweite Arbeit stammt von Bender,[46] Universität Kaiserslautern, und zeigt, daß genaue Tabellen aufgrund einer einzigen empirischen Zustandsgleichung, welche 20 Koeffizienten enthält, innerhalb eines nicht zu ausgedehnten Bereichs erstellt werden können.

Die dritte Studie, die von Juza, Sifner und Hoffer[47] am Institut für Thermomechanik in Prag verfasst wurde, ermöglicht ebenfalls die Erstellung genauer Tabellen in einem begrenzten Bereich; es wird eine einzige Gleichung mit 33 Koeffizienten verwendet.

Auf der Basis dieser beiden Publikationen hat das IUPAC-Projekt-Centre nun Tabellen hergestellt mittels einer Zustandsgleichung, welche den größten Bereich der genauen experimentellen Daten berücksichtigt. Zur Aufstellung der Gleichung wurde von de Reuck und Armstrong[50] ihre Anpassung der von Wagner[49, 49a, 49b, 49c,] entwickelten Regressionstechnik verwendet. Diese Regressionsmethode ermöglichte die Wiedergabe

der Daten über einem erheblich erweiterten Bereich der Variablen mit einer Zustandsgleichung von 22 Koeffizienten.

Thermodynamische Konsistenz zwischen den verschiedenen tabellierten Zustandsgrößen ist dann gewährleistet, wenn aus den vorhandenen Meßwerten eine
Gleichung aufgestellt wird, welche die Helmholtz'sche Energie in Abhängigkeit von der
Dichte und der Temperatur wiedergibt. Alle anderen thermodynamischen Zustandsgrößen können dann durch mathematische Operationen aus dieser Gleichung berechnet
werden. Die freie Energie kann zwar nicht direkt gemessen werden, aber jede aus ihr
ableitbare Zustandsgröße X kann in der Form

$$X(\rho_2, T_2) = X_0 + \int_{T_1}^{T_2} X_1(T)C(T)\,dT + \int_{\rho_1}^{\rho_2} X_2(\rho, T)\,d\rho$$

geschrieben werden. Dabei ist $C(T)$ eine Beziehung für die Temperaturabhängigkeit
einer Wärmekapazität; $X_1(T)$ ist ein Multiplikator, der von der Art der Zustandsgröße
X abhängt; die Funktion $X_2(\rho, T)$ läßt sich aus der thermischen Zustandsgleichung
$P = P(\rho, T)$ rechnen, und X_0 ist eine Konstante. Alle diese Terme sind experimentell
bestimmbar.

Vorteilhafterweise setzt man $C(T)$ gleich der spezifischen isobaren Wärmekapazität im idealen Gaszustand. Dann wird die untere Grenze des Dichteintegrals zu Null,
$\rho_1 = 0$. Die Gleichung für die isobare Wärmekapazität des idealen Gases, die daraus
sich ergebenden Temperaturintegrale und die Wahl von X_0 werden in Abschnitt 2.1
behandelt. Der größbte Teil dieses Buches behandelt die Aufstellung der Zustandsgleichung aus den verfügbaren Meßwerten und die daraus gebildeten Dichteintegrale, die
die verschiedenen $X_2(\rho, T)$ ergeben.

Der größte Teil von Abschnitt 1 dieses Buches betrifft eine Diskussion der
verfügbaren Daten, während Abschnitt 2 die Konstruktion der Zustandsgleichungen
$P(\rho, T)$ von diesen Daten behandelt. Abschnitt 3 erklärt, wie die tabellierten Eigenschaften von den ausgewählten Gleichungen berechnet wurden. Zusätzlich werden die
verfügbaren Daten und die sie korrelierenden Gleichungen für die Dampfdruckkurve
und für die Schmelzdruckkurve diskutiert.

Die vorliegenden Tafeln geben Volumen, Entropie, Enthalpie, isobare Wärmekapazität, Realgasfaktor, das Fugazitäts–Druck-Verhältnis, Joule–Thomson-
Koeffizient, das Verhältnis der Wärmekapazitäten und die Schallgeschwindigkeit als
Funktionen von Druck und Temperatur; außerdem Druck, Entropie, innere Energie
und isochore Wärmekapazität als Funktionen der Dichte und der Temperatur. Die
Tafeln erfassen Zustände im Gas- und Flüssigkeitsgebiet von 0,025 MPa bis 1000 MPa
bei Temperaturen zwischen 90 K und 475 K; bei Temperaturen von 470 K bis 575 K
liegen die Drucke zwischen 0,025 MPa und 10 MPa. Weitere Tafeln geben die
Eigenschaften des Gases bei verschwindendem Druck sowie die Zustandsgrößen
idealen Gaszustand und für die Zustandsgrößen der fluiden Phasen auf der Dampf-
druck- und Schmelzdruckkurve.

Die Gleichungen, die diese Zustandsgrößen innerhalb der genannten Druck- und
Temperaturgrenzen wiedergeben, sind ebenso wie Abschätzungen der Unsicherheit
einiger wesentlicher Zustandsgrößen angegeben.

Введение

В состав рабочей группы по алифатическим углеводородам входят:

Д-р Д. Р. Дуслин   Энергетичиский исследовательский центр
в Бартлесвилле, США

Проф. П. Т. Эубанк   А и М университет в Техасе, США

Д-р Р. Д. Гудвин   Народное бюро по нормам (криогеническое
отделение), Боулдер, США

Д-р Ф. Лазарр   Сосьете Насиональ де Петроль д'Аквитань,
Франция

Проф. Дж. Е. Пауэрс   Мичиганский университет, США

Проф. Д. С. Вишванат Индийский институт науки, Бангалор,
Индия

Проф. В. А. Загорученко Одесский институт инженеров морского
флота, СССР

В 1972 г. в рамках Проекта термодинамических таблиц IUPAC были опубликованы таблицы термодинамических свойств этилена,[51] которые показали, что зкспериментальная база была меньше, чем обсуждалось раньше. Точные таблицы настолько нужны, что в США был начат исследовательский проект, который поддерживают разные фирмы, использующие и производящие этилен, чтобы были найдены точные данные в обширной области. Масштабы знаний о термодинамических свойствах пропилена подобны масштабам знаний об этилен, а так как использование пропилена в промышленности непрерывно возрастает, оказывается, что в настоящее время необходимо внести пропилен в список изучаемых жидкостей.

Недавнс были опубликованы три статьи о пропилене. В одной из них, разработанной группой Высшего технического учебного заведения в Киеве,[1] критически обсуждаются все данные, опубликованные до 1968-го года; в ней были использованы аналитическая и графическая техника для составления термодинамических и транспортных свойств, которые точно представляют доступные экспериментальные данные.

Во второй статье, разработанной БендеромХ[46] из Каизерлаутернского университета, показывается, что точные таблицы могли бы быть составлены из одного уравнения состояния, содержащего 20 коэффициентов, которое действует в сравнительно ограниченном масштабе параметров.

В третьей статье, которую разработали Юза, Шифнер и Хоффер из Пражского института термомеханики, тоже нредставлены точные таблицы В ограниченном масштабе на основе единого уравнения состояния, содержащего 33 коэффициентов.

На основе этих публикаций, центр проекта IUPAC разработал таблицы на основе уравнения состояния, которое охватывает самый широкий диапазон точных экспериментальных данных. Это уравнение было составлено с применением регрессивной техники по Вагнеру,[49, 49а, 49б, 49в] перестроемной Де Рейком и Армстронгом. зтот регрессивхый метод дал возможность охватить данные в значительно широком диапазоне параметров с уравнением состояния, содержащем 22 козффициентов,

Термодинамическая консистенция межлу разными свойствами, приведенными в таблице, обеспечена, если имеющиеся данные используются для подготовки уравнения, представляющего свободную энергию (Гельмгольтз) как функцию густоты и температуры, из которого можно исчислять остальные свойцтва математическим путем. Свободную энергню нельзя измерять прямым путем, но любое свойство, выведенное из нее, может быть зеписано как

$$X(\rho_2, T_2) = X_0 + \int_{T_1}^{T_2} X_1(T)C(T)\,\mathrm{d}T + \int_{\rho_1}^{\rho_2} X_2(\rho, T)\,\mathrm{d}\rho,$$

где $C(T)$–уравнение, представляющее вариацию теплоемкости с температурой, $X_1(T)$–простой множитель, зависящий от свойства X, $X_2(\rho, T)$ выведено из уравнения состояния, $P = P(\rho, T)$, и X_0-константа; все зти члены можно определить зкспериментальным путем.

(C/T) является пригодным в качестве вариации изобарной теплоемкости в идеальном газообразном состоянии, В таком случае нижний предел интеграла гурсоты, ρ_1 становится нулем. У равнение идеальной газовой теплоемкости, из нее бытекающие интегралы температуры н выбор X_0 аналнзнруются в отделе 2.1.

Вольшая часть первой части текста этой публикации посвящается обсуждению доступных данных, в то время как вторая часть посвящена способу, которым были составлены уравнения состояния, $P(\rho, T)$. В третьей части объясняется, каким образом были вычислены табличные величины из выбранного уравнения. Кроме того обсуждаются доступные данные и уравнения, которые их коррелируют, в отношении к кривсй насыщения и кривой таяния.

Приведенные здесь таблицы содержат объем, энтропию, энталпию, изобарную теплоемкость, сжимаемость и фугитивность (отношение давлений), козффициент Джоуль–Томпсона, отношение теплоемкостей и скорости звука как функции давления и температуры, и далее давление, энтропию, внутренюю энергию и изохорную теплоемкость как функции густоты и температуры для газообразного и жидкого состояний от 0,025 MPa до 1000 MPa при температурах от 90 K до 475 K, и от 0,025 MPa до 10 MPa, 475 K до 575 K; приводятся также таблицы для нулевого давления а также таблицы свойств для жидких фаз вдоль кривой насыщения и кривой таяания.

Приводятся уравнения, которые передают эти свойства в приведенных пределах, как и оценки погрешности для некоторых главных приводимых свойств.

Symbols

This list of symbols follows the recommendations of the IUPAC *Manual of Symbols and Terminology for Physicochemical Quantities and Units*.[56] The subscript "m", indicating "molar", has been omitted, since no ambiguity arises.

Symbol	Physical quantity
A	Molar Helmholtz energy
B	Second virial coefficient
C	Molar heat capacity
G	Molar Gibbs energy
H	Molar enthalpy
M	Molar mass
P	Pressure
R	Gas constant
S	Molar entropy
T	Thermodynamic temperature†
U	Molar internal energy
V	Molar volume
Z	Compression factor, Realfaktor
f	Fugacity
w	Speed of sound
γ	Ratio of heat capacities (C_P/C_V)
μ	Joule–Thomson coefficient $(\partial T/\partial P)_H$
ρ	Molar density
τ	T_c/T
ω	V_c/V

Subscripts

P, V, etc.	At constant pressure, volume, etc.
g	In the gas phase
l	In the liquid phase
c	At the critical point
t	At the triple point
σ	Along the saturation curve
m	Along the melting curve
data	Experimental results, or input data derived from calculation
calc	Values calculated from a given equation

Superscripts

id	Ideal gas state
ic	Ideal crystal

†In equations, for T is substituted the numerical value of the temperature on IPTS–68.

Units and Conversion Factors

The units used in these tables are given in the left-hand column; and conversion factors, including a change from molar to specific quantities, are based on the assumption that one mole of propylene molecules has a mass of 0.042 080 4 kg.[57]

To convert from	To	Multiply by
Pressure in MPa	Pressure in Pa	10^6
	Pressure in bar	10
	Pressure in atm	9.869 23
	Pressure in kgf cm^{-2}	10.197 2
	Pressure in lbf in^{-2}	145.038
	Pressure in mmHg	7500.62
Molar volume in $cm^3 mol^{-1}$	Molar volume in $m^3 mol^{-1}$	10^{-6}
	Specific volume in $m^3 kg^{-1}$	0.000 023 764
	Specific volume in $cm^3 g^{-1}$	0.023 764
	Specific volume in $ft^3 lb^{-1}$	0.000 380 66
Density in $mol\, dm^{-3}$	Density in $mol\, m^{-3}$	10^3
	Density in $kg\, m^{-3}$	42.080 4
	Density in $g\, cm^{-3}$	0.042 080 4
	Density in $lb\, ft^{-3}$	2.626 994
Molar energy in $J\, mol^{-1}$	Specific energy in $J\, kg^{-1}$	23.764
	Specific energy in $MPa\, cm^3 g^{-1}$	0.023 764
	Specific energy in $cal_{th}\, g^{-1}$	0.005 679 7
	Specific energy in $cal_{IT}\, g^{-1}$	0.005 675 9
	Specific energy in $BTU\, lb^{-1}$	0.010 217
	Molar energy in $cal_{th}\, mol^{-1}$	0.239 006
Molar entropy in $J\, K^{-1}\, mol^{-1}$	Specific entropy in $J\, K^{-1}\, kg^{-1}$	23.764
	Specific entropy in $MPa\, cm^3\, K^{-1}\, g^{-1}$	0.023 764
	Specific entropy in $cal_{th}\, K^{-1}\, g^{-1}$	0.005 679 7
	Specific entropy in $cal_{IT}\, K^{-1}\, g^{-1}$	0.005 675 9
	Specific entropy in $BTU\, °F^{-1}\, lb^{-1}$	0.005 675 9
	Molar entropy in $cal_{th}\, K^{-1}\, mol^{-1}$	0.239 006
Molar enthalpy in $J\, mol^{-1}$	Conversion factors as for molar energy	
Molar heat capacity in $J\, K^{-1}\, mol^{-1}$	Conversion factors as for molar entropy	

Note:

Units of $P \times V$

In order to avoid the printing of excessively large numbers, the volume (and density) as printed in the various tables differ by multiples or sub-multiples of ten from one another, so users should note the conversions given for $P \times V$ at the beginning of each table.

Nomenclature

Most of the comparisons between experiment and correlation are made in terms of the percentage deviation, $(X(\text{data})-X(\text{calc}))\ 100/X(\text{data})$, where X is the property concerned. If we wish to compare a set of N deviations of the property X, we quote a value for

$$\left[\frac{1}{N}\Sigma\left(\frac{X\ (\text{data})-X(\text{calc})}{X(\text{data})}\times 100\right)^{2}\right]^{1/2},$$

and call this the "relative deviation".

1. Experimental Results

In this section, the available experimental results are discussed and compared with the equations used to generate the tables. The reasons for the choice of the equations are discussed in Section 2 and the equations themselves are given in detail in Section 3. They include an analytic equation of state for the fluid state, an equation for the ideal gas thermodynamic properties, and an equation for the melting curve. For the convenience of users, these are supplemented with auxiliary equations for the variation of the vapour pressure and the saturated liquid densities with temperature, but these are not involved in the comparisons which follow.

1.1 TRIPLE POINTS

In the region covered by these tables, as well as the usual solid–liquid–gas triple point, there also exists a solid–solid–liquid triple point.

The solid–liquid–gas triple point has not been measured but, as is shown in Section 1.2.1, it is possible to obtain an estimate of it from the existing data, the value of which is:

$$T_{t_1} = (87.89 \pm 0.01) \text{ K}, \quad P_{t_1} = (0.95 \pm 0.01) \, 10^{-9} \text{ MPa}.$$

The solid–solid–liquid triple point was measured in 1964 by Reeves, Scott and Babb[2] of the University of Oklahoma, in the course of work discussed in Section 1.2.1, and reported as:

$$T_{t_2} = (-144.2 \pm 1)°C, \quad P_{t_2} = (7150 \pm 10) \text{ bar}.$$

The authors made it clear that this was a difficult point to determine, and the quoted errors were those of the measuring systems rather than of the triple point itself.

However, we have accepted this value and, after correcting to IPTS-68, the values are:

$$T_{t_2} = (129.84 \pm 1) \text{ K}, \quad P_{t_2} = (715.0 \pm 1) \text{ MPa}.$$

1.2 TWO-PHASE REGIONS

1.2.1 Melting Curves

Very few measurements of the melting curves have been reported, and only at the melting point at atmospheric pressure are there replicate measurements. These are:

	T/K	Ref.
Maass and Wright (1921)	88.0	4
Huffman, Parks and Barmore (1931)	88.2 ± 0.5	18
Powell and Giauque (1939)	87.90 ± 0.05	43

In evaluating these results, the important questions are those regarding purity and the temperature scale, since at the dates of these experiments there was no agreement on the determination of temperature below 90.2 K, the boiling point of oxygen (and no internationally accepted scale until 1968). Maass and Wright,[4] whose work is discussed in Section 1.2.2, used a temperature scale which it is now impossible to characterize, and appear to have used propylene which was impure by an indefinite amount. Huffman *et al.*[18] measured the EMF of a copper–constantan thermocouple, and converted this to temperature using an equation by Eastman and Rodebush.[18a] The exact temperature scale used is thus in doubt, but since the authors claim an accuracy of only ±0.5 K, it is not of great importance. They estimated their propylene to contain less than 0.1% of impurities and it had "an extremely sharp melting point", but they did not describe their purification procedures. Powell and Giauque[43] measured the melting point as part of their "third law" experiments, which also included measurements of several other properties, all of which are discussed in the appropriate sections. Their propylene contained 0.02 molar percentage of impurity, and the melting temperature showed an increase of 0.09 K as melting proceeded, the final temperature being taken as the melting point. Powell and Giauque measured the temperature with a thermocouple calibrated against a gas thermometer, described in refs. 43a and 43b, and discussed in Section 2.1.2. Their value, when corrected to IPTS-68, was accepted as the melting point.

The only other measurements on the melting curve are those made in 1964 by Reeves, Scott and Babb.[2] They were examining the suitability of various fluids as pressure-transmitting media and said "no claim of great accuracy is made for these results". They used the plugged capillary method, in which, at constant temperature, the pressure is raised to solidify the sample, and then decreased until melting is detected. The pressure was measured with a manganin resistance gauge, to an accuracy of approximately 0.1 MPa, and the other experimental errors were said to be

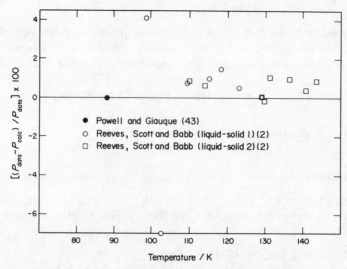

Figure 1. Comparison of experimental melting pressures with the auxiliary melting curve equations (eqns. (8) and (9))

equivalent to about ± 1 K at 273 K, and ± 1.5 K at 77 K. The paper reported only the parameters of a fusion equation, but the individual measurements are given in the Appendix, showing thirteen results in addition to the high-pressure triple point.

From the above results, two equations have been derived in Section 2.3.3 for the two melting curves. A comparison of the values calculated from these equations with the experimental results is shown in Figure 1.

1.2.2 Critical Point

Six experimental investigations of the critical point have been reported, which are summarized in Table A. It is not possible to have complete confidence in any one of them.

Table A. MEASUREMENTS OF THE CRITICAL POINT PARAMETERS

Authors	Date	Method	$\dfrac{T_c}{K}$	$\dfrac{P_c}{MPa}$	$\dfrac{\rho_c}{mol\ cm^3}$
Seibert and Burrell[3]	1915	1	365.7 ± 0.1	4.594	—
Maass and Wright[4]	1921	1	$365.2 \pm$	—	—
Winkler and Maass[5]	1933	—	364.9 ± 0.3	—	0.0053 ± 0.0002
Vaughan and Graves[25]	1940	1	364.5 ± 1.0	—	—
Lu, Newitt and Ruhemann[7]	1941	1	365.2 ± 0.1	4.600 ± 0.005	—
Marchman, Prengle and Motard[23]	1949	5	364.91 ± 0.3	—	—
This work			365.57	4.6646	0.00531

(Estimates of error result from critical review)

Key to method
1   Disappearance of the meniscus.
5   Pressure, volume, temperature relation: $(\partial P/\partial V)_T = 0$.

The earliest investigation was by Seibert and Burrell,[3] of the U.S. Bureau of Mines Laboratory for Gas Investigations, made in 1915, in the course of measurements of the vapour pressures of a number of hydrocarbons. The apparatus is described in detail in ref. 3a. The method used was to observe the disappearance and reappearance of the meniscus as the temperature varied, the two observations differing by 0.1 K. The temperature and pressure measuring systems were adequate for the accuracies claimed, and the thermometer calibration was ultimately referrable to the contemporary NBS standard, which means the temperatures can be taken to be on IPTS-27. The propylene was made by dehydrating propyl alcohol with phosphorous pentoxide, followed by drying and low-temperature distillation. This would appear to be one of the more reliable determinations of the critical parameters, but the account given lacks the details which are now known to be needed for proper appraisal.

In 1921 Maass and Wright[4] of McGill University published some results of work on the vapour pressure of hydrocarbons which includes a value for T_c of propylene. The only reference to this is that the critical temperature was determined "in the usual way". The thermometer used was calibrated at the ice, steam and CO_2/ether points, the temperature assumed for the last not being given. The calibration was checked by measuring the boiling point of liquid air, which was found to be $-188.8°C$: the currently accepted value is $-191.4°C$. The material for investigation was made by heating propyl alcohol to 350°C in the presence of aluminium oxide, then dried, and distilled a number of times. It is impossible to be certain of the purity of the final sample. In view of the uncertainties, this work must be treated with caution.

In 1933 Winkler and Maass[5] published a paper entitled "Density Discontinuities at the Critical Temperature" which includes consideration of propylene. The paper refers readers for numerical values to an immediately preceding paper[6] which does not give the critical point itself. From a graph in the paper it was possible to estimate the values which are given in Table A. Since these papers were written in support of the view that the critical point as such could not be observed, but only a "critical region" in which density differences persisted even though the meniscus could no longer be observed, it is difficult to know what weight to attach to their results.

In 1940 Vaughan and Graves[25] of the Shell Development Company, Emeryville, reported *PVT* measurements which included direct observations of T_c from the disappearance of a meniscus. The propylene used was made by passing isopropyl alcohol over heated alumina, drying, and distillation. The purity was characterized by measuring the difference in vapour pressure between the dew-point and the bubble-point of a sample held at various temperatures between 0°C and 91°C: the difference was always less than $2 \, lbf \, in^{-2}$ which is equivalent to one degree at 0°C. The thermometers were linked to the NBS standards, and the temperature scale is thus IPTS-27. Two capillaries, of differing diameters, were used and it was found that in either of these T_c could be determined to ± 0.05 K, although the absolute values for the two capillaries differed by 0.07 K. Since their preferred value of 91.4°C was said to be in "excellent agreement" with another (unpublished) value of 91.9°C, further analysis seems unnecessary. The values of P_c and ρ_c quoted in their paper were not directly observed. The general standard of the whole paper is not high enough to accept that the value of T_c is any better than that implied by their comment, and since the purity of their sample is in doubt, it is probable that their quoted value for T_c is a lower limit.

In 1941 Lu, Newitt and Ruhemann[7] of Imperial College, London, published a paper on the ethane–propylene system which included observations of the critical points of the mixture in varying proportions, ending with the pure components. The method used was to observe the disappearance and reappearance of the meniscus and the values obtained for both phenomena were plotted against the mixture composition and, it is presumed, were then smoothed. It is known that the thermometers used in this laboratory were linked to the NPL standards, and the temperature is thus on IPTS-27. The authors paid considerable attention to the question of the purity of the material, and describe their techniques in detail, but give no final figure by which the purity of the sample can be estimated. In the same way, they describe in detail the method of observing the appearance and disappearance of the meniscus without quoting the differences found in temperature or pressure. The error of ± 0.01 atm given for the pressure measurement is probably the reproducibility rather than the accuracy, and no error is given for the temperature. The values given for these errors in Table A are our estimates. The results would appear to be reliable estimates for pure propylene, but the account lacks the final details which would give assurance of this.

In 1941 Marchman, Prengle and Motard[23] of the Carnegie Institute of Technology,

Pittsburgh, published the results of *PVT* observations in the form of smoothed tables, and included values of P_c, V_c and T_c. The critical temperature was found by graphing their results as *P* vs. *V* and determining the isotherm on which the finite flat portion (in the two-phase region) became a point. This method is liable to subjective error and, if there is any impurity, must tend to give a low result. The purity of their sample is poorly characterized. The error quoted for T_c (± 0.015 K) appears unrealistic even for a pure sample, and the values P_c and V_c depend on the accuracy of the remainder of their work (for a discussion of this see Section 1.3.1).

In summary, of the six reports, one (Winkler and Maass) must be discarded, three (Maass and Wright, Vaughan and Graves, Marchman *et al.*) are probably low, and two (Seibert and Burrell, Lu *et al.*) may be reliable, but there is not sufficient detail to be certain. The only safe deduction is that T_c probably lies in the region 365.15 ± 0.5 K, P_c lies between 4.5 MPa and 5 MPa, and ρ_c is not known.

In these circumstances, the *PVT* equation was not constrained to a critical point and the values given for this work in Table A are those at which $(\partial P/\partial \rho)_T$ and $(\partial^2 P/\partial \rho^2)_T$ were found to be zero. From earlier experience (see, for example, ref. 52) it is known that when fitting an unconstrained analytic equation, the resulting value of T_c will probably be too high, by up to 0.5 K. Thus it is probable that an accurate experimental investigation of T_c will find a value below that given here, but for the sake of thermodynamic consistency the values given in the table have been used throughout this work.

1.2.3 Saturation Curve

The variation of the properties along the saturation curve and the available experimental data are dealt with in detail in the following subsections.

At any temperature, the analytic equation of state gives a relation between the vapour pressure and the saturated liquid density and between the vapour pressure and the saturated vapour density and, by applying the Maxwell "equal-area" principle, a relation between all three. Simultaneous solution of these three equations will give numerical values of the vapour pressure and the saturated densities of liquid and vapour, and these have been used in constructing the tables. This method gives numbers which are thermodynamically consistent with the equation of state, but does not give analytic expressions for their variation with temperature.

1.2.3.1 *Vapour pressure*

Numerous investigators have made measurements of the vapour pressure but usually only over short temperature ranges and there is no set of results which covers the entire range which runs approximately from 88 K to 366 K. The values of the triple point and critical point temperatures are in some doubt (see Sections 1.1 and 1.2.2). The available results are summarized in Table B.

The results of the various investigators show considerable internal scatter, and differ from one another by significant amounts. They fall roughly into three temperature intervals, which were examined with the intention of choosing a set of results from each interval, if one could be found which was clearly preferable to the others.

The first temperature interval is from about room temperature towards the critical

Table B. AVAILABLE VAPOUR PRESSURE DATA

Authors	Date	No. of points	Temperature range/K
Burrell and Robertson[11]	1915	16	145–225
Seibert and Burrell[3]	1915	7	253–273
Maass and Wright[4]	1921	8	236–288
Powell and Giauque[43]	1939	12	166–226
Vaughan and Graves[25]	1940	6	273–364
Lamb and Roper[12]	1940	7	152–235
Lu, Newitt and Ruhemann[7]	1941	13	243–363
Marchman, Prengle and Motard[23]	1949	3	303–348
Farrington and Sage[24]	1949	14	288–355
Tickner and Lossing[13]†	1951	20	99–169
Michels, Wassenaar, Louwerse, Lunbeck and Wolkers[22]	1953	9	298–363
Voinov, Pavlovich and Timrot[8]	1967	44	213–355
Manley and Swift[26]	1971	10	244–311
Laurance and Swift[9]	1972	8	311–344
Grauso, Fredenslund and Mollerup[10]	1977	2	260–290

†All references in the text are to the unsmoothed results of Tickner and Lossing, which are listed in the Appendix.

point. Here there are seven sets of results from Vaughan and Graves,[25] Lu, Newitt and Ruhermann,[7] Marchman, Prengle and Motard,[23] Farrington and Sage,[24] Michels *et al.*,[22] Voinov *et al.*[8] and Laurance and Swift.[9]

On plotting the deviations of all the results of these experimenters against an arbitrary function, the results of Voinov *et al.* and of Michels *et al.* were found to have the smallest scatter, but within this uncertainty to suggest slightly different trends at the highest temperature. The results of Vaughan and Graves were of similar uncertainty, but pointed towards a critical pressure which was too high to reconcile with other data. Conversely the results of Laurance and Swift tended towards too low a critical pressure. The three points of Marchman *et al.* lay within the band of results by Voinov *et al.* and Michels *et al.*, but showed a different trend. The results of Farrington and Sage and of Lu *et al.* were too scattered to be worth further consideration. A critical review of the experimental methods as reported by the various authors supported the view that the work of Michels *et al.* and of Voinov *et al.* were the most reliable. The work of Voinov *et al.* is reported in a paper which is inaccessible to us, and our use of their results is based on their acceptance by Neduzhii *et al.*[1] The final decision was to use the results both of Voinov *et al.* and of Michels *et al.*

The second temperature interval considered was that from room temperature down to about 230 K. In this interval, into which the results of Voinov *et al.* extend, there are also sets of results from Seibert and Burrell,[3] Maass and Wright,[4] Manley and Swift[26] and Grauso *et al.*[10]

A deviation plot of these results against an arbitrary function showed the Voinov *et al.* results still to have the smallest scatter. The results of the other four investigations in this region have a very large scatter, and there was no doubt that the results of Voinov *et al.* should be chosen to cover this temperature interval.

The third temperature interval extends from about 230 K to 150 K. In this interval, there are sets of results from Burrell and Robertson,[11] Powell and Giauque,[43] and Lamb and Roper[12] and there are also some results from Tickner and Lossing,[13]

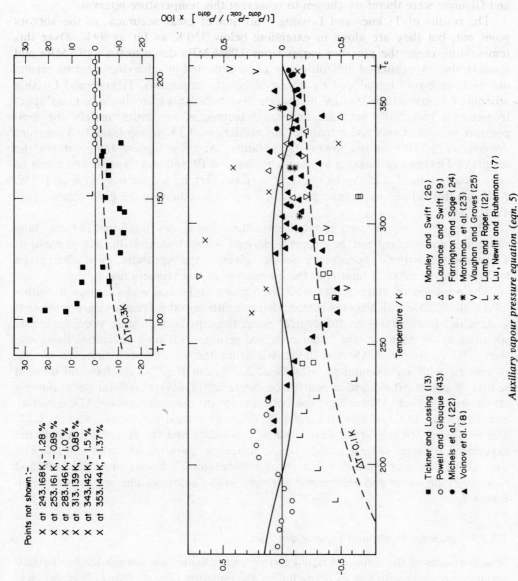

Figure 2. Comparison of experimental vapour pressures with the equation of state (eqn. (10))

which are a special case, and a few points from the extensive results of Voinov *et al.*

A deviation plot shows the results of Powell and Giauque to be very precise, and to join smoothly to the results of Voinov *et al.* The Lamb and Roper results are equally precise, but are systematically lower than those of Powell and Giauque and of Voinov *et al.* The results of Burrell and Robertson are very scattered. The results of Powell and Giauque were therefore chosen to represent this temperature interval.

The results of Tickner and Lossing[13] are not of great accuracy, as the authors point out, but they are alone in extending below 150 K as far as 99 K. Over this temperature range the pressure varies from 0.0036 MPa down to 0.6×10^{-7} MPa and considerable experimental difficulties are to be expected in achieving accurate results due to large errors introduced by traces of volatile impurities. Tickner and Lossing attempted to overcome this by measuring the partial pressure using a mass spectrometer, a potentially accurate and rapid technique, but unfortunately the temperature measurements had a reported uncertainty of 0.3 K which leads to a pressure uncertainty of 10% at the lowest temperature. At their highest temperatures the results of Tickner and Lossing overlap with those of Powell and Giauque and it can be seen from Figure 2 that the two sets of results differ by as much as 11% at 173 K. Although this is large in percentage terms, it represents an absolute difference of only 4×10^{-4} MPa.

The results were not used in initial correlation, but it was found that the very long saturation curve could not be properly defined without some data, and the use of Tickner and Lossing's experimental results, given in the Appendix, even when given low weight, significantly improved the correlation at low temperature.

The equation of state agrees with the results of Powell and Giauque to within -0.1% to $+0.25\%$ which is considered adequate although the deviations are somewhat systematic. From 300 K to the critical point the equation of state predicts values which lie in the middle of the rather scattered results, with most deviations being less than $\pm 0.2\%$. Below 300 K the deviations from the results of Voinov *et al.* are systematic with a maximum of -0.48% at 273 K, but their values have an internal scatter of about $\pm 0.4\%$ and appear to be the result of several different experiments taken over a period of time. Near the critical point the measurements of Michels *et al.* and Vaughan and Graves are diverging to different critical points, see Section 1.2.2. The results of Michels *et al.* were used as input data and the equation predicts the experimental values closest to the chosen critical temperature to within $\pm 0.1\%$. The data of Seibert and Burrell,[3] Burrell and Robertson,[11] Maass and Wright,[4] and Grauso, Fredenslund and Mollerup[10] are very widely scattered and are not shown on Figure 2.

1.2.3.2 *Saturated liquid and vapour densities*

Measurements of the saturated liquid and vapour densities are summarized in Table C and a comparison with values predicted by the equation of state (eqn. 10) is shown in Figures 3 and 4. Neither liquid nor vapour densities have been measured at low temperatures, the lowest values being at 173 K for the liquid, 213 K for the vapour, both measurements by Voinov, Pavlovich and Timrot,[8] whose results extend to within 2 K of the critical temperature for both liquid and vapour. Because of the inconsistencies between the results of various workers and, in some cases, the large internal scatter of

Table C. AVAILABLE SATURATED DENSITY DATA

Authors	Date	No. of points	Temperature range/K
Liquid			
Maass and Wright[4]	1921	9	195–292
Winkler and Maass[6]	1933	8	341–365
Pall and Maass[14]	1936	15	293–362
Lu, Newitt and Ruhemann[7]	1941	12	243–353
Vaughan and Graves[25]	1940	6	273–364
Farrington and Sage[24]	1949	7	278–361
Morecroft[15]	1958	8	273–348
Voinov, Pavlovich and Timrot[8]	1967	62	173–363
Vapour			
Winkler and Mass[6]	1933	8	341–365
Vaughan and Graves[25]	1940	6	273–364
Lu, Newitt and Ruhemann[7]	1941	12	253–363
Farrington and Sage[24]	1949	7	278–361
Marchmann, Prengle and Motard[23]	1949	3	303–348
Voinov, Pavlovich and Timrot[8]	1967	60	213–363

results, none of the experimental results were used in constructing the equation of state, and in the comparisons, the predicted values are those obtained by solving the equation of state to find the densities where $\Delta G = 0$.

Figure 3 shows that, considering the results for the liquid density as a whole, there is a small systematic difference between their general trend and the predicted values at temperatures away from the critical. The large negative deviations close to the

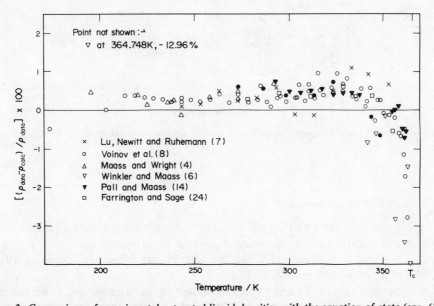

Figure 3. Comparison of experimental saturated liquid densities with the equation of state (eqn. (10))

critical point of the results of Voinov *et al.*, Winkler and Maass, and Pall and Maass are a consequence of their values of the critical temperature.

Although the equation differs in its predictions from directly measured saturated liquid densities, it agrees well with the results Golovskii and Tsimarnii[16] obtained by extrapolation of their liquid-phase densities to a selected vapour pressure curve. The differences between these results and those predicted lie within ±0.3%.

The results of Voityuk,[31] listed in Neduzhii *et al.*,[1] were obtained by a similar procedure, but were not considered since Golovskii and Tsimarnii have pointed out that there were errors in the processing of the data.

In Figure 4 the deviations of the vapour density results from the equation of state predictions show agreement which is as good as can be expected at low temperatures, and close to the critical again shows that the results of Voinov *et al.* and of Winkler and Maass require a lower value of the critical temperature.

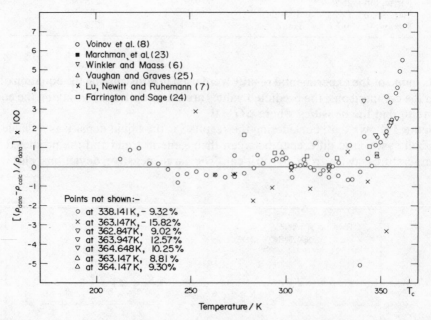

Figure 4. Comparison of experimental saturated vapour densities with the equation of state (eqn. (10))

1.2.3.3 *Enthalpy of evaporation*

The experimental results consist of a single value by Powell and Giauque,[43] of the University of California, at the normal boiling point, published in 1939, and four values by Clusius and Konnertz,[17] of the Universität München, between 193.132 K and 288.394 K, published in 1949.

The experimental results and values predicted by the equation of state are given in Table D. The maximum difference of 166 J mol⁻¹ at 288.394 K is equivalent to 1.1%.

Table D. COMPARISON OF EXPERIMENTAL DATA FOR
THE ENTHALPY OF EVAPORATION WITH PREDICTION

	$\dfrac{T}{K}$	$\dfrac{\Delta H_{data}}{J\,mol^{-1}}$	$\dfrac{\Delta H_{calc}}{J\,mol^{-1}}$	$\dfrac{\Delta(\Delta H)}{J\,mol^{-1}}$
Powell and Giauque[43]	225.43	18417 ± 13	18481	-64
Clusius and Konnertz[17]	193.13	19816 ± 42	19875	-59
	225.43	18473 ± 38	18481	-8
	268.20	16294 ± 42	16211	83
	288.39	15014 ± 50	14848	166

1.2.3.4 *Isobaric heat capacity*

Experimental measurements of the isobaric heat capacity at saturated liquid densities have been made by Huffman, Parks and Barmore[18] in 1931 at Stanford University, at ten temperatures between 93 K and 210 K and over approximately the same temperature range by Powell and Giauque[43] in 1939 at the Chemical Laboratory of the University of California at eighteen temperatures between 94 K and 224 K; these two sets of results differ from each other by no more than 1%. The difference between the results and values calculated from the equation of state (eqn. 10) is shown in Figure 5 where it can be seen that the shallow minimum at 150 K in the experimental data is not reproduced by the equation of state which instead shows a much sharper minimum shifted to 110 K, where the deviations have increased to -15% from $\pm 1\%$ at 223 K.

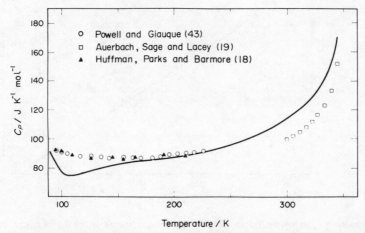

Figure 5. *Comparison of experimental isobaric heat capacities in the liquid phase under saturation conditions with the equation of state (eqn. (10))*

At higher temperatures Auerbach, Sage and Lacey,[19] of the Californian Institute of Technology, in 1950 reported nine values between 300 K and 344 K. The measurements made were of the apparent heat capacity, and extensive corrections had to be made to them, amounting in the worst case to 35% of the measured value. Some of the data used for these corrections are now superseded, but the calculations cannot be revised because insufficient information was given by the authors. Their values are some 11% to 16% lower than those predicted by the equation of state, and a similar discrepancy was found by Neduzhii *et al.*[1] when comparing them with values produced by extrapolating to the saturation curve the measurements by Vaschenko[35] of liquid heat capacities.

1.2.3.5 *Speed of sound*

Blagoi, Butko, Mikhailenko and Yakuba,[20] of the Physicotechnical Institute of Low Temperatures, Ukrainian SSR Academy of Sciences, Kharkov, in 1968 published thirty-six results along the saturated liquid curve between 88 K and 282 K. A comparison of these with the values calculated from the equation of state shows good agreement down to 95 K, with the predicted values generally lying 2% below the experimental results. At 88 K the deviation has increased to +5% because the equation of state predicts an incorrect maximum at about 95 K, as can be seen in Figure 6.

Soldatenko and Dregulyas,[21] of the Kiev Institute of Technology, in 1968 published eight results on the saturated vapour curve. A comparison of these results with values predicted by the equation of state shows deviations which vary systematically from +0.6% at their lowest temperature of 193.15 K, to −4% at 363.15 K which is less than 2.5 K below the critical temperature.

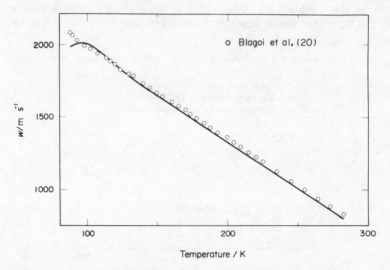

Figure 6. Comparison of experimental speed of sound measurements in the liquid phase under saturation conditions with the equation of state (eqn. (10))

1.3 SINGLE-PHASE REGION

1.3.1 Density

Experimental measurements of density extend from approximately 90 K to 573 K and have been made by a large number of workers using a variety of techniques, with which it is possible to deal only briefly. Many of the sets of results are not of high precision, use propylene of uncertain purity or are presented in the form of smoothed tables with no indication as to the extra errors introduced by interpolation of the original experimental results.

In some regions, the only data available have one or more of these defects, and in order to construct a table at all it has been necessary to relax somewhat the critical standards used in earlier books in this series regarding the acceptance of data. An attempt has been made to allow for these data defects in the weightings used, and we believe the final results justify our practice.

The available results are listed by author in Table E and their distribution is shown in Figure 7.

Table E. AVAILABLE $P\rho T$ DATA

Authors	Date	No. of points	Temperature range/K	Pressure range/MPa	Density range/mol dm^{-3}
Vaughan and Graves[25]	1940	281	273–573	0.23–8.28	0.1–14.9
Roper[27]	1940	8	223–343	0.05–0.15	0.02–0.06
Marchman, Prengle and Motard[23]	1949	317	303–523	0.5–21.8	0.12–10.9
Farrington and Sage [24]	1949	408	278–511	0.1–69	0.02–14.7
Michels, Wassenaar, Louwerse, Lunbeck and Wolkers[22]	1953	255	298–423	0.66–232.7	0.3–15.5
Pfennig and McKetta[28]	1957	24	305–339	0.08–0.16	0.03–0.06
Robertson and Babb[33]	1969	50	308–473	96.5–974	14.3–19.4
Dittmar, Schultz and Strese[32]	1962	219	273–413	2.0–98	7.0–15.0
Voityuk[31]	1968	96	183–312	0.5–5.9	11.4–15.7
Manley and Swift[26]	1971	37	244–333	0.15–11	0.07–14.3
Golovskii and Tsimarnii[29]	1975	120	190–300	1.3–60	13.0–15.4
Golovskii and Tsimarnii[30]	1975	176	90–200	1.6–60	15.2–18.3

After all the available data had been reviewed, it was decided that the results of Michels et al.[22] should be used as the standard against which other work would be assessed since not only does it appear to be reliable when considered alone, but also it forms part of the long experimental programme of the Van Der Waals Laboratory the value of which has been proven for many substances. The work occupies a central position, including both the gas and liquid phases and extending from well below the critical pressure to well above it, and all their results were used in constructing the equation of state.

The equation predicts all the results of Michels et al. with a relative deviation of $\pm 0.18\%$ in density and pressure and deviates by less than $\pm 0.1\%$ in density for 218 of the 255 experimental measurements. Many of the deviations in the liquid or at high pressure are less than $\pm 0.05\%$. The highest density deviations are found in a critical region defined by $0.66\rho_c < \rho < 1.33\rho_c$, $0.8T_c < T < 1.25T_c$, where they reach a maximum of -1.17% at 4.3504 MPa and 361.34 K. The relative deviations for the fifty-two results which lie in this critical region are $\pm 0.20\%$ in pressure and $\pm 0.48\%$ in density.

The results of Marchman, Prengle and Motard[23] of the Carnegie Institute of Technology, Pittsburgh, overlap with the gas phase results of Michels et al. and extend 100 K higher in temperature. The experimental results were not given, but the smoothed table of values presented show systematic deviations from the results of

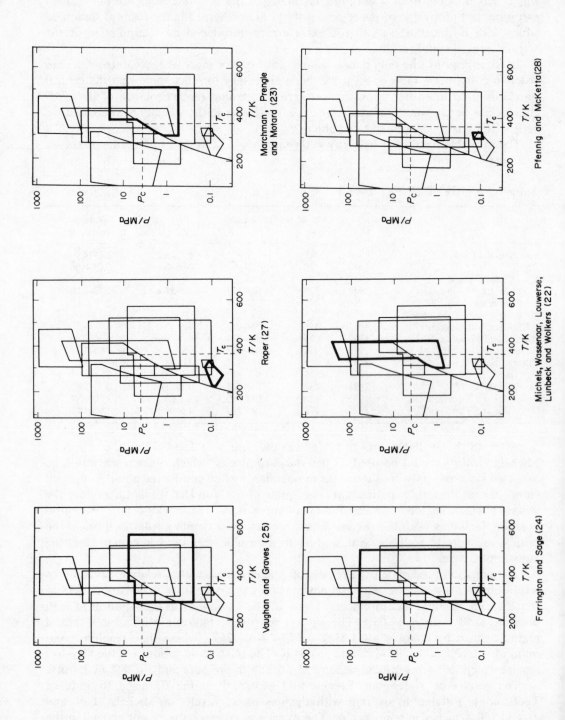

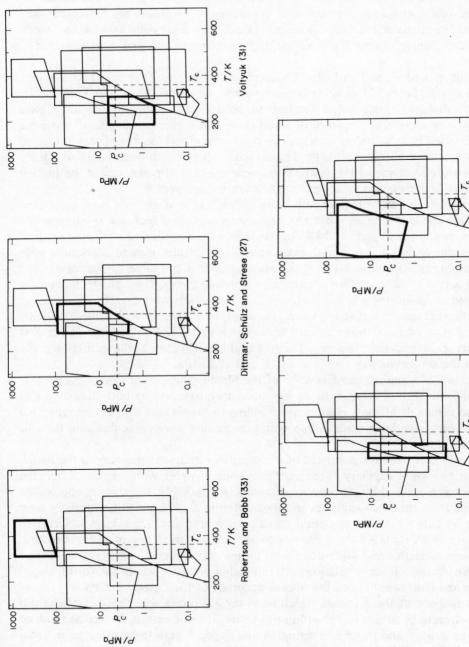

Figure 7. *Available PρT data*

Michels *et al.* These smoothed results were therefore not used in constructing the equation of state, and comparison of its predictions with the 317 tabulated values shows, as would be expected, similar deviations to the comparison of Michels *et al.* with their table. At higher temperatures, the deviations increase as the temperature increases. The relative deviation in density is ±0.33% and in pressure ±0.82%, the largest deviation in density being 0.8% at the highest temperature and pressure (523 K, 21.8 MPa).

Farrington and Sage,[24] of the California Institute of Technology, Pasadena, California, also published only smoothed values, covering the whole temperature range of Michels *et al.*, but extending to both higher temperatures and lower pressures. The apparatus consisted of a stainless-steel vessel whose effective volume could be varied by introducing or withdrawing mercury, and the vessel was immersed in a temperature-controlled oil-bath. The authors estimated an uncertainty of ±0.3% in the measured volume of the fluid except near the critical point, and at the highest temperature and pressures where polymerization was suspected.

A comparison of these results with those of Michels *et al.* showed good agreement at both high and low pressures, but a systematic increase and decrease at intermediate pressures, reaching as much as 0.8%. In this region the equation of state also predicts values of the speed of sound and heat capacity which are in good agreement with experimental results and it would therefore appear that the table of Farrington and Sage has a systematic error, but whether this is due to the experimental techniques or the method of smoothing it is impossible to say. The results were therefore not used in constructing the equation of state. A comparison of the 408 results with values predicted by the equation of state shows a relative deviation of ±0.37% in density. The largest deviations are in the critical region and in the gas phase very close to the saturation curve of which the maximum was +3.7% at 411 K and 10.3 MPa.

The results of Vaughan and Graves[25] of the Shell Development Company, Emeryville, California, are at comparatively low pressures, overlapping both the liquid and gas phase results of Michels *et al.*, and extending to the highest temperature at which measurements have been made. Once again, the results were presented as a table of smoothed values.

The apparatus basically consisted of a compressor unit, with mercury as the liquid, on which two glass capillary tubes containing the samples were mounted, and the pressure was measured by nitrogen manometers. Remarkably, the authors considered the results from the two capillaries to be of different accuracy and accordingly gave differing weights to the experimental results when graphical smoothing was carried out. Above 200°C (473 K) the authors considered that their low-pressure measurements were unreliable and discarded them. In this region therefore the values in their table were obtained by interpolation between higher pressure points and the ideal gas. Vaughan and Graves estimated the overall accuracy of their table as ±1%.

A comparison of those results which lie in the gas phase with those of Michels *et al.* show them to be in agreement within this figure. If these results, as well as those of Marchman *et al.*[23] and those of Farrington and Sage,[24] were to be rejected as being composed of smoothed values only, there would be no data at all above 423 K.

In order to avoid this, and because the Vaughan and Graves results show agreement to within their claimed accuracy with Michels *et al.*, sixty-three of their results on the 523.2 K, 548.2 K and 573.2 K isotherms were used in constructing the equation of state, but given low weights. The remaining 218 points which lie in the region of overlap with Michels *et al.* were rejected.

A comparison of these sixty-three results with the values predicted by the equation of state shows systematic deviations in density ranging from −0.2% at the lowest pressures to −1% at the highest pressures with a relative deviation of ±0.44%.

A comparison of the 148 values in the gas region which were not used shows that in general they do not deviate by more than ±0.5% in density from the values predicted by the equation.

In the liquid and critical regions deviations in density between the published values which were not used and the predictions of the equation of state are as high as 11% on the lowest isotherm of 273.15 K and in general not less than ±5%.

In 1971 Manley and Swift[26] of the University of Kansas, Lawrence, published thirty-seven somewhat scattered experimental points at seven temperatures between 244.26 K and 333.1 K in both the gas and liquid phases at pressures up to 11 MPa. The experimental apparatus was described by Manley.[26a] Manley and Swift estimated the probable errors to be ±0.1% for the liquid specific volume and ±0.1% for the compressibility factor in the gas, although the apparatus was designed to measure the specific volumes of liquid mixtures with an accuracy of ±0.5%. Their results do not agree with any of the other authors with whom comparison is possible, and so they were not used in constructing the equation of state. A comparison with values calculated from it gave relative density deviations of ±0.52%, the highest deviations being in the gas phase with a maximum of 2.1% at 260.92 K, 0.396 MPa.

Turning now to work in the gas phase at low pressures, there are two sets of results, obtained as part of a programme which included measurements of ethylene. In our study of that fluid (ref. 51) there was a considerable amount of low-pressure data and we were able to regard them as measurements giving information on the second virial coefficient. In the case of propylene, they are the only measurements of the density at low pressure, and have been used as such.

In 1940 Roper,[27] at Harvard University, published eight points between 223 K and 343 K at low pressures. The method consisted of filling an evacuated globe of known internal volume with gas under measured conditions of temperature and pressure. The work was carried out with care, the precision being estimated to be ±0.02% and the results were used in constructing the equation of state. Comparisons of these results with values predicted by the equation show relative deviations of ±0.05% in both pressure and density.

A study of the compression factor at low pressures in the gas phase was carried out by Pfennig and McKetta,[28] of the University of Texas, Austin, in 1957 who reported twenty-four results on three isotherms up to a maximum pressure of 0.16 MPa. An Edwards-type gas density balance was used and the overall precision was estimated to be ±0.1%. The work was used in constructing the equation of state. A comparison of their values with those predicted by the equation shows a relative deviation of ±0.06% in both pressure and density.

In the liquid phase, the most extensive results were reported by Golovskii and Tsimarnii[29,30] of the Odessa Institute of Marine Technology, whose two papers together deal with a region extending from near that of Michels *et al.* almost to the melting curve and cover a reasonable range of pressure. The experimental results were found to be internally consistent with very little scatter. They were used in the construction of the equation of state, and a comparison with its predictions shows that the relative deviation in density of the 296 points is ±0.06%, the maximum deviation being 0.14%.

Still in the liquid region, but at lower pressures and at temperatures closer to the saturation curve, experiments were made by Voityuk[31] of the Kiev Institute of Technology, which were reported by Neduzhii *et al.*[1] in the form of a smoothed table. Experimental details of the hydrostatic weighing technique were given by Boiko and Voityuk[31a] and the accuracy of the specific volumes were reported to be ±0.15%.

In their region of mutual overlap the results agree well with those of Golovskii and Tsimarnii, and the rest of them lie in an important region not covered by other experimenters, so although they are smoothed values, they were all used in constructing the equation of state.

A comparison of the ninety-six results with the values predicted by the equation gives a relative deviation of ±0.09%, the largest being −0.2%. Deviations of less than ±0.1% are found for sixty-seven of the ninety-six experimental points.

In the same paper, Voityuk also reported results for ethylene in a roughly comparable region. These were used by us in our study of that fluid (ref. 51), and those results were found to differ systematically from the ethylene equation of state by about −0.15% at low temperatures, with large deviations on the two isotherms at approximately 268 K and 273 K. The improvement in fitting the propylene results probably reflects the improvement in methods of constructing equations of state since 1972.

The results of Dittmar, Schulz and Strese[32] form a valuable bridge between those of Golovskii and Tsimarnii and of Michels *et al.*, as well as overlapping with parts of the work of Voityuk, Vaughan and Graves, Farrington and Sage, Marchman *et al.*, and reach the lower limit of the high-pressure work of Robertson and Babb. Unfortunately, their results are systematically about ±0.4% different from what are regarded as the most accurate results, those of Golovskii and Tsimarnii and of Michels *et al.* The experiments were carried out using a constant-volume piezometer, and all the necessary precautions and corrections appear to have been made. The fact that the systematic difference found is a constant percentage of the density suggests that the possible cause lies in the initial calibration of the piezometer. The results were not used in constructing the equation of state. A comparison of their 219 smoothed tabular values with the predictions of the equation shows, as expected, a systematic deviation of about +0.4%, except near the critical point where the deviations become much larger, reaching a maximum −6.9% at 373.15 K, 6.08 MPa.

The highest pressure results yet reported were published in 1969 by Robertson and Babb,[33] of the University of Oklahoma, and lay along three isotherms at 308.14 K, 373.17 K and 473.09 K at pressures from 96.5 MPa to 974 MPa. The results were relative measurements and in their paper were reduced to absolute terms using the results of Michels *et al.* The authors reported that polymerization occurred at 473 K and that the reaction rate increased with pressure making the density values uncertain along this isotherm. In case this effect should be present but unnoticed at the lower isotherms, their results were not used in constructing the equation of state. A comparison of their results with the values predicted by the equation of state shows at the lowest temperature that the relative deviation is ±0.75% in pressure and ±0.13% in density. At the intermediate temperature the relative deviations are ±0.84% in pressure and ±0.14% in density. At the highest temperature the relative deviations are ±2.16% in pressure and ±0.38% in density. If the two highest pressure points are omitted, the deviations fall to ±1.06% and ±0.20%, respectively.

It would appear that Robertson and Babb were correct in their diagnosis that polymerization occurred only at the highest temperature.

1.3.2 Second Virial Coefficient

The coefficients of the first power of density of the equation of state (eqn. 10) represent the second virial coefficient:

$$B = \frac{1}{\rho_c}(N_1\tau + N_2\tau^2 + N_3\tau^3).$$

In 1964 McGlashan and Wormald[34] published from Reading University fifteen experimental measurements between 303.7 K and 413.6 K. The apparatus and differential piezometer technique were described in ref. 34a. The results were not used in constructing the equation of state but values predicted from it are in good agreement, as shown in Figure 8. The mean deviation of the calculated values from the experimental is ±5.8 cm³ mol⁻¹ with a maximum of −11 cm³ mol⁻¹ at 343.4 K.

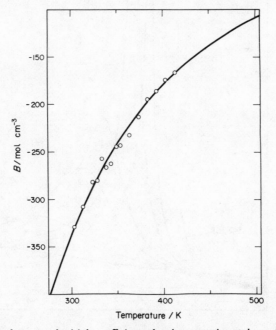

Figure 8. Variation of the second virial coefficient showing experimental results of McGlashan and Wormald[34]

1.3.3 Isobaric Heat Capacity

The distribution of the experimental values of the isobaric heat capacity is shown on Fig. 15.

   In 1974 Bier, Ernst, Kunze and Maurer,[36] of the Universität Karlsruhe (TH), published 152 experimental results in the gas phase from 298 K to 473 K. At super-critical temperatures the pressure range extended up to 12 MPa and the investigation included a number of points in the near neighbourhood of the critical point, 35 of which were within a region defined by $0.8 T_c < T < 1.25 T_c$, $0.66 \rho_c < \rho < 1.33 \rho_c$. The apparatus, which was described fully in ref. 36a, was capable of high precision and great care was taken with the measurements: the results are considered to be of high accuracy. Values predicted by the equation of state deviate by less than $\pm 1\%$ for 93 of the 152 results, and by less than $\pm 2\%$ for a further 49 results. Large differences were only found for the remaining 10 points which are near the critical point where the sharp "peak" associated with the isobaric heat capacity is qualitatively reproduced but slightly shifted in temperature. Figure 9 shows the behaviour of seven representative isotherms together with the appropriate experimental points.

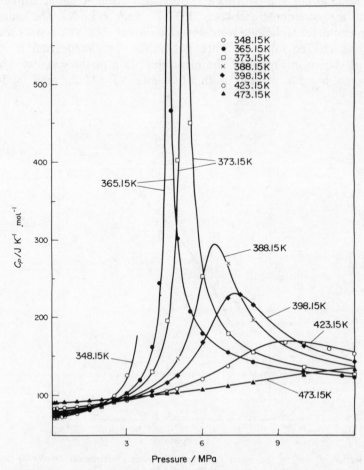

Figure 9. Variation of the isobaric heat capacity showing experimental results of Bier, Ernst, Kunze and Maurer[36]

Vaschenko,[35] at the Odessa Institute of Marine Technology in 1971, reported experimental measurements in the liquid phase from 164 K to 361 K at pressures from below 1 MPa to 6 MPa. A graph and a smoothed table extracted from ref. 35 were given by Neduzhii *et al.* in ref. 1. The unsmoothed measurements are listed in the Appendix, and a comparison of values predicted by the equation of state with these 112 measurements gives a relative deviation of ±5.6% but there were several misprints; if the most obvious cases were omitted the relative deviation fell to ±4.8%. The deviations are fairly evenly scattered except at both high temperatures and high pressures where the equation consistently predicts values lower than the experimental results. The maximum deviation is at 358.65 K, 5.702 MPa where the difference is 10.2%.

In the light of the information currently available to us on this work, the agreement is considered to be satisfactory.

1.3.4 Joule–Thomson Coefficient

Bier, Ernst, Kunze and Maurer,[36] in 1974, as well as publishing the isobaric heat capacity measurements discussed in Section 1.3.3 also published seventy-six experi-

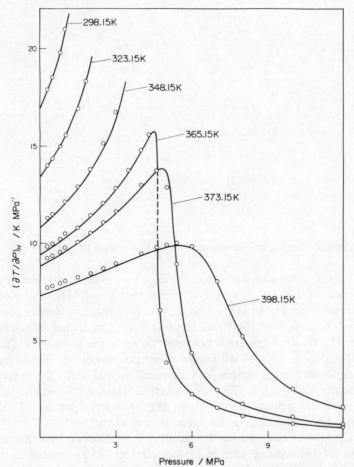

Figure 10. Variation of the Joule–Thomson coefficient $(\partial T/\partial P)_H$ showing experimental results of Bier, Ernst, Kunze and Maurer[36]

mental measurements of the isenthalpic Joule–Thomson coefficient, $(\partial T/\partial P)_H$, at temperatures between 298 K and 398 K at pressures up to 12 MPa. The maximum error was estimated by Bier *et al.* to be ±5%. Values predicted by the equation of state deviated by less than this amount for all but six of the experimental results, and the mean deviations for all the results was ±3.15%.

The six results whose deviations were greater than ±5% lay either close to the critical point or in the liquid region where the error in the experimental measurements was assessed to be at its greatest and the maximum deviation is −13.9% at 365.15 K, 5 MPa ($T/T_c = 0.9986$, $P/P_c = 1.0682$). Figure 10 shows the Joule–Thomson coefficient calculated from the equation of state on seven isotherms with the appropriate data points.

The distribution of their experimental measurements is the same as that of their heat capacity measurements, which is shown on Figure 15.

The Joule–Thomson inversion curve has been calculated from the equation of state and is illustrated in Figure 11.

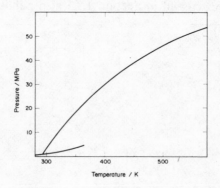

Figure 11. Joule–Thomson inversion curve

1.3.5 Speed of Sound

The distribution of the experimental values of the speed of sound is shown on Figure 16.

In 1957 Terres, Jahn and Reissmann,[37] of Universität Karlsruhe (TH), published speed of sound results on seven isotherms between 293 K and 448 K, up to a maximum pressure of 11.768 MPa. The majority of the ninety-two points are in the gas phase but several at sub-critical temperatures and higher pressures are in the liquid phase. The relative deviation between the equation of state predictions and the experimental results is ±1.65% with sixty-seven points having a deviation of less than ±1%, with a further eight points less than ±2%, and only three points having a deviation greater than ±5%. The maximum deviation of −6.9% is at 373.15 K, 5.149 MPa, near the critical point where the analytic equation of state does not reproduce the sharp minimum in the speed of sound surface.

A similar region is covered by the results of Soldatenko and Dregulyas[38] of the Kiev Institute of Technology who in 1968 published 223 experimental measurements at pressures up to 9.87 MPa. The temperature range is somewhat greater than that of

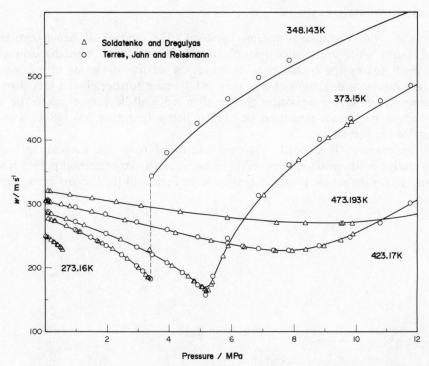

Figure 12. Variation of the speed of sound showing experimental results of Soldatenko and Dregulyas[38] and Terres, Jahn and Reissmann[37]

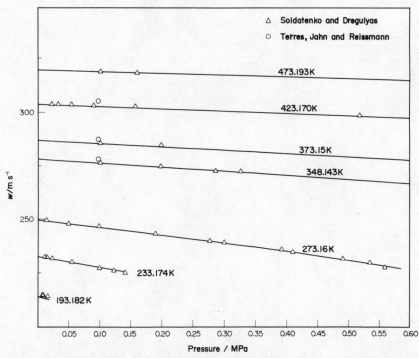

Figure 13. Variation of the speed of sound at low pressures showing the experimental results of Soldatenko and Dregulyas[38] and Terres, Jahn and Reissmann[37]

Terres *et al.*, with eighteen isotherms between 193 K and 473 K being reported and several results which are at very low pressures below 0.1 MPa. A comparison with the values predicted by the equation of state gives a relative deviation of ±2.24%, with 197 points having a deviation of less than ±1% and a further eleven less than ±2%. Six points which have a deviation greater than ±5% all lie very close to the critical point with a maximum deviation of −19% being found at 365.148 K, 4.661 MPa ($T/T_c = 0.9985$, $P/P_c = 0.9957$).

The variation of the speed of sound calculated from the equation of state on representative isotherms together with the appropriate experimental points is shown in Figure 12 for the whole pressure range and in Figure 13 for the low-pressure range.

2. Correlating Equations

2.1 IDEAL GAS EQUATIONS

In 1975 Chao and Zwolinski[39] of Texas A & M University published a set of tables of the ideal gas properties. These were calculated from the most recent fundamental and molecular spectroscopic constants and were therefore chosen for use in these tables.

2.1.1 Isobaric Heat Capacity

The ideal gas isobaric heat capacity table of Chao and Zwolinski,[39] was represented by an equation of the form

$$\frac{C_P^{id}}{R} = \sum_{i=1}^{5} f_i \tau^{(1-i)} + f_6 \tau^2 + f_7 \frac{u^2 e^u}{(e^u - 1)^2} \tag{1}$$

where $u = f_8 \cdot \tau$ and $\tau = T_c/T$. (The critical temperature as such plays no part in the ideal gas heat capacity function, but it is convenient to use the same variable as that of the equation for the real fluid.) The numerical values of the coefficients are given in Table J, and values of the isobaric heat capacity are listed in Table 1 from 90 K to 1500 K.

Chao and Zwolinski compared their calculated values with the experimental measurements of Kistiakowsky and Rice,[40] Kistiakowsky, Lacher and Ransom,[41] Telfair[42] and Bier, Ernst, Kunze and Maurer[36] all extrapolated to zero pressure. The maximum deviation between the calculated and measured values was $\pm 0.50 \, \text{J K}^{-1} \, \text{mol}^{-1}$.

Over the temperature range 50 K to 1500 K the maximum deviations between equation (1) and the tables of Chao and Zwolinski are $+0.18 \, \text{J K}^{-1} \, \text{mol}^{-1}$ and $-0.09 \, \text{J K}^{-1} \, \text{mol}^{-1}$. The comparison was made with a more extended table than that given in ref. 39, kindly supplied by Chao and Zwolinski.

2.1.2 Entropy

For the calculation of entropy in this book, the point at which the entropy was assumed to be zero was that of the ideal gas at 298.15 K and 1 atm (0.101 325 MPa). The entropy at any other temperature thus requires the appropriate integration of the isobaric heat capacity equation from 298.15 K to the required temperature, i.e.

$$S^{id}(T) = \int_{298.15}^{T} \frac{C_P^{id}}{T} \, dT$$

which written in full is

$$\frac{S^{id}(T)}{R} = \left[f_1 \ln T + \sum_{i=2}^{5} f_i \frac{\tau^{(1-i)}}{(i-1)} - f_6 \frac{\tau^2}{2} + f_7 \left\{ \frac{u e^u}{e^u - 1} - \ln(e^u - 1) \right\} \right]_{298.15}^{T} \tag{2}$$

25

The maximum difference between equation (2) and the extended tables of Chao and Zwolinski is $-0.45\,\text{J K}^{-1}\,\text{mol}^{-1}$ at 50 K, but between 100 K and 1500 K the maximum differences are less than $\pm0.05\,\text{J K}^{-1}\,\text{mol}^{-1}$.

If the entropy of the crystalline solid at 0 K were to be taken as zero, then the entropy difference between this point and that of the ideal gas at 298.15 K and 1 atm would need to be added to the values given in this book. The value given by Chao and Zwolinski is $266.62\,\text{J K}^{-1}\,\text{mol}^{-1}$ and this is recommended as the best value.

Calorimetric measurements by Powell and Giauque[43] are available from which this value can also be calculated. The work of Powell and Giauque provides results for several different properties: the heat capacity of the solid, the heat of fusion, the heat of evaporation, and the vapour pressure of the liquid. Individually they are of importance, and together they provide the essential data for calculating the enthalpy difference between the real fluid and the ideal crystal at zero temperature.

The work was part of a programme carried out by Giauque and his collaborators and it is necessary to consult several papers to find full details of the experimental procedures, refs. 43a, 43b, 43c, 43d, 43e, 43f. The main instrument was a low-temperature calorimeter of a form, and operated in a manner, which has by now become standard, the modern version owing much to this early work of Giauque. The calorimeter was a gold vessel with a heating coil wound on the outside, and well insulated by a vacuum and by radiation shields. Energy was supplied electrically, and the quantity measured in terms of the voltage, current and time. The temperature was measured absolutely by a copper–constantan thermocouple and changes in temperature by a platinum resistance thermometer, both having been calibrated against a hydrogen gas thermometer. The pressure was measured by a mercury manometer, the connection to which required adjustments for the heat loss it caused in the calorimeter. The quantity of sample was determined as a volume of gas at low pressure by transferring the contents of the calorimeter to equipment at room temperature. The many corrections needed were carefully set out, and though some of the constants assumed need to be changed, the corrections are themselves so small that changes in them will not affect the overall accuracy, with one exception.

Throughout his work, Giauque used an ice-point of 273.10°C, which raises the question of the temperature scale to which his temperatures should be referred. We have taken the view that since his temperatures were ultimately referred to a gas thermometer, then the values quoted are as nearly as possible on a thermodynamic temperature scale in which the size of a degree is 1/273.10 that of the interval from zero temperature to the ice-point. Our first correction to Giauque's temperature was therefore to multiply them all by (273.15/273.10). At low temperatures, the fixed points used in calibration were given values by Giauque differing from those used today, and his whole temperature scale has been recalculated in terms of IPTS-68.

Using their data, the calculation of the entropy above zero temperature gives the contributions shown in Table F. The difference between the real and the ideal gas values at the boiling point were calculated from $-P(\mathrm{d}B/\mathrm{d}T)$ using the second virial coefficient from equation (24). It can be seen that the total difference of $S^{\text{id}}(298.15\,\text{K},\ 1\,\text{atm}) - S^{\text{ic}}(0\,\text{K})$ is in good agreement with that found by Chao and Zwolinski, the former being only $0.148\,\text{J K}^{-1}\,\text{mol}^{-1}$ higher.

Soldatenko, Morozov, Dregulyas and Vaschenko[44] calculate the difference $S^{\text{id}}(225.45\,\text{K},\ 1\,\text{atm}) - S^{\text{ic}}(0\,\text{K})$ to be $168.532\,\text{J K}^{-1}\,\text{mol}^{-1}$, whilst the value derived from Table F is $168.146\,\text{J K}^{-1}\,\text{mol}^{-1}$, which is $0.386\,\text{J K}^{-1}\,\text{mol}^{-1}$ lower.

Table F. INCREMENTS OF ENTROPY AND ENTHALPY FROM THE PERFECT CRYSTAL AT ZERO TEMPERATURE TO THE IDEAL GAS AT 298.15 K AND 1 atm PRESSURE

Change	Source of values	$\dfrac{\Delta S}{\text{J K}^{-1}\,\text{mol}^{-1}}$	$\dfrac{\Delta H}{\text{J mol}^{-1}}$
0–15 K	Debye function[45]	2.015	22.3
15–87.90 K	Experimental C_P[43]	48.243	2544.1
Fusion at 87.90 K	Experimental ΔH[43]	34.156	3002.3
87.90–225.43 K	Experimental C_P[43]	83.730	12 208.9
Evaporation at 225.43 K	Experimental ΔH[43]	81.696	18 416.9
Real to ideal gas at 225.43 K	See Sections 2.1.2 and 2.1.3	0.556	187.7
225.43–298.15 K	Equation (1)	16.375	4277.6
Total change		266.771	40 659.8
0–298.15 K, ideal gas	Chao and Zwolinski[39]	266.623	

Tables of the variation of the ideal gas entropy with temperature are given in Table 1 from 90 K to 1500 K.

2.1.3 Enthalpy and Internal Energy

In the calculation of enthalpy and internal energy in this book, the zero reference point was taken to be the enthalpy of the ideal gas at 298.15 K. The enthalpy at any other point in the ideal gas state is given by

$$H^{\text{id}}(T) = \int_{298.15}^{T} C_P^{\text{id}}\, dT,$$

which in terms of equation (1) becomes

$$\frac{H^{\text{id}}(T)}{R} = \left[T_c \left\{ \sum_{i=1}^{5} \frac{f_i \tau^{-i}}{i} - f_6 \tau + \frac{f_7 \cdot f_8}{e^u - 1} \right\} \right]_{298.15}^{T}. \tag{3}$$

Tables of the variation of the ideal gas enthalpy with temperature are given in Table 1 from 50 K to 1500 K.
Since

$$H = U + P/\rho$$

the internal energy of the ideal gas at any temperature becomes

$$U^{\text{id}}(T) = H^{\text{id}}(T) - RT. \tag{4}$$

For calculations of the enthalpy and internal energy, there are three different states in common use at which the property may be assumed to be zero.

The first state is that of the ideal gas at 298.15 K, and this is the one used for the numbers printed in all the tables in this book.

The second commonly used state is that of the ideal gas at 0 K, and this may be derived from any of the enthalpies printed in this book by adding to them the value 13546 J mol^{-1} which is the value found by Chao and Zwolinski.[39]

The third commonly used state is that of the ideal crystal at 0 K, and this may be found by adding the value 40 660 J mol^{-1} to the enthalpies printed in this book. The calculation of this value is shown in Table F and was derived from the same data of Powell and Giauque,[43] as was used for calculating the ideal gas entropies in the table, with the same corrections applied. The difference between the ideal and real enthalpies at the boiling point was calculated from

$$P\left[B - T\frac{dB}{dT}\right],$$

using the second virial coefficient from equation (24).

The agreement found between the entropy calculated from spectroscopic data and that calculated from Powell and Giauque's calorimetric data gives confidence that the enthalpy calculation is correct.

Soldatenko, Morozov, Dregulyas and Vaschenko[44] give for $H^{id}(225.45 \text{ K}) - H^{ic}(0 \text{ K})$ the value 17 895 J mol^{-1} which is 81 J mol^{-1} higher than the value calculated here.

2.2 EQUATION OF STATE OF THE REAL FLUID

In 1971 the State Committee of Standards of the Council of Ministers, USSR, published a monograph on the thermodynamic and transport properties of ethylene and propylene. All the available literature prior to 1969 was critically examined and tables of properties compiled which made obsolete all formulations produced before this date. The work was carried out at the Kiev Institute of Technology and this monograph, the equation of Bender, the equation of Juza, Sifner and Hoffer, and the present equation of the IUPAC Project Centre are the only correlations considered here.

2.2.1 The Correlation of the Kiev Institute of Technology[1]

Twelve authors, with Neduzhii as the principal author, assessed the existing data with the object of producing thermodynamically concordant tables of properties. Where necessary additional experimental measurements were made at the Kiev Institute, and in some cases the results and experimental accounts are only to be found in ref. 1.

The thermodynamic tables for propylene were constructed using a combination of graphical and analytical techniques so that a comprehensive equation of state was not presented. The tables contain: properties on the liquid–vapour equilibrium curve in the range 160 K to the critical point; specific volume, enthalpy, entropy and isobaric heat capacity of the single-phase at pressure up to 6 MPa in the temperature range 180 K to 370 K and up to 200 MPa in the temperature range 370 K to 450 K. Speed of sound, heat-capacity ratio and adiabatic indices are also given for the gas in the 210 K to 470 K temperature range at pressures up to 10 MPa.

An equation of state for the gas phase only is given in the form,

$$Z = 1 + \sum_{i=1}^{m} \frac{1}{V^i} \sum_{k=0}^{n} n_{ik} \left(\frac{1}{T}\right)^k.$$

Two sets of coefficients are given and their use depends upon the region of interest. Region 1 covers the range 270 K to 450 K, 0.1 MPa to 3 MPa (30 coefficients) and region 2 covers the range 370 K to 450 K, 0.1 MPa to 300 MPa (27 coefficients).

Tables of the speed of sound were produced by graphical smoothing of experimental data, and heat capacity and adiabatic indices by combining the equation of state with experimental speed of sound data using known thermodynamic relationships and graphically smoothing the results. The expressions used for calculating the heat capacities are,

$$C_P = \frac{T^2(\partial\rho/\partial T)^2_\rho w^2}{\rho^2\left(\frac{\partial P}{\partial \rho}\right)_T\left[w^2 - \left(\frac{\partial P}{\partial \rho}\right)_T\right]}$$

and

$$C_V = \frac{T(\partial\rho/\partial T)^2_\rho}{\rho^2\left[w^2 - \left(\frac{\partial P}{\partial \rho}\right)_T\right]}.$$

The adiabatic indices were calculated from

$$\gamma = \frac{C_P}{C_V} = \frac{w^2}{(\partial P/\partial \rho)_T}$$

and

$$k_V = \frac{w^2}{PV}.$$

In the liquid phase the experimental isobaric heat capacities and densities were graphically smoothed and enthalpies and entropies derived from graphical integration of the heat capacities.

Auxiliary equations are given for the second virial coefficient and ideal gas heat capacity as a function of temperature, but equations are not presented for the vapour pressure or orthobaric density curves.

It can be seen that though these methods ensure a good representation of the available experimental data, they cannot guarantee thermodynamic consistency between the various tabulated properties as can the use of a single equation of state.

2.2.2 The Equation of Bender[46]

The first equation of state for propylene which can be used to calculate *PVT* and derived properties in both gas and liquid phases was published by Bender in 1975. The author has published coefficients for several fluids[46a, 46b] retaining the same form of equation in each case;

$$P = \rho T[R + B\rho + C\rho^2 + D\rho^3 + E\rho^4 + F\rho^5 + (G + H\rho^2)\rho^2 \exp(-n_{20}\rho^2)]$$

where

$$B = n_1 + n_2/T + n_3/T^2 + n_4/T^3 + n_5/T^4,$$
$$C = n_6 + n_7/T + n_8/T^2,$$
$$D = n_9 + n_{10}/T,$$
$$E = n_{11} + n_{12}/T,$$
$$F = n_{13}/T,$$
$$G = n_{14}/T^3 + n_{15}/T^4 + n_{16}/T^5,$$
$$H = n_{17}/T^3 + n_{18}/T^4 + n_{19}/T^5.$$

The saturation properties can be calculated directly from the equation of state by solving the equilibrium condition (Maxwell relation):

$$G(\rho_{\sigma_g}, T_\sigma) - G(\rho_{\sigma_l}, T_\sigma) = -\int_{\rho_{\sigma_g}}^{\rho_{\sigma_l}} \frac{P(\rho, T)}{\rho^2} \cdot d\rho + P_\sigma\left(\frac{1}{\rho_{\sigma_g}} - \frac{1}{\rho_{\sigma_l}}\right) = 0.$$

A consequence of the requirement that the equation of state satisfies the above condition is that properties such as heat capacities, enthalpy, entropy, fugacity, etc., can be calculated in both gas and liquid phases without making special provision for the discontinuity at the saturation curve of the real fluid.

Nine hundred and fifty-one experimental *PVT* results of Roper,[27] Manley and Swift,[26] Michels *et al.*,[22] Marchman *et al.*[23] and Farrington and Sage[24] were used as input data in fitting the equation as well as ninety-nine points on the saturation curve to enable the Maxwell "equal area" principle to be applied. The latter points were calculated from auxiliary equations for the vapour pressure and orthobaric densities as a function of temperature which the author also presents.

In addition the equation of state was constrained to pass through a selected critical point and to satisfy the conditions $(\partial P/\partial \rho)_{T_c} = 0$, $(\partial^2 P/\partial \rho^2)_{T_c} = 0$ at the critical density.

The equation accurately represents the input data and predicts the experimental heat capacities of Bier *et al*,[36] and speeds of sound of Soldatenko and Dregulyas[38] and is valid over the temperature range 200 K to 523 K at pressures up to 50 MPa. Extrapolation outside these limits is not recommended as deviations from existing accurate measurements increase rapidly to unacceptable values.

2.2.3 The Equation of Juza, Sifner and Hoffer[47]

These authors recently produced an equation state for propylene in the form

$$P = \rho R T\left[1 + \sum \sum a_{ij} \tau^{-i} \rho^j\right]$$

which covers the liquid and gaseous phases, there being thirty-three coefficients.

The experimental *PVT* results used were those of Michels *et al.*,[22] Farrington and Sage[24] and Voityuk,[31] and by a process of numerical and graphical analysis, the results of speed of sound by Soldatenko[38] and of C_P by Bier[36] were also incorporated in the fit. Saturation-curve data was used together with an auxiliary vapour-pressure equation to ensure a good representation of the two-phase region. The authors specifically exclude the near-critical region from the equation.

The equation, which gives an accurate representation of the data used, is valid over the temperature range 180 K to 460 K at pressures up to 200 MPa. The *PVT* equation is also given in integrated form as an $A\rho T$ equation, using the C_P^{id} values from Neduzhii *et al.*[1] It is of interest that a second equation is given of the form $\rho = f(P, T)$. This equation has nineteen coefficients, and is said to give almost identical values to the first equation over the restricted range of temperature from 220 K to 460 K and pressure from 0.01 MPa to 6 MPa.

The authors report, in a private communication, that their tables agree well with the IUPAC tables.

2.2.4 The IUPAC Equation

The correlation of the Kiev Institute of Technology while giving accurate tables does not include an equation of state which can be used over the range of the tables. The equations of Bender and of Juza *et al.* represent the experimental results in a single form, but over a somewhat limited range. The recent accurate *PVT* data of Golovskii and Tsimarnii[29, 30] in the liquid extends to near the melting curve and is represented by neither formulation. It was therefore felt that an equation taking in the widest range of accurate experimental data should be constructed.

Initially an attempt was made to fit an equation of the form first developed by Jacobsen[48] and successfully used for a number of fluids.[53, 54, 55] Difficulties were encountered in reproducing the Maxwell "equal area" principle at low temperatures and it was decided to construct a new equation using a technique due to Wagner[49] which, briefly, is a search procedure based upon least-squares methods which selects from an initial set of terms only those which are statistically important. The method was developed by Wagner for determining vapour-pressure equations and the adaptation of the method for work on equations of state is described in ref. 50.

The initial assembly of terms, from which the most significant were found, were members of the equation

$$P - \rho RT = \sum \sum n_a \rho^i T^j + E \sum \sum n_b \rho^k T^l$$

where

$$E = \exp{(\gamma \rho^2)}$$

and the relation of n_a to i and j, and of n_b to k and 1 is shown below, the assembly being substantially the same as that used by Jacobsen[48] in constructing an equation for nitrogen by a forward selection least-squares technique.

				j				
i	3	2	1	$\frac{1}{2}$	0	-1	-2	-3
2	n_1	n_2	n_3	n_4	n_5	n_6	n_7	
3			n_8		n_9	n_{10}	n_{11}	n_{12}
4			n_{13}		n_{14}	n_{15}	n_{16}	n_{17}
5		n_{18}	n_{19}		n_{20}	n_{21}	n_{22}	
6						n_{23}	n_{24}	
7						n_{25}	n_{26}	
8						n_{27}	n_{28}	
9						n_{29}	n_{30}	
11						n_{31}	n_{32}	

	l		
k	-2	-3	-4
3	n_{33}	n_{34}	n_{35}
5	n_{36}		n_{37}
7		n_{38}	n_{39}
9	n_{40}	n_{41}	n_{42}
11	n_{43}	n_{44}	n_{45}
13	n_{46}	n_{47}	n_{48}
15	n_{49}	n_{50}	

The equation of state resulting from this search procedure is

$$P - \rho RT = \rho^2(n_5 + n_6 T^{-1} + n_7 T^{-2}) + \rho^3(n_8 T + n_9 + n_{10}T^{-1})$$
$$+ \rho^4(n_{15}T^{-1} + n_{16}T^{-2})$$
$$+ \rho^5(n_{18}T^2 + n_{19}T + n_{20} + n_{22}T^{-2})$$
$$+ \rho^6(n_{24}T^{-2}) + \rho^7(n_{26}T^{-2})$$
$$+ \rho^8(n_{28}T^{-2})$$
$$+ \rho^3 E(n_{35}T^{-4}) + \rho^5 E(n_{37}T^{-4}) + \rho^7 E(n_{38}T^{-2})$$
$$+ \rho^9 E(n_{40}T^{-2}) + \rho^{11} E(n_{43}T^{-2}) + \rho^7 E(n_{38}T^{-2})$$

where
$$E = \exp(-\rho^2/\rho_c^2).$$

The critical point was taken to be the point where $(\partial P/\partial \rho)_T = 0$ and $(\partial^2 P/\partial \rho^2)_T = 0$, and a final least-squares fit was then made constraining the equation to the values of P_c, ρ_c and T_c thus found.

The tables printed here have been calculated from this equation. The coefficients were determined using the assembly of data shown in Table G.

Table G. SELECTED INPUT DATA FOR DETERMINING THE COEFFICIENTS OF THE EQUATION OF STATE

Type	Data	No. of points	Source
1	$P\rho T$	255	Michels et al.[22]
		8	Roper[27]
		14	Pfennig and McKetta[28]
		63	Vaughan and Graves[25]
		296	Golovskii and Tsimarnii[29, 30]
		96	Voityuk[31]
2	$T_\sigma, P_\sigma, V_{\sigma_1}, V_{\sigma_g}$	20	Tickner and Lossing[13]
		9	Michels et al.[22]
		12	Powell and Giauque[43]
		44	Voinov et al.[8]
3	$C_V \rho T$	30	see text
4	$(\partial P/\partial \rho)_T, \rho, T$	16	see text

The distribution of the data of type (1) is shown in Figure 14 and this data was discussed in detail in Section 1, but some consideration of types (2), (3) and (4) is relevant here.

The extended method of least squares fitting to include the Maxwell "equal area" principle was discussed by Bender.[46b] The Maxwell relation is:

$$P_\sigma(V_{\sigma_g} - V_{\sigma_1}) - \int_{V_{\sigma_g}}^{V_{\sigma_1}} P(V, T)\, dV = 0$$

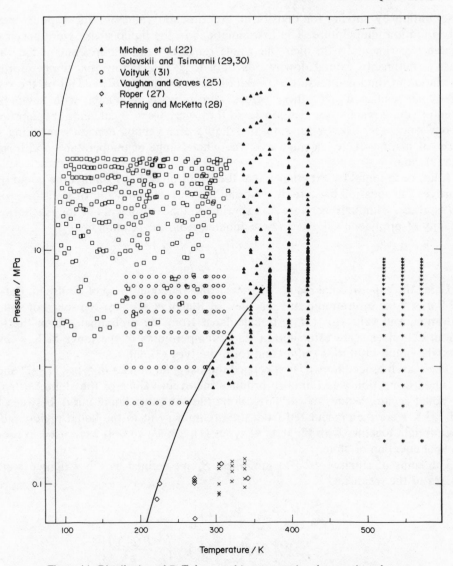

Figure 14. Distribution of PρT data used in constructing the equation of state

from which it can be seen that, to make use of it, the vapour pressure and both coexistence densities must be known or estimated at the same saturation temperature.

The technique used by Bender was unsatisfactory because experimental vapour and liquid saturation densities for propylene do not extend much below the temperature of the normal boiling point leaving a temperature range of 100 K above the triple point where there are no experimental results. Experimental measurements of the vapour pressure extend from 99 K (10 K above the triple point) to the critical point but below 165 K are of doubtful accuracy.

Therefore, for this work, it was decided to fit equations to the single-phase data, one in the vapour region and one in the liquid region, and extrapolate their isochores to saturation by solving for density at experimental values of the vapour pressure and saturation temperature. The extrapolations in the liquid phase were based upon reliable experimental data over the whole range of temperature, but in the vapour phase experiments extend downwards only to 223 K. Below this temperature, the calculation of the low-pressure $P\rho T$ data is almost entirely dependent on the second virial coefficient, and since these values were in good agreement with low-pressure speed of sound results (see Section 1.3.2) they were used to calculate the appropriate vapour isochores. Thus both saturated liquid and vapour densities resulting from these calculations were accepted over the whole range of temperature, and form the type (2) data.

It has been found by experience that the inclusion of $C_V\rho T$ data in the liquid region improves the equation of state by ensuring the correct behaviour of the slopes of the $P\rho T$ surface. Although there are no experimental measurements of the isochoric heat capacity of propylene values can be estimated from the relation

$$C_V = \frac{C_P}{w^2}\left(\frac{\partial P}{\partial \rho}\right)_T.$$

Using the experimental results of Powell and Giauque[43] and of Huffman, Barmore and Parks[18] for saturation heat capacities, of Blagoi *et al.*[20] for the speed of sound at saturation, and values of $(\partial P/\partial \rho)_T$ calculated from the preliminary fit in the liquid region, C_v values were estimated at thirty temperatures in the range 88 K $-$ 225.2 K along the liquid saturation curve, and form the type (3) data.

The search procedure was first carried out using input data of types (1), (2) and (3), but upon inspection was found to produce incorrect values of the slope $(\partial P/\partial \rho)_T$ in the liquid region. Values of $(\partial P/\partial \rho)_T$ along the saturated liquid curve between 88 K and 155 K were then calculated from the preliminary fit to the liquid region and this type (4) data together with the data of types (1), (2), (3) and (4) were used to produce the final equation of state.

The sums of squares used to minimize, S, are defined as the weighted sums of squares of the residues:

$$S = \sum_{n=1}^{N} W_n \left[P_{\text{data}} - P(\rho, T) \right]_n^2$$
$$+ \sum_{j=1}^{J} W_j \left[P_{\text{data}} \left(\frac{1}{\rho_{\sigma_g}} - \frac{1}{\rho_{\sigma_l}} \right) + \int_{\rho_{\sigma_g}}^{\rho_{\sigma_l}} \frac{P(\rho, T)}{\rho^2} \right]_j^2$$
$$+ \sum_{k=1}^{K} W_k \left[C_{V_{\text{data}}} - C_V(\rho, T) \right]_k^2$$
$$+ \sum_{l=1}^{L} W_l \left[\left(\frac{\partial P}{\partial \rho} \right)_{T_{\text{data}}} - \left(\frac{\partial P(\rho, T)}{\partial \rho} \right)_T \right]_l^2.$$

The weights W were calculated from the equation:

$$\frac{1}{W} = \epsilon_Y^2 + \left(\frac{\partial Y}{\partial \rho} \right)_T \epsilon_\rho^2 + \left(\frac{\partial Y}{\partial T} \right)_\rho \epsilon_T^2$$

where Y denotes the relevant property and the ϵ are estimates of the experimental uncertainties deduced from a critical appraisal of the experimental details. Theoretically ϵ should in every case be the standard deviation of the measured value, but due to the difficulties encountered in assessing them, there has been no hesitation in changing values of ϵ to improve the general representation of the $P\rho T$ surface, subject only to the requirement that the values of ϵ should be reasonable estimates of the standard deviation, having regard to the experimenters' account. Section 1 contains a detailed discussion of the accuracy of the equation of state and comparisons with all available experimental data.

The good agreement between experimental values of the heat capacity and of the speed of sound and their calculated values shows the equation to be well behaved in regions where no $P\rho T$ data were used as input, so the distribution of these measurements relative to the *PVT* data are shown in Figures 15 and 16.

It will have been noticed in reading Section 1 that the largest disagreements

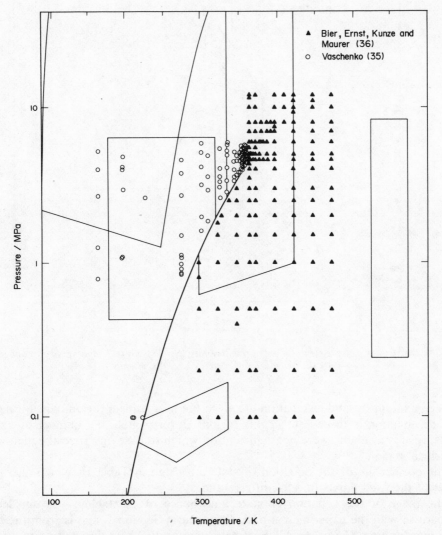

Figure 15. Distribution of experimental isobaric heat capacity results. (Symbols are individual heat capacity results: areas are the boundaries of main PρT results, see figs 7, 14).

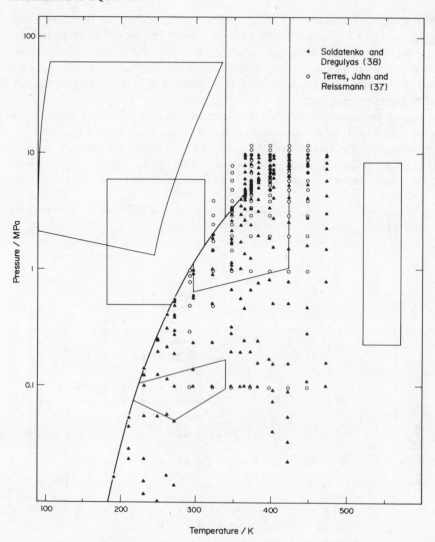

Figure 16. Distribution of experimental speed of sound results. (Symbols are individual speed of sound results: areas are the boundaries of main PρT results, see figs 7, 14).

between experiment and calculation are all in the near critical region, which suggests that the equation is inaccurate in this region. Unfortunately the number of experimental results available were not sufficient on which to base any special equation for the critical region.

The coefficients of the equation of state are listed in Table H, which also gives values of the fixed points for ease of reference.

The range of the equation of state, and hence of the tables, was decided by comparison with the experimental data discussed in Section 1, and is from the triple point temperature, 87.85 K to 475 K at pressures up to 1000 MPa, and up to 10 MPa between 475 K and 575 K.

2.3 AUXILIARY EQUATIONS

2.3.1 Vapour Pressure

To provide an auxiliary vapour-pressure equation, 277 values of the temperature and pressure at saturation were calculated at 1 K intervals from the main equation of state. An empirical equation relating the vapour pressure to the temperature was then found using Wagner's technique.[49] The pressure and temperature variables were chosen so that the equation must pass through the critical point, and the equation was constrained to pass through the triple point.

The final equation is of the form

$$\ln \left(\frac{P}{P_c}\right) = \left[b_1\theta + b_2\theta^{1.5} + b_3\theta^4 + b_4\theta^{4.5} \right] (1-\theta)^{-1} \tag{5}$$

where $\theta = (1 - (T/T_c))$ and P_c and T_c take the values selected in Section 1.2.2.

The numerical coefficients are given in Table K. When solved for temperature at any pressure the temperature given by this equation never differs by more than 0.06 K from that given by the equation of state and conversely, when solved at a given temperature, the pressure never differs by more than 0.0037 MPa.

The final equation closely resembles those found by Wagner[49, 49a] for nitrogen, oxygen and argon, and like them predicts that at the critical point d^2P/dT^2 varies as the square root of θ.

2.3.2 Saturated Densities

To provide auxiliary saturated density equations, the saturated liquid density and saturated vapour density were each calculated at 1 K intervals between the triple point and the critical point from the main equation of state. Each were then fitted to empirical equations relating the density to the temperature using Wagner's technique.[49] The density and temperature variables were chosen so that the equations must pass through the selected critical point, and the equations were constrained to pass through their respective triple point values.

The final equation selected for the saturated liquid density is of the form

$$\ln \left(\frac{\rho_l}{\rho_c}\right) = \sum_{i=1}^{6} c_i\theta^{(i+2)/6} + c_7\theta^{13/6} \tag{6}$$

where $\theta = 1 - (T/T_c)$.

The numerical coefficients are given in Table K. When solved for temperature at any density the temperature given by this equation never differs by more than 0.001 K from that given by the equation of state, and, conversely, when solved at a given temperature the density never differs by more than 6×10^{-6} mol cm^{-3}.

The final equation selected for the saturated vapour density is of the form

$$\ln \left(\frac{\rho_g}{\rho_c}\right) = \left[d_1\theta^{0.5} + d_2\theta + d_3\theta^{1.5} + d_4\theta^2 + d_5\theta^4 + d_6\theta^{5.5} + d_7\theta^9 \right] (1-\theta)^{-1} + d_8 \ln (1-\theta) \tag{7}$$

where $\theta = 1 - (T/T_c)$.

The numerical coefficients are given in Table K. When solved for temperature at any density the temperature given by this equation never differs by more than 0.003 K from that given by the equation of state and conversely when solved at a given temperature the density never differs by more than 1.5×10^{-6} mol cm^{-3}.

2.3.3 Melting Pressure

As was explained in Section 1.2.1, there is a need for two melting curve equations. Reeves *et al.*[2] had observed that the angle of intersection of these two curves was very small, and preliminary fits of the data showed that small changes in the fitting parameters could lead to widely different values for this intersection. It was therefore considered preferable to constrain both curves to pass through the measured triple point value, even though this has a large uncertainty attached to it. For the first melting curve, extending from the gas–liquid–solid triple point to the solid–solid–liquid triple point, that is, approximately from 87.9 K to 129 K, there are only eight data points available. It has been found in the past that in the absence of any theoretically based equation, that due to Watson:

$$\ln \left(\frac{P}{P_t} \right) = \sum_{i=1}^{I} a_i \left(\frac{T}{T_t} - 1 \right)^{i/10}$$

was capable of fitting helium, nitrogen and methane, as described in our previous books (refs. 53, 54, 55).

The triple point needed for this equation is not known, but cannot be far below the melting temperature of 87.90 K, at which the corresponding vapour pressure of the liquid calculated from the equation of state (10) is 0.958×10^{-9} MPa. Accordingly, the least-squares fit was made using several values of pressure and temperature calculated from the liquid–gas equilibrium given by the equation of state. The goodness of fit of the equation proved to be fairly insensitive to the choice of triple point, but the values of $T_{t_1} = 87.89$ K, $P_{t_1} = 0.95 \times 10^{-9}$ MPa were the best. The equation itself was found using Wagner's method[49] to select the most significant terms, and constraining it to pass through the selected solid–solid–liquid triple point. The final equation is:

$$\ln \left(\frac{P}{P_{t_1}} \right) = a_1 \left(\frac{T}{T_{t_1}} - 1 \right)^{1/10} + a_2 \left(\frac{T}{T_{t_1}} - 1 \right)^{2/10} + a_3 \left(\frac{T}{T_{t_1}} - 1 \right)^{5/10} \tag{8}$$

The numerical coefficients are listed in Table K.

From near the solid–solid–liquid triple point and up to the limits of experiment, that is approximately from 129 K to 144 K, there are seven results, two of which are in the super-cooled state. These data were used to fit, by Wagner's method, an equation of the form

$$\ln P = \sum_{i=1}^{I} a_i T^{i/10}$$

which was also constrained to pass through the solid–solid-liquid triple point. The final equation is

$$\ln \left(\frac{P}{P_{t_2}} \right) = 6.1664 \left[\left(\frac{T}{T_{t_2}} \right)^{1/2} - 1 \right]. \tag{9}$$

3. The IUPAC Tables

3.1 PHYSICAL CONSTANTS

In the sections which follow, the numerical values given below are to be used for these quantities:

Quantity	Value	Ref.
Molar mass of C_3H_6	0.042 080 4 kg mol^{-1}	57
Gas constant	8.314 34 J K^{-1} mol^{-1}	58
Critical temperature	365.57 K	
Critical pressure	4.6646 MPa	Section 1.2.2
Critical density	0.005 3086 mol cm^{-3}	
$S^{id}(298.15\ K, 1\ atm) - S^{ic}(0\ K)$	266.62 J K^{-1} mol^{-1}	
$H^{id}(298.15\ K) - H^{id}(0\ K)$	13.546 kJ mol^{-1}	Section 2.1.2
$U^{id}(298.15\ K) - U^{id}(0\ K)$	11.067 kJ mol^{-1}	
$H^{id}(298.15\ K) - H^{ic}(0\ K)$	40.660 kJ mol^{-1}	Section 2.1.3
$U^{id}(298.15\ K) - U^{ic}(0\ K)$	38.224 kJ mol^{-1}	

3.2 CONSTRUCTION OF THE TABLES

The main tables were produced from the equation of state whose genesis was described in Section 2. It may be written in two forms, the general shape being best shown by writing it as

$$Z = 1 + \omega(N_1\tau + N_2\tau^2 + N_3\tau^3) + \omega^2(N_4 + N_5\tau + N_6\tau^2)$$
$$+ \omega^3(N_7\tau^2 + N_8\tau^3) + \omega^4(N_9\tau^{-1} + N_{10} + N_{11}\tau + N_{12}\tau^3)$$
$$+ \omega^5 N_{13}\tau^3 + \omega^6 N_{14}\tau^3 + \omega^7 N_{15}\tau^3$$
$$+ \omega^2 e^{-\omega^2}[N_{16}\tau^5 + \omega^2 N_{17}\tau^5 + \omega^4 N_{18}\tau^3 + \omega^6 N_{19}\tau^3$$
$$+ \omega^8 N_{20}\tau^3 + \omega^{12}N_{21}\tau^4], \tag{10}$$

where $\omega = \rho/\rho_c$, $\tau = T_c/T$ and ρ_c and T_c take the values given in Section 3.1.

In computing the tables it was more useful to write the equation as

$$P = \rho RT\left[1 + \sum_{i=1}^{21} N_i(X)_i\right]. \tag{11}$$

The N_i are listed in Table H, and the $(X)_i$ are given in Table I.

In the calculation of other properties, the partial derivatives of the pressure were required. Writing them as

$$(\partial P/\partial \rho)_T = RT\left[1 + \sum_{i=1}^{21} N_i(X)_i\right]. \tag{12}$$

$$\text{and } (\partial P/\partial T)_\rho = R\rho\left[1 + \sum_{i=1}^{21} N_i(XT)_i\right] \tag{13}$$

they were evaluated from the N_i in Table H and the $(X\rho)_i$ and $(XT)_i$ in Table I.

**Table H. NUMERICAL VAL-
UES OF THE COEFFICIENTS
N_i OF EQUATION 10**

Coef.	Value
N_1	0.186 248 290 035
N_2	−1.292 611 016 62
N_3	−0.054 101 609 741 5
N_4	1.013 803 406 54
N_5	−2.121 229 224 61
N_6	1.526 272 166 48
N_7	−0.255 219 915 870
N_8	1.314 787 724 75
N_9	−0.045 653 388 880 9
N_{10}	0.092 659 828 640 8
N_{11}	0.102 014 965 320
N_{12}	−2.293 103 240 48
N_{13}	1.251 447 761 24
N_{14}	−0.281 035 528 699
N_{15}	0.022 765 984 902 0
N_{16}	−0.235 159 642 461
N_{17}	0.220 999 857 935
N_{18}	0.336 805 009 198
N_{19}	−0.021 024 854 178 5
N_{20}	0.029 849 352 904 4
N_{21}	0.000 285 153 473 859

Values of the molar volume and compression factor are tabulated as a function of pressure and temperature in Table 2 and the pressure as a function of density and temperature in Table 3.

3.2.1 Entropy

The entropy at any point $S(\rho, T)$ may be represented by

$$S(\rho, T) = S^{id}(T) - R \ln \frac{\rho RT}{P_a} + S_1 \tag{14}$$

The arbitrary reference pressure P_a is taken as 1 atm (0.101 325 MPa), and $S^{id}(T)$, defined by equation (2) in Section 2.1.2 becomes

$$\frac{S^{id}(T)}{R} = \left[f_1 \ln T + \sum_{i=2}^{5} f_i \frac{\tau^{(1-i)}}{(i-1)} - f_6 \frac{\tau^2}{2} + f_7 \left\{ \frac{ue^u}{e^u - 1} - \ln(e^u - 1) \right\} \right]_{298.15}^{T} \tag{2}$$

where $u = f_8 \tau$

and

$$S_1 = \int_0^\rho \left[\frac{R}{\rho} - \frac{1}{\rho^2} \left(\frac{\partial P}{\partial T} \right)_\rho \right] d\rho.$$

After integration, this may be written, in terms of the general equation of state, as

$$S_1 = -R \left[\sum_{i=1}^{21} N_i (XS)_i \right]_0^\omega \tag{15}$$

Table I. CONTRIBUTORY TERMS OF THE EQUATION OF STATE FOR THE CALCULATION OF PROPERTIES

i	$(X)_i$	$(X_\rho)_i$	$(XT)_i$	$(XS)_i$	$(XU)_i$	$(XC)_i$
1	$\omega\tau$	$2\omega\tau$	0	0	$\omega\tau$	0
2	$\omega\tau^2$	$2\omega\tau^2$	$-\omega\tau^2$	$-\omega\tau^2$	$\omega\tau^2$	$2\omega\tau^2$
3	$\omega\tau^3$	$2\omega\tau^3$	$-2\omega\tau^3$	$-2\omega\tau^3$	$\omega\tau^3$	$6\omega\tau^3$
4	ω^2	$3\omega^2$	ω^2	$\omega^2/2$	$\omega^2/2$	0
5	$\omega^2\tau$	$3\omega^2\tau$	0	0	$\omega^2\tau/2$	0
6	$\omega^2\tau^2$	$3\omega^2\tau^2$	$-\omega^2\tau^2$	$-\omega^2\tau^2/2$	$\omega^2\tau^2/2$	$\omega^2\tau^2$
7	$\omega^3\tau^2$	$4\omega^3\tau^2$	$-\omega^3\tau^2$	$-\omega^3\tau^2/3$	$\omega^3\tau^2/3$	$2\omega^3\tau^2/3$
8	$\omega^3\tau^3$	$4\omega^3\tau^3$	$-2\omega^3\tau^3$	$-2\omega^3\tau^3/3$	$\omega^3\tau^3/3$	$2\omega^3\tau^3$
9	$\omega^4\tau^{-1}$	$5\omega^4\tau^{-1}$	$2\omega^4\tau^{-1}$	$\omega^4\tau^{-1}/2$	$\omega^4\tau^{-1}/4$	$\omega^4\tau^{-1}/2$
10	ω^4	$5\omega^4$	ω^4	$\omega^4/4$	$\omega^4/4$	0
11	$\omega^4\tau$	$5\omega^4\tau$	0	0	$\omega^4\tau/4$	0
12	$\omega^4\tau^3$	$5\omega^4\tau^3$	$-2\omega^4\tau^3$	$-2\omega^4\tau^3/4$	$\omega^4\tau^3/4$	$3\omega^4\tau^3/2$
13	$\omega^5\tau^3$	$6\omega^5\tau^3$	$-2\omega^5\tau^3$	$-2\omega^5\tau^3/5$	$\omega^5\tau^3/5$	$6\omega^5\tau^3/5$
14	$\omega^6\tau^3$	$7\omega^6\tau^3$	$-2\omega^6\tau^3$	$-2\omega^6\tau^3/6$	$\omega^6\tau^3/6$	$\omega^6\tau^3$
15	$\omega^7\tau^3$	$8\omega^7\tau^3$	$-2\omega^7\tau^3$	$-2\omega^7\tau^3/7$	$\omega^7\tau^3/7$	$6\omega^7\tau^3/7$
16	$\omega^2\tau^5 E$	$\omega^2(3+a)\tau^5 E$	$-4\omega^2\tau^5 E$	$-4A_1\tau^5 E/2\alpha$	$A_1\tau^5 E/2\alpha$	$10A_1\tau^5 E/\alpha$
17	$\omega^4\tau^5 E$	$\omega^4(5+a)\tau^5 E$	$-4\omega^4\tau^5 E$	$-4A_2\tau^5 E/2\alpha$	$A_2\tau^5 E/2\alpha$	$10A_2\tau^5 E/\alpha$
18	$\omega^6\tau^3 E$	$\omega^6(7+a)\tau^3 E$	$-2\omega^6\tau^3 E$	$-2A_3\tau^3 E/2\alpha$	$A_3\tau^3 E/2\alpha$	$3A_3\tau^3 E/\alpha$
19	$\omega^8\tau^3 E$	$\omega^8(9+a)\tau^3 E$	$-2\omega^8\tau^3 E$	$-2A_4\tau^3 E/2\alpha$	$A_4\tau^3 E/2\alpha$	$3A_4\tau^3 E/\alpha$
20	$\omega^{10}\tau^3 E$	$\omega^{10}(11+a)\tau^3 E$	$-2\omega^{10}\tau^3 E$	$-2A_5\tau^3 E/2\alpha$	$A_5\tau^3 E/2\alpha$	$3A_5\tau^3 E/\alpha$
21	$\omega^{14}\tau^4 E$	$\omega^{14}(15+a)\tau^4 E$	$-3\omega^{14}\tau^4 E$	$-3A_7\tau^4 E/2\alpha$	$A_7\tau^4 E/2\alpha$	$6A_7\tau^4 E/\alpha$

j	A_j	
1	1	
2	$\omega^2 - A_1/\alpha$	$E = \exp(\alpha\omega^2)$
3	$\omega^4 - 2A_2/\alpha$	$a = 2\alpha\omega^2$
4	$\omega^6 - 3A_3/\alpha$	$\alpha = -1.0$
5	$\omega^8 - 4A_4/\alpha$	
6	$\omega^{10} - 5A_5/\alpha$	
7	$\omega^{12} - 6A_6/\alpha$	

Table J. NUMERICAL VALUES OF THE COEFFICIENTS f_i OF EQUATIONS (1), (2), (3), (17) and (20)

Coef.	Value
f_1	0.655 913 81
f_2	16.216 554
f_3	−4.898 063 3
f_4	0.821 854 68
f_5	−0.058 314 808
f_6	0.025 252 519
f_7	−4.703 242 0
f_8	1.684 494 4

where the N_i are listed in Table H and the functions $(XS)_i$ in Table I. The f_i in equation (2) are listed in Table J.

Since the equation of state satisfies the Maxwell relation, there is no term for the entropy change on evaporation, and it may be used for the direct calculation of changes in entropy which involve a path through the saturation region. If the entropy change on evaporation is needed, the saturated liquid and vapour densities are inserted in equation (14).

Values of entropy are tabulated as a function of pressure and temperature in Table 2, and as a function of density and temperature in Table 3. Ideal gas values are tabulated as a function of temperature in Table 1.

3.2.2 Enthalpy and Internal Energy

The internal energy at any point $U(\rho, T)$ may be represented by

$$U(\rho, T) = U^{id}(T) + U_1 \tag{16}$$

where $U^{id}(T)$ is defined by equation (4) in Section 2.1.3 as

$$\frac{U^{id}(T)}{R} = \left[T_c \left\{ \sum_{i=1}^{5} f_i \frac{\tau^{-i}}{i} - f_6\tau + \frac{f_7 f_8}{e^u - 1} \right\} \right]_{298.15}^{T} - T \tag{17}$$

where $u = f_8\tau$

and

$$U_1 = \int_0^\rho \left[\frac{P}{\rho^2} - \frac{T}{\rho^2} \left(\frac{\partial P}{\partial T} \right)_\rho \right] d\rho$$

This integral may be written as

$$U_1 = \int_0^\rho \left[\frac{P}{\rho^2} - \frac{RT}{\rho} \right] d\rho + T \int_0^\rho \left[\frac{R}{\rho} - \frac{1}{\rho^2} \left(\frac{\partial P}{\partial T} \right)_\rho \right] d\rho$$

which in terms of the general equation of state becomes

$$U_1 = RT \left[\sum_{i=1}^{21} N_i(XU)_i - \sum_{i=1}^{21} N_i(XS)_i \right]_0^\omega$$

where the N_i are listed in Table H and the functions $(XU)_i$ and $(XS)_i$ in Table I. The f_i in equation (17) are listed in Table J.

The enthalpy was obtained from the internal energy by the relation

$$H = U + P/\rho. \tag{18}$$

Since the equation of state satisfies the Maxwell relation, there are no terms for the internal energy or enthalpy change on evaporation, and the equation was integrated direct when calculating changes in these properties which involved a path through the saturation region. The change in enthalpy on evaporation was calculated by inserting the saturated liquid and vapour densities into equation (18).

Values of the enthalpy are tabulated as a function of pressure and temperature in Table 2. Values of the internal energy are tabulated as a function of density and temperature in Table 3. Ideal gas values of the enthalpy are tabulated as a function of temperature in Table 1.

3.2.3 Isochoric Heat Capacity

The heat capacity at constant volume at any point $C_v(\rho, T)$ may be represented by

$$C_V(\rho, T) = C_V^{id} + C_{V1} \tag{19}$$

where C_V^{id} is defined from equation (1) for C_P^{id} in Section 2.1.1 as

$$C_V^{id} = C_P^{id} - R$$

$$= R\left[\sum_{i=1}^{5} f_i\tau^{(1-i)} + f_6\tau^2 + f_7\frac{u^2e^u}{(e^u-1)^2} - 1\right] \tag{20}$$

where $u = f_8\tau$ and

$$C_{V1} = -\int_0^\rho \frac{T}{\rho^2}\left(\frac{\partial^2 P}{\partial T^2}\right)_\rho d\rho$$

In terms of the general equation of state,

$$C_{V1} = -R\left[\sum_{i=1}^{21} N_i(XC)_i\right]_0^\omega \tag{21}$$

where the N_i are listed in Table H and the functions $(XC)_i$ in Table I. The f_i in equation (20) are listed in Table J.

Since the equation of state satisfies the Maxwell relation, it was integrated direct when calculating properties which involved a path through the saturation region.

Values of the isochoric heat capacity are tabulated as a function of density and temperature in Table 3. Note that values have been omitted in that part of the liquid region, where the equation of state gives values which are believed to be increasingly in error.

3.2.4 Isobaric Heat Capacity

The heat capacity at constant pressure at any point $C_P(\rho, T)$ is related to the heat capacity at constant volume by the expression

$$C_P(\rho, T) = C_V(\rho, T) + C_{P1}$$

where

$$C_{P1} = \frac{T}{\rho^2}\frac{(\partial P/\partial T)_\rho^2}{(\partial P/\partial \rho)_T},$$

which in terms of the general equation of state, and using expressions from Section 3.2 and 3.2.3, becomes

$$C_P(\rho, T) = C_P^{id} - R - R\left[\sum_{i=1}^{21} N_i(XC)_i\right]_0^\omega + R\frac{\left[1 + \sum_{i=1}^{21} N_i(XT)_i\right]^2}{\left[1 + \sum_{i=1}^{21} N_i(X\rho)_i\right]} \tag{22}$$

where the N_i are listed in Table H and the functions $(XC)_i$, $(XT)_i$ and $(X\rho)_i$ in Table I.

Values of the isobaric heat capacity and the ratio C_P/C_V are tabulated as a function of pressure and temperature in Table 2. Note that values have been omitted in that part of the liquid region where the equation of state gives values which are believed to be increasingly in error. Ideal gas values are tabulated as a function of temperature in Table 1.

3.2.5 Fugacity

The fugacity, f, was not calculated direct, since it is more convenient to tabulate, and frequently to use, the function (f/P). At any point (ρ, T), this function may be calculated from

$$\ln\left(\frac{f}{P}\right) = -\ln Z + Z - 1 + \frac{1}{RT}\int_0^\rho \left[\frac{P}{\rho^2} - \frac{RT}{\rho}\right]d\rho.$$

In terms of the general equation of state, this becomes

$$\ln\left(\frac{f}{P}\right) = -\ln\left[1 + \sum_{i=1}^{21} N_i(X)_i\right] + \sum_{i=1}^{21} N_i(X)_i + \left[\sum_{i=1}^{21} N_i(XU)_i\right]_0^\infty \tag{23}$$

where the N_i are listed in Table H, and the functions $(X)_i$ and $(XU)_i$ are in Table I.

Values of the fugacity/pressure ratio are tabulated as a function of pressure and temperature in Table 2.

3.2.6 Speed of Sound

The speed of sound at any point $w(\rho, T)$ may be written as

$$[w(\rho, T)]^2 = \frac{C_P(\rho, T)}{C_V(\rho, T)}\left(\frac{\partial P}{\partial\rho}\right)_T \frac{1}{M}$$

and it was calculated from the components of this equation, where $C_P(\rho, T)$ is given by equation (22), $C_V(\rho, T)$ by equation (19) and $(\partial P/\partial\rho)_T$ by equation (12).

Values of the speed of sound at zero pressure are given in Table 1 and are tabulated as a function of pressure and temperature in Table 2. Note that values have been omitted in that part of the liquid region, where the equation of state gives values which are believed to be increasingly in error.

3.2.7 Joule–Thomson Coefficient

The Joule–Thomson coefficient at any point $\mu(\rho, T)$ is usually written as

$$\mu(\rho, T) = (\partial T/\partial P)_H$$

but is more conveniently calculated from

$$\mu(\rho, T) = \left[\frac{C_P(\rho, T) - C_V(\rho, T)}{(\partial P/\partial T)_\rho} - \frac{1}{\rho}\right]\bigg/ C_P(\rho, T)$$

where $C_P(\rho, T)$ is given by equation (22), $C_V(\rho, T)$ by equation (19) and $(\partial P/\partial T)_\rho$ by equation (13).

Values of the Joule–Thomson coefficient are tabulated as a function of pressure and temperature in Table 2.

3.2.8 Virial Coefficients

The second virial coefficient is represented by the coefficients of the first power of density in the equation of state, that is

$$B = \frac{1}{\rho_c}(N_1\tau + N_2\tau^2 + N_3\tau^3) \tag{24}$$

where ρ_c and T_c take the values given in Section 3.1 and the N_i are listed in Table H.

Values of the second virial coefficient are tabulated as a function of temperature in Table 1.

3.2.9 Saturation Properties

The saturation curve consists of the locus along which $G_l(P, T) = G_g(P, T)$. The Maxwell relation,

$$P(V_1 - V_g) = \int_{V_g}^{V_1} P \, dV$$

along an isotherm, is equivalent to equating the Gibbs free energies and this relation enables saturation curve values to be calculated; in terms of the general equation of state it may be written as:

$$P_\sigma = \frac{\rho_l\rho_g}{(\rho_g - \rho_l)} RT \left\{ \ln\frac{\rho_g}{\rho_l} + \left[\sum_{i=1}^{21} N_i(XU)_i\right]_{\rho_l}^{\rho_g} \right\} \tag{25}$$

where the N_i are listed in Table H and the function $(XU)_i$ in Table I. Similarly the equation of state gives $P = P_\sigma$ when $\rho = \rho_g$ and also when $\rho = \rho_l$. At any given T there are thus three equations in three unknowns: P_σ, ρ_g and ρ_l. If P_σ is eliminated, the resulting equations can be solved for ρ_g and ρ_l by an iterative technique, and hence, P_σ found.

This procedure was followed at each temperature where the saturation curve values were needed for Tables 2 and 3. Values of other properties at saturation were found by inserting ρ_g or ρ_l as required, into the appropriate equations.

The auxiliary equations given for vapour pressure in Section 2.3.1 and for saturated liquid and gaseous densities in Section 2.3.2, the coefficients of which are given in Table K, were not used in constructing the tables, but are recommended to readers interested only in properties at saturation. The differences between the vapour pressures calculated from the auxiliary equation and from the main equation of state are shown in Fig. 2. The corresponding differences for the auxiliary densities are too small to be shown on Figs. 3 and 4.

Using the values of density given in these tables, the rectilinear diameter can be represented with a maximum error of 0.5% from 90 K to 350 K by the equation

$$\frac{1}{2}\frac{(\rho_{l\sigma} + \rho_{g\sigma})}{\text{mol dm}^{-3}} = 10.3014 - 1.344\left(\frac{T}{100 \text{ K}}\right).$$

From 350 K to the critical point the rectilinear diameter is sharply curved. Since there are no experimental data for this region, this curvature may not be correct for the real fluid.

Table K. VALUES OF THE COEFFICIENTS OF AUXILIARY EQUATIONS (5), (6), (7) AND (8), REPRESENTING PROPERTIES ALONG THE TWO-PHASE BOUNDARIES

Boundary	Saturation curve			Melting curve
Equation	(5)	(6)	(7)	(8)
Dependent variable	Vapour pressure $\ln(P/P_c)$	Liquid density $\ln(\rho/\rho_c)$	Vapour density $\ln(\rho/\rho_c)$	Melting pressure $\ln(P/P_t)$
Independent variable	$1-(T/T_c)$	$1-(T/T_c)$	$1-(T/T_c)$	$(T/T_c)-1$
Coefficient	b_i	c_i	d_i	a_i
k_1	$b_1 = -6.553\ 517\ 503\ 34$	$c_1 = -6.744\ 667\ 589\ 58$	$d_1 = -4.492\ 388\ 452\ 10$	$a_1 = 62.634\ 036\ 459\ 1$
k_2	$b_2 = 0.957\ 645\ 928\ 738$	$c_2 = 104.170\ 660\ 384$	$d_2 = -217.978\ 872\ 730$	$a_2 = -42.174\ 533\ 442\ 8$
k_3	$b_3 = -4.747\ 026\ 376\ 50$	$c_3 = -361.099\ 213\ 639$	$d_3 = -22.758\ 163\ 639\ 3$	$a_3 = 8.080\ 210\ 187\ 58$
k_4	$b_4 = 1.931\ 420\ 850\ 86$	$c_4 = 570.399\ 232\ 365$	$d_4 = 143.832\ 778\ 695$	
k_5		$c_5 = -439.539\ 212\ 314$	$d_5 = 53.117\ 942\ 146\ 0$	
k_6		$c_6 = 135.508\ 176\ 771$	$d_6 = -1.098\ 555\ 881\ 05$	
k_7		$c_7 = -1.323\ 118\ 818\ 17$	$d_7 = 26.013\ 652\ 606\ 5$	
k_8			$d_8 = -223.581\ 566\ 935$	

3.2.10 Melting Curves

In Section 2.3.3 the pressure–temperature relations for the melting curves were discussed, and two equations derived to fit the data. These are

$$\ln(P/p_t) = a_1(T/T_{t1}-1)^{1/10} + a_2(T/T_{t1}-1)^{2/10} + a_3(T/T_{t1}-1)^{5/10}$$

and

$$\ln(P/P_{t2}) = 6.1664[(T/T_{t2})^{1/2} - 1],$$

where $P_{t1} = 0.95 \times 10^{-9}$ MPa, $T_{t1} = 87.89$ K,

$P_{t2} = 715.0$ MPa, $T_{t2} = 129.84$ K

and values of a_1, a_2 and a_3 are given in Table K.

Values of pressure and temperature when put into the equation of state (10) give values for other properties in the liquid phase at melting. These are tabulated as a function of temperature in Table 6 and as a function of pressure in Table 7.

3.3 LIMITS OF THE TABLES

The lowest temperature entries in the two-phase tables are at the triple point, 87.89 K, and in the main tables are 90 K. Despite the absence of data for the liquid in the low-pressure region, we have sufficient confidence in the equation of state to use the melting curve as the low-temperature limit for density values, but have omitted some derived properties where their behaviour suggests that the slopes $(\partial P/\partial \rho)_T$ and $(\partial P/\partial T)_\rho$ are not correct. The high-pressure limit has been taken as 1000 MPa, despite the gap in the data above 60 MPa and below 298 K, largely because the successful

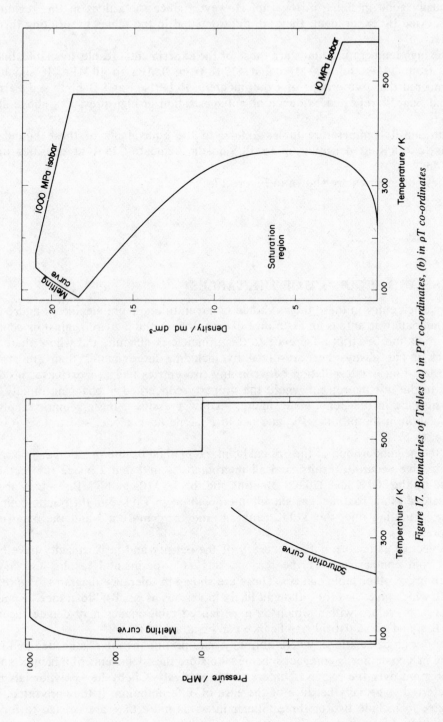

Figure 17. Boundaries of Tables (a) in PT co-ordinates, (b) in ρT co-ordinates

prediction of the results of Robertson and Babb[33] suggests that the equation should be equally good in filling in the gap. However, since the values in this region are unsupported by experiment, they are differentiated in the tables by printing them in brackets.

The high-temperature limits are those of the experimental results available, that is, 475 K from 0 MPa to 1000 MPa and 575 K from 0 MPa to 10 MPa. It should be remembered that two groups of experimenters, Robertson and Babb[33] and Farrington and Sage,[24] reported evidence of polymerization at high pressures above about 360 K.

The density–temperature tables extend to the equivalents of these boundaries that is to 475 K at densities up to 18.5 mol dm^{-3} and to 575 K at densities up to 2 mol dm^{-3}.

These boundaries are shown in Figure 17.

3.4 USE AND ACCURACY OF THE TABLES

Ideally, each entry in these tables should be given to one more significant figure than the input data warrant, as an assurance of smoothness and a guard against inaccuracy in interpolation. A strict adherence to this principle is difficult, and where there is a conflict, it has always been resolved by including more figures than are strictly necessary. Linear interpolation between any two entries will in most cases produce errors in the last figure; but where the property concerned is changing rapidly, the error may be in the penultimate figure. At low pressures, these remarks apply to interpolation in the product PV and not in P alone or V alone, and to $(S + R \ln P)$ and not to S.

In the neighbourhood of the critical point, no method of interpolation can be relied upon to give accurate results, and all interpolations in Table 2 inside the rectangle formed by the 360 K and 370 K isotherms and the 4.5 MPa and 5 MPa isobars should be regarded as approximate, as should interpolations in Table 3 in the region bounded by the saturation line, the 370 K isotherm and the 5 mol dm^{-3} and the 6 mol dm^{-3} isochores.

Subjective estimates of the accuracy of the density and heat capacity have been made from comparisons of the various sets of experimental results and of the predictions of other table makers. These are shown in tolerance diagrams, Figures 18 and 19, which give the errors thought likely in various areas. By "tolerance" is meant the range of values within which an experienced table maker may decide the best value lies, and no statistical significance can be attached to it.

For design calculations, where normally only part of the table is needed and is to be held in a computer, a compact scheme is to store the coefficients of a bicubic spline interpolation over the region of interest. Alternatively, Chebyshev polynomials may be preferred, especially because of the ease of differentiation. If they are fitted, it is advisable to include two or three tabular intervals more than are needed to form a border around the region of interest.

3.5 RECOMMENDATIONS FOR IMPROVEMENT OF TABLES

It is thought that the representation by these tables of the experimental data is capable of only marginal improvements which are not worth undertaking until further experimental measurements are carried out.

The most urgent need is for accurate measurements to be made of the vapour pressure, saturated liquid density and, if possible, saturated vapour density over the entire range from the triple point to the critical point. The critical point has not been experimentally determined with any reliability and the triple point temperature has not been measured.

The $P\rho T$ data of Michels *et al.*,[22] Golovskii and Tsimarnii,[29, 30] Voityuk[31] and Robertson and Babb,[33] the speed of sound data of Soldatenko and Dregulyas[38] and the heat capacity and Joule–Thomson coefficient data of Bier *et al.*[36] will endure for many years. In the liquid region $P\rho T$ measurements are required from 300 K to the critical point and at pressures from the saturation pressure to say 100 MPa, and also at low pressures and temperatures below 200 K. In the gas phase above 473 K the $P\rho T$ data of Vaughan and Graves[25] was used to construct the equation but is not of the same accuracy as the rest of the data. If more accurate tables are desired in this region new experimental measurements will be required.

A number of well-spaced reliable measurements of heat capacities and speed of sound in the liquid phase would be most useful in assessing the reliability of tables produced from the equation of state. Measurements of the isochoric heat capacity are especially useful in constructing equations of state.

In the critical region there exists accurate measurements of $C_P P T$ and $\omega P T$ but these properties are difficult to use in constructing equations of state and $P\rho T$ and $C_V \rho T$ measurements are needed if an accurate representation of the surface in this region is to be produced.

Acknowledgements

We should like to thank Professor Zwolinski, Dr. Lossing and Professor Babb for supplying us with unpublished data, and Professor Juza for helpful comments.

We also wish to record our appreciation of those organizations whose financial support has enabled the Project Centre to co-ordinate the work of this Project, including the preparation of this volume. These include the Academy of Sciences of the USSR, Air Liquide, BOC Ltd., Imperial Chemical Industries Ltd., Shell Research Ltd. and Sulzer Brothers Ltd. We are particularly grateful to the Science Research Council of the United Kingdom, who provided the major share of the finance, and to the Imperial College of Science and Technology of London for providing accommodation and services for the Project Centre without charge.

Appendix

EXPERIMENTAL RESULTS HITHERTO UNPUBLISHED OR INACCESSIBLE

1. Tickner and Lossing[13] published smoothed values for the vapour pressure of propylene at low pressures. The experimental results, kindly communicated by Dr. Lossing, on which the smoothed values are based, are

T/K	P/MPa	T/K	P/MPa
99.460	0.5733×10^{-7}	135.638	0.5999×10^{-4}
103.156	0.1293×10^{-6}	135.838	0.5986×10^{-4}
103.856	0.1733×10^{-6}	142.442	0.1466×10^{-3}
106.453	0.2840×10^{-6}	145.545	0.2213×10^{-3}
109.949	0.5893×10^{-6}	150.150	0.4053×10^{-3}
115.143	0.1973×10^{-5}	158.959	0.1037×10^{-2}
115.543	0.1906×10^{-5}	163.564	0.1627×10^{-2}
121.039	0.6253×10^{-5}	167.568	0.2426×10^{-2}
121.339	0.6053×10^{-5}	169.069	0.2746×10^{-2}
128.537	0.1973×10^{-4}	173.073	0.3573×10^{-2}

2. Reeves, Scott and Babb[2] published smoothed values of the melting pressures of propylene. The experimental results, kindly communicated by Professor Babb, on which the smoothed values are based, are

Solid phase	$t_{48}/°C$	P/bar
I	−174.9	1671
I	−170.6	2096
I	−163.9	3325
I	−158.3	4313
I	−155.0	4959
I	−150.3	5847
II	−163.6	4450
II	−159.5	4936
II	−143.6	7244
II	−142.3	7563 + 15, −0†
II	−136.9	8580 ± 15†
II	−132.7	9400
II	−130.1	10023

†The authors have indicated that these two measurements have larger errors than the rest.

3. Vaschenko,[35] referred to in Neduzhii *et al.*[1] published his results in full only in his thesis, and so they are listed below

Point no.	$\dfrac{T}{K}$	$\dfrac{P}{\text{bar}}$	$\dfrac{C_P}{\text{J K}^{-1}\text{g}^{-1}}$	Point no.	$\dfrac{T}{K}$	$\dfrac{P}{\text{bar}}$	$\dfrac{C_P}{\text{J K}^{-1}\text{g}^{-1}}$
1	164.59	7.82	2.068	57	352.36	49.48	4.112
2	164.70	12.5	2.063	58	352.12	38.05	4.926
3	164.68	26.7	2.060	59	352.12	38.05	4.947
4	164.63	39.5	2.037	60	352.10	49.95	4.050
5	164.66	51.95	2.060	61	352.30	38.15	5.052
6	197.72	47.86	2.075	62	351.12	37.99	5.153
7	197.78	47.86	2.070	63	355.14	55.90	4.025
8	196.83	39.47	2.075	64	355.15	42.27	5.020
9	197.92	29.32	2.080	65	355.39	50.99	4.328
10	196.74	10.79	2.080	66	355.31	45.70	4.720
11	196.13	10.69	2.085	67	355.33	50.31	4.282
12	197.32	40.99	2.129	68	355.28	47.17	4.492
13	197.11	10.80	2.108	69	355.08	43.74	4.725
14	197.02	10.86	2.143	70	355.03	39.62	5.599
15	274.84	14.87	2.379	71	358.65	57.02	4.321
16	274.85	11.13	2.391	72	358.35	42.89	5.950
17	274.85	14.91	2.389	73	358.74	57.12	4.292
18	274.86	10.77	2.381	74	358.66	45.71	5.372
19	274.84	9.02	2.401	75	358.34	42.10	6.002
20	274.85	45.53	2.379	76	358.58	44.52	5.703
21	274.91	8.39	2.405	77	358.79	47.83	5.115
22	274.90	9.76	2.505	78	358.74	50.99	4.805
23	274.87	37.45	2.348	79	358.85	55.11	4.367
24	274.86	37.27	2.356	80	360.23	44.33	6.597
25	274.85	26.97	2.315	81	360.25	46.29	5.597
26	301.38	15.95	2.675	82	360.49	49.03	5.292
27	301.72	19.63	2.632	83	360.30	51.68	4.902
28	301.63	31.65	2.601	84	360.18	45.51	6.327
29	325.66	30.40	3.075	85	360.32	48.44	5.258
30	325.92	35.30	3.050	86	360.21	54.23	4.602
31	325.81	35.55	3.025	87	360.27	50.64	5.002
32	325.89	40.26	3.045	88	360.41	56.68	4.505
33	325.88	33.05	3.085	89	301.75	41.02	2.573
34	325.44	54.18	2.875	90	301.55	50.97	2.553
35	326.00	40.21	3.015	91	301.59	58.55	2.532
36	325.66	30.40	3.080	92	361.54	46.92	6.402
37	325.86	35.55	3.025	93	361.49	48.69	5.446
38	325.88	33.05	3.052	94	361.64	48.68	5.582
39	325.92	35.30	3.049	95	347.36	36.58	4.412
40	326.75	26.09	3.240	96	347.69	48.93	3.820
41	335.07	40.80	3.220	97	347.59	45.31	3.672
42	335.06	33.64	3.602	98	354.11	40.01	5.052
43	335.16	28.34	3.662	99	354.15	56.39	3.955
44	335.18	49.25	3.182	100	354.29	52.96	4.052
45	335.17	57.62	3.092	101	354.29	52.91	3.753
46	335.88	59.45	3.087	102	344.00	44.82	3.620
47	334.91	42.17	3.733	103	223.00	1.00	2.145
48	344.94	51.97	3.448	104	223.06	1.00	2.148
49	344.80	51.48	3.446	105	211.50	1.00	2.112
50	344.75	35.70	3.152	106	209.72	1.00	2.102
51	344.20	33.73	4.202	107	327.15	26.09	2.735
52	344.47	33.73	4.200	108	274.85	8.55	2.409
53	344.40	39.52	3.852	109	310.45	35.66	2.707
54	352.23	48.05	4.102	110	310.41	48.67	2.667
55	352.23	45.01	4.252	111	310.41	18.49	2.757
56	352.33	50.94	3.994	112	310.43	29.15	2.723

References

General

1. Vashchenko, D. M., Voinov, Yu. F., Voityuk, B. V., Dregulyas, E. K., Kolomiets, A. Ya., Labinov, S. D., Morozov, A. A., Neduzhi, I. A. (Principal Author), Provotar, V. P., Soldatenko, Yu. A., Storozhenko, E. I., and Khmara, Yu. I., *Термодинамические и Транспортные Своиства Этилена и Пропилена. GS SSD Monograph No. 8, Moscow (1971)*. English translation: "Thermodynamic and Transport Properties of Ethylene and Propylene", NBSIR 75-763, U.S. Dept. of Commerce (1972).

Triple Point

See ref. 2.

Melting Curve

2. Reeves, L. E., Scott, G. J. and Babb, S. E., Melting curves of pressure transmitting fluids. *J. Chem. Phys.* **40**, 3662–6 (1964).
   See also refs. 4, 18, and 43.

Critical Point

3. Seibert, F. M. and Burrell, G. A., The critical constants of normal butane, iso-butane and propylene and their vapor pressures at temperatures between 0°C and 120°C. *J. Amer. Chem. Soc.* **37**, 2683–91 (1915).
3a. Burrell, G. A. and Robertson, I. W., The vapor pressure of ethane and ethylene at temperatures below their normal boiling points. *J. Amer. Chem. Soc.* **37**, 1892–1902 (1915).
4. Maass, O. and Wright, C. H., Some physical properties of hydrocarbons containing two and three carbon atoms. *J. Amer. Chem. Soc.* **43**, 1098–1111 (1921).
5. Winkler, C. A. and Maass, O., Density discontinuities at the critical temperature. *Can. J. Res.* **9**, 613–29 (1933).
6. Winkler, C. A. and Maass, O., The density of propylene in the liquid and vapor phases near the critical temperature. *Can. J. Res.* **9**, 610–12 (1933).
7. Lu. H., Newitt, D. M. and Ruhemann, M., Two-phase equilibrium in binary and ternary systems. IV. The system ethane—propylene. *Proc. Roy. Soc. (London)*, **178**, 506–25 (1941).
   See also refs. 23 and 25.

Saturation Curve

8. Voinov, Yu. F., Pavlovich, N. V. and Timrot, D. L., see ref. 1, pp. 43–50. (English translation, pp. 47–55.)
8a. Voinov, Yu. F., Pavlovich, N. V. and Timrot, D. L., Тензометриеский метод измерения плотностей равновесных фаз на линии насыщения. (A tensometric method of measuring the phase equilibrium density on the saturation curve.) *Khimicheskaya Promyshlennost' Ukrainy, No.* 3, pp. 51–54, 1967. (In Russian.)
9. Laurance, D. R. and Swift, G. W., Relative volatility of propane–propene system from 100–160 degrees F. *J. Chem. Engng. Data* **17** (3), 333–7 (1972).
10. Grauso, L., Fredenslund, A. and Mollerup, J., Vapour–liquid equilibrium data for the systems $C_2H_6 + N_2$, $C_2H_4 + N_2$, $C_3H_8 + N_2$, and $C_3H_6 + N_2$. *Fluid Phase Equilibria*, **1**, 13–26 (1977).
11. Burrell, G. A. and Robertson, J. W., The vapor pressures of propane, propylene and normal butane at low temperature. *J. Amer. Chem. Soc.* **37**, 2188–93 (1915).
12. Lamb, A. B. and Roper, E. E., The vapour pressures of certain unsaturated hydrocarbons. *J. Amer. Chem. Soc.* **62**, 806–14 (1940).
13. Tickner, A. W. and Lossing, F. P., The measurement of low vapour pressures by means of a mass spectrometer. *J. Phys. Colloid Chem.* **55**, 733–40 (1951).
14. Pall, D. B. and Maass, O., The liquid densities of propylene and methyl ether as determined by a modified dilatometer method. *Can. J. Res.* **14B**, 96–104 (1936).
15. Morecroft, D. W., The densities of liquid C_3 and C_4 alkenes at temperatures above their boiling points. *J. Inst. Petroleum*, **44**, 433–4 (1958).
16. Golovskii, E. A. and Tsimarnii, V. A., Плотность жидкого пропилена на кривыхǀсосуществования

жидкость–пар и жидкость–твердое тело. (The density of liquid propylene on the coexistence curve for liquid–vapour and liquid–solid.) *Izv. Vyssh. Uchebn. Zaved., Neft. Gaz*, **20**, 70–2 (1977). (Ih Russian.)

17. Clusius, K. and Konnertz, F., Ergebnisse der Tieftemperaturforschung: VI, Kalorimetrische Messungen der Verdampfungswärme des Sauerstoffs bei Normalen Druck sowie des Äthylens und Propylens unterhalb und oberhalb vom Atmosphärendruck. (Results of low temperature research: VI, Calorimetric measurements of heat of vaporisation of oxygen at normal pressure and of ethylene and propylene below and above atmospheric pressure.) *Z. Naturf.* **4a**, 117–24 (1949). (In German.)
18. Huffman, H. M., Parks, G. S. and Barmore, M., Thermal data on organic compounds. X. Further studies on the heat capacities, entropies and free energies of hydrocarbons. *J. Amer. Chem. Soc.* **53**, 3876–88 (1931).
18a. Eastman, E. D. and Rodebush, W. H., The specific heats at low temperatures of sodium, potassium, magnesium and calcium metals, and of lead sulfide. *J. Amer. Chem. Soc.* **40**, 489–500 (1918).
19. Auerbach, C. E., Sage, B. H. and Lacey, W. N., Isobaric heat capacities at bubble point. Propene, neohexane, cyclohexane, and iso-octane. *Ind. Engng Chem.* **42**, 110–13 (1950).
20. Blagoi, Yu. P., Butko, A. E., Mikhailenko, S. A., and Yakuba, V. V., Скорость звука в жидких пропилене, четырехфтористом углероде и аммиаке. *Zh. Fiz. Khim.* **42**, 1075–8 (1968). English translation: "Speed of sound in liquid propylene, carbon tetrachloride, and ammonia." *Russ. J. Phys. Chem.* **42**, 564–5 (1968).
21. Soldatenko, Yu. A. and Dregulyas, E. K., see ref. 1, p. 69. (English translation, p. 77.) See also refs. 3, 4, 6, 7, 22, 23, 24, 25, 26, 31, 35, and 43.

Density
22. Michels, A., Wassenaar, T., Louwerse, P., Lunbeck, R. J. and Wolkers, G. J., Isotherms and thermodynamical functions of propene at temperatures between 25° and 150°C and at densities up to 340 amagat (pressures up to 2800 atm). Com. Van der Waals–Fund. 132. *Physica* **19**, 287–97 (1953).
23. Marchman, H., Prengle, H. W. and Motard, R. L., Compressibility and critical constants of propylene vapour. *Ind. Engng. Chem.* **41**, 2658–60 (1949).
24. Farrington, P. S. and Sage, B. H., Volumetric behaviour of propene. *Ind. Engng Chem.* **41**, 1734–7 (1949).
25. Vaughan, W. E. and Graves, N. R., *P–V–T* relations of propylene. *Ind. Engng Chem.* **32**, 1252–6 (1940).
26. Manley, D. B. and Swift, G. W., Relative volatility of propane–propene system by integration of general coexistence equation. *J. Chem. Engng Data*, **16**, 301–7 (1971).
26a. Manley, D. B., Relative volatility of the propane–propene system. Ph.D. Thesis, Univ. of Kansas (1970).
27. Roper, E. E., Gas imperfection. I. Experimental determination of second virial coefficients for seven unsaturated aliphatic hydrocarbons. *J. Phys. Chem.* **44**, 835–47 (1940).
28. Pfenning, H. W. and McKetta, J. J., Compressibility factors at low pressures. *Petroleum Refiner*, **36**, 309–12 (1957).
29. Golovskii, E. A. and Tsimarnii, V. A., Экспериментальное исследование сжимаемости жидкого пропилена в интервале температур 90–200 K при давлениях до 600 ьар. (Compressibility of liquid propylene in the 90–200 K range at pressures to 600 bar.) *Izv. Vyssh. Uchebn. Zaved., Neft Gaz*, **18**, 60 and 64 (1975). (In Russian.)
30. Golovskii, E. A. and Tsimarnii, V. A., Экспериментальное исследование сжимаемости жидкого пропилена в ингервале температур 190–300 K при давлениях до 600 ьар. (Compressibility of liquid propylene in the 190–300 K range at up to 600 bar.) *Izv. Vyssh. Uchevn. Zaved., Neft Gaz*, **18**, 96 and 108 (1975). (In Russian.)
31. Voityuk, B. V., see ref. 1, p. 41. (English translation, pp. 44–45.)
31a. Boiko, N. V. and Voityuk, B. V., Экспериментальное исследование илотности жидких углеводородов методом гидростатического взвешивания на тензометрических весах. *Teplofiz. Charakteristiki Veshtchestv*, **1**, 33–37 (1968). English translation: Experimental investigation of the density of liquid hydrocarbons by hydrostatic weighing on a tensometric balance. *Thermoph. Prop. of Gases and Liquids*, **1**, 26–30 (1968).
32. Dittmar, P., Schultz, F. and Strese, G., Druck/Dichte/Temperatur-Werte für Propan und Propylen. (Pressure/density/temperature values for propane and propylene.) *Chemie-Ing. Techn.* **34**, 437–41 (1962). (In German.)
33. Robertson, S. L. and Babb, S. E., Jr., PvT properties of methane and propene to 10 kbar and 200°C. *J. Chem. Phys.* **53**, 1097–9, (1970).
33a. Robertson, S. L., Babb, S. E. and Scott, G. J., Isotherms of argon to 10 000 bars and 400°C. *J. Chem. Phys.* **50**, 2160–6 (1969).

Second Virial Coefficient
34. McGlashan, M. L. and Wormald, C. J., Second virial coefficients of some alk-l-enes, and of a mixture of propene + hept-l-ene. *Trans. Farad. Soc.* **60**, 646–52 (1964).
34a. McGlashan, M. L. and Potter, D. J. B., An apparatus for the measurement of the second virial coefficients of vapours; the second virial coefficients of some n-alkanes and of some mixtures of n-alkanes. *Proc. Roy. Soc.* A, **267**, 478–500 (1962).
35. Vashchenko, D. M., Исследование изоъарной теплоеткосьи жидких углеводоподов. (An investigation of the isobaric heat capacity of hydrocarbons.) Doctoral Thesis, Odessa (1971). (In Russian.)

Isobaric Heat Capacity
36. Bier, K., Ernst, G., Kunze, J. and Maurer, G., Thermodynamic properties of propylene from calorimetric measurements. *J. Chem. Thermodynamics*, **6**, 1039–52 (1974).
36a Bier, K., Ernst, G. and Maurer, G., Flow apparatus for measuring the heat capacity and the Joule–Thomson coefficient of gases. *J. Chem. Thermodynamics*, **6**, 1027–37 (1974).

Joule–Thomson Coefficients
See ref. 36.

Speed of Sound
37. Terres, Von E., Jahn, W. and Reissmann, H., Zur Kenntnis der Bestimmung von adiabatischen Exponenten leicht siedender Kohlenwasserstoffe bei verschiedenen Drucken und Temperaturen durch Messung der Ultraschallgeschwindigkeiten. (Concerning the determination of the adiabatic exponents of low boiling point hydrocarbons at various pressures and temperatures by measuring the ultra sonic velocity.) *Brennstoff–Chemie*, **38**, 129–60 (1957). (In German.)
38. Soldatenko, Yu. A. and Dregulyas, E. K., Экспериментальное исследование скорости распространения звука и расчет калорнческих свойств этилена и пропилена в сверхкритической области состояния, (Speed of sound and calculation of the caloric properties of ethylene and propylene in the supercritical region.) *Proc. 3rd All-Union Scientific-Technical Thermodynamics Conf., 1968*, pp. 344–52 (1970). (In Russian.)

Ideal Gas Properties
39. Chao, J. and Zwolinski, B. J., Ideal gas thermodynamic properties of ethylene and propylene. *J. Phys. Chem. Ref. Data* **4**, 251–61 (1975).
40. Kistiakowsky, G. B. and Rice, W. W., Gaseous heat capacities. II. *J. Chem. Phys.* **8**, 610–18 (1940).
41. Kistiakowsky, G. B., Lacher, J. R. and Ransom, W. M., The low temperature gaseous heat capacities of certain C_3 hydrocarbons. *J. Chem. Phys.* **8**, 970–7 (1940).
42. Telfair, D., Supersonic measurement of the heat capacity of propylene. *J. Chem. Phys.* **10**, 167–71 (1942).
43. Powell, T. M. and Giauque, W. F., Propylene. The heat capacity, vapor pressure, heats of fusion and vaporization. The Third Law of Thermodynamics and orientation equilibriums in the solid. *J. Amer. Chem. Soc.* **61**, 2366–70 (1939).
43a. Giauque, W. F., Buffington, R. M. and Schulze, W. A., Copper-constantan thermocouples and the hydrogen thermometer compared from 15 to 283° absolute. *J. Amer. Chem. Soc.* **49**, 2343–54 (1927).
43b. Stephenson, C. C. and Giauque, W. F., A test of the third law of thermodynamics by means of two crystalline forms of phosphine. The heat capacity, heat of vaporization and vapor pressure of phosphine. Entropy of the gas. *J. Chem. Phys.* **5**, 149–58 (1937).
43c. Giauque, W. F. and Wiebe, R., The entropy of hydrogen chloride. Heat capacity from 16°K to the boiling point. Heat of vaporization. Vapor pressures of solid and liquid. *J. Amer. Chem. Soc.* **50**, 101–22 (1928).
43d. Giauque, W. F. and Johnston, H. L., The heat capacity of oxygen from 12°K to its boiling point and its heat of vaporization. The entropy from spectroscopic data. *J. Amer. Chem. Soc.* **51**, 2300–21 (1929).
43e. Giauque, W. F. and Egan, C. J., Carbon dioxide. The heat capacity and vapor pressure of the solid. The heat of sublimation. Thermodynamic and spectroscopic values of the entropy. *J. Chem. Phys.* **5**, 45–54 (1937).
43f. Giauque, W. F. and Powell, T. M., Chlorine. The heat capacity, vapor pressure, heats of fusion and vaporization, and entropy. *J. Amer. Chem. Soc.* **61**, 1970–4 (1939).
44. Soldatenko, Yu. A., Morozov, A. A., Dregulyas, E. K. and Vashchenko, D. M., see ref. 1, p. 87. (English translation p. 98.)
45. Furukawa, G. T., Douglas, T. B. and Pearlman, N., *Amer. Inst. Phys. Handbook*, Zemansky, M. W. (ed.), Chap. 4e (1972).
See also ref. 36.

Equations of State

46. Bender, E., Equations of state for ethylene and propylene. *Cryogenics*, **15**, 667–73 (1975.)

46a. Bender, E., Equations of state representing the phase behaviour of pure substances. *Proc. 5th Symposium on Thermophysical Properties ASME, New York* (1970).

46b. Bender, E., *Die Berechnung von Phasengleichgewichten mit der thermischen Zustandsgleichung-dargestellt an den reinen Fluiden Argon, Stickstoff, Sauerstoff und an ihren Gennischen.* Verlag C. F. Müller, Karlsruhe (1973). English translation: "The calculation of Phase Equilibria from a Thermal Equation of state—Applied to the Pure Fluids Argon, Nitrogen, Oxygen and their Mixtures." Verlag C. F. Müller, Karlsruhe (1973).

47. Juza, J., Sifner, O. and Hoffer, V., Thermodynamic properties of propylene including liquid and vapour phases from 180 to 460 K with pressures to 200 MPa. *Acta Technica CSAV*, **23**, 425–45 (1978).

48. Jacobsen, R. B., The thermodynamic properties of nitrogen from 65 K to 2000 K with pressures to 10,000 Atmospheres. Ph.D. Thesis, Washington State Univ., Pullman, Washington (1972).

49. Wagner, W., Eine mathematisch-statistische Methode zum Aufstellen thermodynamischer Gleichungen-gezeigt am Beispiel der Dampfdruckkurve reiner fluider Stoffe. *Fortschr.-Ber. VDI-Z*, Reihe 3, Nr. 39 (1974). (In German.) Shortened English translation: "A new correlation method for thermodynamic data applied to the vapour pressure curve of argon, nitrogen and water." Report PC/T15, IUPAC TTPC, London (1977).

49a. Wagner, W., Neue Dampfdruckmessungen für die reinen Fluide Argon und Stickstoff Sowie Methoden zur Korrelation von Dampfdruckmesswerten. (New vapour pressure measurements for the pure fluids argon and nitrogen and methods for correlation of vapour pressure measurements.) Paper in the Thermodynamik-Kolloquium of the VDI in Bad Kissingen, 1972. Cf. also *Brennst–Wärme–Kraft*, **25** (1973), 16/18.

49b. Wagner, W., A method to establish equations of state representing all saturated state variables—applied to nitrogen. *Cryogenics*, **12**, 214–21 (1972).

49c. Wagner, W., New vapour pressure measurements for argon and nitrogen and a new method to establish rational vapour pressure equations. *Cryogenics*, **13**, 470–82 (1973).
For Corrigenda, see *Cryogenics*, **14**, 63 (1974).

49d. Pentermann, W. and Wagner, W., New pressure–density–temperature measurements and new rational equations for the saturated liquid and vapour densities of oxygen. *J. Chem. Thermodynamics*, **10**, 1161–72 (1978).
See also ref. 1.

Miscellaneous

50. de Reuck, K. M. and Armstrong, B., A method of correlation using a search procedure, based on a step-wise least-squares technique, and its application to an equation of state for propylene. *Cryogenics.* **19**, 505–12, (1979).

51. Angus, S., Armstrong, B., de Reuck, K. M., Featherstone, W. and Gibson, M. R., *International Thermodynamic Tables of the Fluid State, Ethylene.* Butterworths, London (1972).

52. Angus, S., Armstrong, B., de Reuck, K. M., Altunin, V. V., Gadetskii, O. C., Chapela, G. A. and Rowlinson, J. S., *International Thermodynamic Tables of the Fluid State, Carbon Dioxide.* Pergamon, Oxford (1976).

53. Angus, S., de Reuck, K. M. and McCarty, R. D., *International Thermodynamic Tables of the Fluid State, Helium.* Pergamon, Oxford, (1977).

54. Angus, S., Armstrong, B. and de Reuck, K. M., *International Thermodynamic Tables of the Fluid State, Methane.* Pergamon, Oxford (1978).

55. Angus, S., de Reuck, K. M., Armstrong, B., Stewart, R. B. and Jacobsen, R. T. *International Thermodynamic Tables of the Fluid State, Nitrogen.* Pergamon, Oxford (1979).

56. McGlashan, M. L., Paul, M. A., and Whiffen, D. H. (Eds.), *IUPAC Manual of Symbols and Terminology for Physico-Chemical Quantities and Units.* Pergamon, Oxford (1979).

57. Atomic Weights of the Elements 1975. Report from the Commission on Atomic Weights. *Pure and Appl. Chem.* **47**, 75–95 (1976).

58. Le Neindre, B. and Vodar, B., VI. Values of the fundamental constants. *Experimental Thermodynamics.* Vol. II, 1. General Introduction, pp. 25–28. Butterworths, London (1975).

Table 1

THE VARIATION OF
MOLAR ISOBARIC HEAT CAPACITY,
MOLAR ENTROPY AND
MOLAR ENTHALPY
IN THE IDEAL GAS STATE
AND OF
THE ISOTHERMAL ENTHALPY–PRESSURE COEFFICIENT,
SECOND VIRIAL COEFFICIENT
AND
SPEED OF SOUND
AT THE ZERO-PRESSURE LIMIT

Notes:
1. In this table the molar entropy is given a value of zero at 298.15 K and 1 atm (0.101 325 MPa) in the ideal gas state.

$$S^{id}(298.15 \text{ K}, 1 \text{ atm}) - S^{ic}(0 \text{ K}) = 266.62 \text{ J K}^{-1} \text{mol}^{-1}.$$

2. In this table the molar enthalpy is given a value of zero at 298.15 K in the ideal gas state.

$$H^{id}(298.15 \text{ K}) - H^{id}(0 \text{ K}) = 13\,546 \text{ J mol}^{-1},$$
$$H^{id}(298.15 \text{ K}) - H^{ic}(0 \text{ K}) = 40\,660 \text{ J mol}^{-1},$$

Table 1

$\dfrac{T}{K}$	$\dfrac{C_P^{id}}{J\,K^{-1}\,mol^{-1}}$	$\dfrac{S^{id}}{J\,K^{-1}\,mol^{-1}}$	$\dfrac{H^{id}}{J\,mol^{-1}}$	$\dfrac{(\partial H/\partial P)_T}{cm^3\,mol^{-1}}$	$\dfrac{B}{cm^3\,mol^{-1}}$	$\dfrac{w}{m\,s^{-1}}$
90	37.78	−57.02	−10425.1	−14499	−4557	151.0
95	38.45	−54.96	−10234.5	−12870	−4050	154.8
100	39.08	−52.97	−10040.7	−11497	−3623	158.4
105	39.69	−51.05	−9843.8	−10331	−3259	162.0
110	40.26	−49.19	−9643.9	−9331	−2946	165.5
115	40.82	−47.39	−9441.2	−8468	−2675	168.9
120	41.35	−45.64	−9235.8	−7718	−2440	172.3
125	41.87	−43.94	−9027.7	−7062	−2234	175.5
130	42.38	−42.29	−8817.0	−6486	−2052	178.8
135	42.89	−40.68	−8603.9	−5976	−1892	181.9
140	43.39	−39.11	−8388.2	−5523	−1749	185.0
145	43.89	−37.58	−8170.0	−5120	−1622	188.0
150	44.39	−36.08	−7949.3	−4758	−1507	191.0
155	44.91	−34.62	−7726.1	−4433	−1404	193.9
160	45.43	−33.18	−7500.2	−4139	−1312	196.7
165	45.96	−31.78	−7271.7	−3874	−1227	199.5
170	46.51	−30.40	−7040.6	−3632	−1151	202.2
175	47.07	−29.04	−6806.6	−3413	−1081	204.9
180	47.64	−27.71	−6569.9	−3212	−1017	207.6
185	48.22	−26.39	−6330.2	−3028	−960.1	210.2
190	48.82	−25.10	−6087.6	−2860	−906.5	212.7
195	49.44	−23.82	−5842.0	−2704	−857.2	215.2
200	50.07	−22.56	−5593.2	−2561	−811.6	217.7
205	50.71	−21.32	−5341.3	−2429	−769.6	220.1
210	51.36	−20.09	−5086.1	−2307	−730.6	222.5
215	52.03	−18.87	−4827.6	−2193	−694.4	224.9
220	52.71	−17.67	−4565.8	−2087	−660.8	227.2
225	53.40	−16.48	−4300.5	−1989	−629.5	229.5
230	54.10	−15.30	−4031.8	−1898	−600.3	231.7
235	54.81	−14.12	−3759.5	−1812	−573.0	234.0
240	55.53	−12.96	−3483.7	−1732	−547.5	236.2
245	56.26	−11.81	−3204.2	−1657	−523.6	238.3
250	57.00	−10.67	−2921.1	−1587	−501.2	240.5
255	57.74	−9.53	−2634.2	−1521	−480.2	242.6
260	58.50	−8.40	−2343.6	−1459	−460.4	244.7

Table 1—*continued*

$\dfrac{T}{\text{K}}$	$\dfrac{C_P^{\text{id}}}{\text{J K}^{-1}\text{mol}^{-1}}$	$\dfrac{S^{\text{id}}}{\text{J K}^{-1}\text{mol}^{-1}}$	$\dfrac{H^{\text{id}}}{\text{J mol}^{-1}}$	$\dfrac{(\partial H/\partial P)_T}{\text{cm}^3\text{mol}^{-1}}$	$\dfrac{B}{\text{cm}^3\text{mol}^{-1}}$	$\dfrac{w}{\text{m s}^{-1}}$
265	59.26	−7.28	−2049.2	−1400	−441.7	246.8
270	60.02	−6.17	−1751.0	−1345	−424.2	248.8
273.15	60.50	−5.47	−1561.2	−1312	−413.6	250.1
275	60.79	−5.06	−1449.0	−1293	−407.6	250.9
280	61.56	−3.95	−1143.1	−1244	−391.9	252.9
285	62.34	−2.86	−833.3	−1198	−377.1	254.9
290	63.12	−1.77	−519.7	−1154	−363.1	256.9
295	63.91	−.68	−202.1	−1112	−349.8	258.9
298.15	64.41	0	0	−1087	−341.8	260.1
300	64.70	.40	119.4	−1073	−337.3	260.8
305	65.49	1.48	444.9	−1036	−325.3	262.7
310	66.28	2.55	774.3	−1000	−314.0	264.6
315	67.07	3.61	1107.7	−966	−303.2	266.5
320	67.86	4.68	1445.0	−934	−292.9	268.4
325	68.66	5.73	1786.3	−903	−283.1	270.3
330	69.45	6.79	2131.6	−874	−273.8	272.2
335	70.25	7.84	2480.8	−846	−264.9	274.0
340	71.04	8.89	2834.0	−820	−256.4	275.8
345	71.83	9.93	3191.2	−794	−248.3	277.6
350	72.63	10.97	3552.4	−770	−240.6	279.5
355	73.42	12.00	3917.5	−747	−233.2	281.2
360	74.21	13.04	4286.5	−725	−226.1	283.0
365	74.99	14.06	4659.5	−703	−219.4	284.8
370	75.78	15.09	5036.4	−683	−212.9	286.6
380	77.34	17.13	5802.1	−645	−200.7	290.0
390	78.90	19.16	6583.3	−610	−189.5	293.5
400	80.44	21.18	7380.0	−577	−179.1	296.9
410	81.97	23.18	8192.0	−547	−169.5	300.3
420	83.49	25.18	9019.3	−519	−160.7	303.6
430	84.99	27.16	9861.8	−493	−152.4	306.9
440	86.48	29.13	10719.1	−469	−144.8	310.1
450	87.95	31.09	11591.3	−447	−137.7	313.4
475	91.56	35.94	13835.5	−397	−121.9	321.3
500	95.07	40.73	16168.6	−355	−108.5	329.0
525	98.46	45.45	18587.9	−319	−97.1	336.6

Table 1—*continued*

$\dfrac{T}{\text{K}}$	$\dfrac{C_P^{\text{id}}}{\text{J K}^{-1}\text{mol}^{-1}}$	$\dfrac{S^{\text{id}}}{\text{J K}^{-1}\text{mol}^{-1}}$	$\dfrac{H^{\text{id}}}{\text{J mol}^{-1}}$	$\dfrac{(\partial H/\partial P)_T}{\text{cm}^3\,\text{mol}^{-1}}$	$\dfrac{B}{\text{cm}^3\,\text{mol}^{-1}}$	$\dfrac{w}{\text{m s}^{-1}}$
550	101.74	50.11	21090.6	− 288	− 87.2	344.0
575	104.91	54.70	23674.0	− 261	− 78.7	351.3
600	107.97	59.23	26335.2			
650	113.76	68.10	31880.3			
700	119.15	76.73	37704.8			
750	124.15	85.13	43788.7			
800	128.79	93.29	50113.6			
850	133.11	101.23	56662.5			
900	137.14	108.95	63419.8			
950	140.89	116.47	70371.5			
1000	144.40	123.78	77504.8			
1050	147.69	130.91	84808.0			
1100	150.78	137.85	92270.6			
1150	153.67	144.62	99882.5			
1200	156.39	151.22	107634.8			
1250	158.94	157.65	115518.6			
1300	161.32	163.93	123525.7			
1350	163.53	170.06	131647.6			
1400	165.58	176.05	139876.0			
1450	167.44	181.89	148202.2			
1500	169.12	187.60	156616.9			

Table 2

THE VARIATION OF

MOLAR VOLUME,

COMPRESSION FACTOR,

MOLAR ENTROPY,

MOLAR ENTHALPY,

MOLAR ISOBARIC HEAT CAPACITY,

SPEED OF SOUND,

HEAT CAPACITY RATIO,

FUGACITY/PRESSURE RATIO

AND

ISENTHALPIC JOULE–THOMSON COEFFICIENT

WITH

TEMPERATURE AND PRESSURE

IN THE

SINGLE-PHASE REGION

Notes:

1. Expressions such as 1.0 E-9 are to be read as 1.0×10^{-9}.
2. Interpolation in the rectangle formed by the isobars 4.5 MPa and 5 MPa and the isotherms 365 K and 370 K can only be approximate.
3. Numbers in parentheses are interpolations into regions unsupported by experiment.
4. In the units used in this table,

$$\frac{P}{MPa} \times \frac{V}{cm^3 \, mol^{-1}} = \frac{PV}{J \, mol^{-1}}.$$

5. In this table the molar entropy is given a value of zero at 298.15 K and 1 atm (0.101 325 MPa) in the ideal gas state.

$$S^{id}(298.15 \text{ K, 1 atm}) - S^{ic}(0 \text{ K}) = 266.62 \text{ J K}^{-1} \text{ mol}^{-1}.$$

6. In this table the molar enthalpy is given a value of zero at 298.15 K in the ideal gas state.

$$H^{id}(298.15 \text{ K}) - H^{id}(0 \text{ K}) = 13\,546 \text{ J mol}^{-1}.$$

$$H^{id}(298.15 \text{ K}) - H^{ic}(0 \text{ K}) = 40\,660 \text{ J mol}^{-1}.$$

Table 2. PRESSURE–TEMPERATURE CO-ORDINATES

P/MPa									0.025
T_σ/K									198.807
Liq	V_σ 65.63	Z_σ .000993	S_σ −110.25	H_σ −25360	C_P 88.0	w_σ 1336	γ_σ 1.571	$(f/P)_\sigma$ .9876	μ_σ −.474
Vap	65284	.98738	−11.45	−5719	50.6	215.2	1.210	.9876	52.8
$\dfrac{T}{K}$	$\dfrac{V}{cm^3\,mol^{-1}}$	Z	$\dfrac{S}{J\,K^{-1}\,mol^{-1}}$	$\dfrac{H}{J\,mol^{-1}}$	$\dfrac{C_P}{J\,K^{-1}\,mol^{-1}}$	$\dfrac{w}{m\,sec^{-1}}$	γ	(f/P)	$\dfrac{\mu}{K\,MPa^{-1}}$
90	54.90	.001834	−174.45	−34248				.8200E−07	
95	55.32	.001751	−169.72	−33811				.4356E−06	
100	55.75	.001676	−165.70	−33419				.1925E−05	
105	56.18	.001609	−162.04	−33045				.7294E−05	
110	56.62	.001548	−158.60	−32674	74.2	1947	1.585	.2413E−04	−.631
115	57.07	.001492	−155.28	−32302	75.1	1904	1.583	.7045E−04	−.622
120	57.53	.001441	−152.06	−31923	76.3	1861	1.575	.1956E−03	−.610
125	57.98	.001395	−148.92	−31538	77.7	1819	1.566	.4840E−03	−.598
130	58.45	.001352	−145.85	−31146	78.9	1779	1.558	.1110E−02	−.587
135	58.92	.001312	−142.85	−30749	80.1	1741	1.552	.2378E−02	−.577
140	59.39	.001276	−139.92	−30346	81.1	1705	1.547	.4796E−02	−.568
145	59.87	.001242	−137.06	−29938	81.9	1670	1.544	.9164E−02	−.559
150	60.36	.001210	−134.27	−29527	82.7	1637	1.543	.1668E−01	−.552
155	60.86	.001181	−131.54	−29112	83.3	1605	1.543	.2906E−01	−.545
160	61.36	.001153	−128.89	−28694	83.9	1574	1.544	.4862E−01	−.538

0.025 MPa

165	61.87	.001128	−126.30	−28273	84.4	1543	1.546	.7839E−01	−.531
170	62.39	.001104	−123.77	−27850	85.0	1512	1.549	.1220	−.524
175	62.93	.001081	−121.30	−27424	85.5	1481	1.552	.1836	−.516
180	63.47	.001060	−118.89	−26995	86.0	1451	1.556	.2776	−.508
185	64.02	.001041	−116.53	−26564	86.5	1420	1.559	.4007	−.500
190	64.59	.001022	−114.21	−26131	87.0	1390	1.563	.5655	−.491
195	65.17	.001005	−111.95	−25694	87.5	1359	1.567	.7818	−.482
200	65692	.98762	−11.15	−5658	50.8	215.8	1.209	.9878	52.0
205	67398	.98857	−9.89	−5403	51.4	218.4	1.205	.9887	48.6
210	69101	.98941	−8.64	−5144	52.0	220.9	1.201	.9895	45.5
215	70801	.99018	−7.41	−4883	52.6	223.3	1.198	.9903	42.7
220	72499	.99088	−6.20	−4619	53.2	225.7	1.194	.9910	40.1
225	74194	.99151	−4.99	−4351	53.9	228.1	1.191	.9916	37.7
230	75886	.99208	−3.80	−4080	54.6	230.4	1.187	.9921	35.4
235	77577	.99261	−2.62	−3805	55.2	232.7	1.184	.9927	33.4
240	79266	.99309	−1.45	−3527	55.9	235.0	1.181	.9931	31.5
245	80953	.99353	−.29	−3246	56.6	237.2	1.178	.9936	29.7
250	82639	.99393	.86	−2961	57.3	239.4	1.175	.9940	28.1
255	84323	.99430	2.00	−2672	58.1	241.6	1.172	.9943	26.5
260	86006	.99465	3.14	−2380	58.8	243.8	1.169	.9947	25.1
265	87688	.99496	4.26	−2084	59.5	245.9	1.166	.9950	23.8
270	89369	.99525	5.38	−1785	60.3	248.0	1.164	.9953	22.5
275	91048	.99552	6.50	−1481	61.0	250.1	1.161	.9955	21.4
280	92727	.99577	7.60	−1174	61.8	252.1	1.159	.9958	20.3
285	94405	.99601	8.70	−863	62.6	254.2	1.156	.9960	19.3

Table 2—continued

P/MPa					0.025				
T_σ/K					198.807				
Liq	V_σ 65.63	Z_σ .000993	S_σ -110.25	H_σ -25360	C_P 88.0	w_σ 1336	γ_σ 1.571	$(f/P)_\sigma$ .9876	μ_σ -.474
Vap	65284	.98738	-11.45	-5719	50.6	215.2	1.210	.9876	52.8
$\dfrac{T}{K}$	$\dfrac{V}{cm^3\,mol^{-1}}$	Z	$\dfrac{S}{J\,K^{-1}\,mol^{-1}}$	$\dfrac{H}{J\,mol^{-1}}$	$\dfrac{C_P}{J\,K^{-1}\,mol^{-1}}$	$\dfrac{w}{m\,sec^{-1}}$	γ	(f/P)	$\dfrac{\mu}{K\,MPa^{-1}}$
290	96082	.99622	9.80	-549	63.3	256.2	1.154	.9962	18.4
295	97758	.99642	10.89	-230	64.1	258.2	1.152	.9964	17.5
300	99434	.99661	11.97	92	64.9	260.1	1.149	.9966	16.7
305	101109	.99678	13.05	419	65.7	262.1	1.147	.9968	15.9
310	102783	.99695	14.13	749	66.5	264.0	1.145	.9970	15.1
315	104457	.99710	15.20	1083	67.2	266.0	1.143	.9971	14.5
320	106130	.99724	16.26	1422	68.0	267.9	1.141	.9972	13.8
325	107803	.99737	17.32	1764	68.8	269.8	1.139	.9974	13.2
330	109475	.99750	18.38	2110	69.6	271.6	1.137	.9975	12.6
335	111147	.99762	19.43	2460	70.4	273.5	1.136	.9976	12.1
340	112818	.99773	20.48	2813	71.2	275.4	1.134	.9977	11.6
345	114489	.99783	21.52	3171	72.0	277.2	1.132	.9978	11.1
350	116160	.99793	22.57	3533	72.7	279.0	1.130	.9979	10.6
355	117830	.99802	23.60	3899	73.5	280.8	1.129	.9980	10.2
360	119500	.99811	24.64	4268	74.3	282.6	1.127	.9981	9.79
365	121170	.99819	25.67	4642	75.1	284.4	1.126	.9982	9.40
370	122839	.99827	26.69	5019	75.9	286.2	1.124	.9983	9.03
380	126177	.99841	28.74	5786	77.4	289.7	1.121	.9984	8.35
390	129514	.99854	30.77	6568	79.0	293.2	1.119	.9985	7.74
400	132850	.99865	32.79	7366	80.5	296.6	1.116	.9987	7.19

410	136186	.99876	34.80	8178	82.0	300.0	1.113	.9988	6.68
420	139520	.99885	36.79	9006	83.6	303.3	1.111	.9989	6.23
430	142854	.99893	38.77	9849	85.1	306.6	1.109	.9989	5.81
440	146188	.99901	40.75	10707	86.5	309.9	1.107	.9990	5.43
450	149520	.99908	42.71	11580	88.0	313.1	1.105	.9991	5.09
475	157851	.99923	47.56	13826	91.6	321.1	1.100	.9992	4.34
500	166178	.99935	52.35	16160	95.1	328.9	1.096	.9993	3.74
525	174504	.99944	57.07	18580	98.5	336.4	1.092	.9994	3.24
550	182828	.99952	61.73	21083	101.8	343.9	1.089	.9995	2.83
575	191151	.99959	66.33	23667	104.9	351.1	1.086	.9996	2.49

Table 2—continued

P/MPa					0.050				
T_σ/K					211.056				
Liq	V_σ 67.15	Z_σ .001913	S_σ −104.95	H_σ −24272	C_P 89.5	w_σ 1260	γ_σ 1.581	$(f/P)_\sigma$ .9794	μ_σ −.446
Vap	34356	.97891	−14.34	−5149	52.7	219.7	1.209	.9794	45.5
$\dfrac{T}{\mathrm{K}}$	$\dfrac{V}{\mathrm{cm^3\,mol^{-1}}}$	Z	$\dfrac{S}{\mathrm{J\,K^{-1}\,mol^{-1}}}$	$\dfrac{H}{\mathrm{J\,mol^{-1}}}$	$\dfrac{C_P}{\mathrm{J\,K^{-1}\,mol^{-1}}}$	$\dfrac{w}{\mathrm{m\,sec^{-1}}}$	γ	(f/P)	$\dfrac{\mu}{\mathrm{K\,MPa^{-1}}}$
90	54.90	.003669	−174.45	−34247				.4115E−07	
95	55.32	.003502	−169.73	−33810				.2182E−06	
100	55.75	.003352	−165.70	−33418				.9642E−06	
105	56.18	.003218	−162.05	−33044				.3656E−05	
110	56.62	.003096	−158.60	−32673	74.2	1947	1.585	.1214E−04	−.631
115	57.07	.002984	−155.29	−32300	75.1	1904	1.583	.3578E−04	−.622
120	57.52	.002883	−152.06	−31922	76.3	1861	1.575	.9794E−04	−.610
125	57.98	.002790	−148.92	−31537	77.7	1819	1.566	.2423E−03	−.598
130	58.45	.002704	−145.85	−31145	78.9	1779	1.558	.5556E−03	−.587
135	58.92	.002624	−142.85	−30748	80.1	1741	1.552	.1190E−02	−.577
140	59.39	.002551	−139.92	−30345	81.1	1705	1.547	.2401E−02	−.568
145	59.87	.002483	−137.06	−29937	81.9	1671	1.544	.4588E−02	−.559
150	60.36	.002420	−134.27	−29526	82.7	1637	1.543	.8353E−02	−.552
155	60.85	.002361	−131.55	−29111	83.3	1605	1.543	.1455E−01	−.545
160	61.36	.002306	−128.89	−28693	83.9	1574	1.544	.2437E−01	−.538

165	61.87	.002255	−126.30	−28272	84.4	1543	1.546	.3935E−01	−.531
170	62.39	.002207	−123.77	−27849	85.0	1512	1.549	.6143E−01	−.524
175	62.92	.002162	−121.30	−27423	85.5	1482	1.552	.9289E−01	−.516
180	63.47	.002120	−118.89	−26994	85.9	1451	1.556	.1389	−.508
185	64.02	.002081	−116.53	−26563	86.5	1421	1.559	.2006	−.500
190	64.59	.002044	−114.22	−26129	87.0	1390	1.563	.2831	−.491
195	65.17	.002010	−111.95	−25693	87.5	1359	1.567	.3913	−.482
200	65.77	.001978	−109.73	−25254	88.1	1329	1.571	.5306	−.472
205	66.38	.001947	−107.54	−24812	88.7	1297	1.576	.7070	−.460
210	67.02	.001919	−105.40	−24367	89.4	1266	1.580	.9267	−.448
215	35042	.98014	−13.36	−4940	53.2	221.7	1.205	.9806	43.2
220	35909	.98156	−12.13	−4672	53.8	224.2	1.201	.9819	40.5
225	36773	.98285	−10.91	−4402	54.4	226.7	1.197	.9832	38.1
230	37635	.98403	−9.71	−4128	55.0	229.1	1.193	.9843	35.8
235	38495	.98510	−8.52	−3852	55.7	231.5	1.190	.9853	33.7
240	39353	.98607	−7.34	−3572	56.3	233.8	1.186	.9863	31.7
245	40209	.98697	−6.17	−3288	57.0	236.1	1.183	.9871	29.9
250	41064	.98779	−5.01	−3001	57.7	238.4	1.179	.9879	28.3
255	41917	.98854	−3.86	−2711	58.4	240.6	1.176	.9887	26.7
260	42769	.98923	−2.72	−2417	59.1	242.8	1.173	.9893	25.3
265	43620	.98987	−1.59	−2120	59.8	245.0	1.170	.9900	23.9
270	44469	.99046	−.47	−1819	60.6	247.1	1.167	.9906	22.7
275	45318	.99101	.65	−1514	61.3	249.2	1.164	.9911	21.5
280	46165	.99151	1.76	−1206	62.1	251.3	1.161	.9916	20.4
285	47012	.99198	2.87	−894	62.8	253.4	1.159	.9920	19.4

Table 2—continued

P/MPa						0.050				
T_σ/K						211.056				
	V_σ 67.15 34356	Z_σ .001913 .97891	S_σ −104.95 −14.34	H_σ −24272 −5149	C_P 89.5 52.7	w_σ 1260 219.7	γ_σ 1.581 1.209	$(f/P)_\sigma$ .9794 .9794	μ_σ −.446 45.5	
Liq Vap										
$\dfrac{T}{\text{K}}$	$\dfrac{V}{\text{cm}^3\,\text{mol}^{-1}}$	Z	$\dfrac{S}{\text{J K}^{-1}\,\text{mol}^{-1}}$	$\dfrac{H}{\text{J mol}^{-1}}$	$\dfrac{C_P}{\text{J K}^{-1}\,\text{mol}^{-1}}$	$\dfrac{w}{\text{m sec}^{-1}}$	γ	(f/P)	$\dfrac{\mu}{\text{K MPa}^{-1}}$	
290	47857	.99241	3.97	−578	63.6	255.4	1.156	.9925	18.5	
295	48702	.99282	5.06	−258	64.3	257.5	1.154	.9929	17.6	
300	49547	.99319	6.15	65	65.1	259.5	1.152	.9932	16.7	
305	50390	.99355	7.23	393	65.9	261.5	1.149	.9936	15.9	
310	51233	.99387	8.31	724	66.6	263.4	1.147	.9939	15.2	
315	52075	.99418	9.38	1059	67.4	265.4	1.145	.9942	14.5	
320	52917	.99447	10.45	1398	68.2	267.3	1.143	.9945	13.9	
325	53759	.99474	11.51	1741	69.0	269.2	1.141	.9948	13.2	
330	54600	.99499	12.57	2088	69.7	271.1	1.139	.9950	12.7	
335	55440	.99522	13.62	2438	70.5	273.0	1.137	.9952	12.1	
340	56280	.99545	14.67	2793	71.3	274.9	1.135	.9955	11.6	
345	57120	.99565	15.72	3151	72.1	276.7	1.133	.9957	11.1	
350	57959	.99585	16.76	3514	72.9	278.6	1.131	.9959	10.7	
355	58798	.99604	17.80	3880	73.6	280.4	1.130	.9961	10.2	
360	59636	.99621	18.84	4250	74.4	282.2	1.128	.9962	9.81	
365	60475	.99637	19.87	4624	75.2	284.0	1.127	.9964	9.42	
370	61313	.99653	20.90	5002	76.0	285.8	1.125	.9965	9.05	
380	62988	.99682	22.95	5770	77.5	289.3	1.122	.9968	8.37	
390	64662	.99707	24.98	6553	79.1	292.8	1.119	.9971	7.75	
400	66335	.99730	27.00	7351	80.6	296.3	1.117	.9973	7.20	

410	68008	.99751	29.01	8165	82.1	299.7	1.114	.9975	6.69
420	69680	.99770	31.01	8993	83.6	303.0	1.112	.9977	6.24
430	71351	.99787	32.99	9837	85.1	306.4	1.109	.9979	5.82
440	73021	.99802	34.97	10696	86.6	309.6	1.107	.9980	5.44
450	74691	.99816	36.93	11569	88.1	312.9	1.105	.9982	5.09
475	78864	.99846	41.79	13816	91.7	320.9	1.101	.9985	4.35
500	83035	.99869	46.58	16151	95.1	328.7	1.096	.9987	3.74
525	87203	.99889	51.30	18572	98.5	336.3	1.093	.9989	3.24
550	91371	.99905	55.96	21076	101.8	343.7	1.089	.9990	2.83
575	95536	.99918	60.56	23661	105.0	351.0	1.086	.9992	2.49

Table 2—*continued*

P/MPa	0.075								
T_σ/K	219.059								
	V_σ	Z_σ	S_σ	H_σ	C_P	w_σ	γ_σ	$(f/P)_\sigma$	μ_σ
Liq	68.21	.002808	−101.60	−23550	90.7	1209	1.588	.9725	−.424
Vap	23596	.97165	−15.91	−4779	54.3	222.2	1.210	.9725	41.5
$\dfrac{T}{K}$	$\dfrac{V'}{\text{cm}^3\,\text{mol}^{-1}}$	Z	$\dfrac{S}{\text{J K}^{-1}\,\text{mol}^{-1}}$	$\dfrac{H}{\text{J mol}^{-1}}$	$\dfrac{C_P}{\text{J K}^{-1}\,\text{mol}^{-1}}$	$\dfrac{w}{\text{m sec}^{-1}}$	γ	(f/P)	$\dfrac{\mu}{\text{K MPa}^{-1}}$
90	54.90	.005503	−174.45	−34246				.2750E−07	
95	55.32	.005253	−169.73	−33809				.1457E−06	
100	55.75	.005029	−165.70	−33417				.6439E−06	
105	56.18	.004827	−162.05	−33042				.2442E−05	
110	56.62	.004643	−158.60	−32672	74.2	1947	1.585	.8118E−05	−.631
115	57.07	.004477	−155.29	−32299	75.1	1904	1.583	.2401E−04	−.622
120	57.52	.004324	−152.07	−31921	76.3	1861	1.575	.6539E−04	−.610
125	57.98	.004184	−148.92	−31536	77.7	1819	1.566	.1618E−03	−.598
130	58.45	.004055	−145.85	−31144	78.9	1779	1.558	.3709E−03	−.587
135	58.91	.003937	−142.85	−30747	80.1	1741	1.552	.7946E−03	−.577
140	59.39	.003827	−139.92	−30344	81.1	1705	1.547	.1603E−02	−.568
145	59.87	.003725	−137.06	−29936	81.9	1671	1.544	.3063E−02	−.559
150	60.36	.003630	−134.27	−29525	82.7	1638	1.543	.5576E−02	−.552
155	60.85	.003541	−131.55	−29110	83.3	1605	1.543	.9716E−02	−.545
160	61.36	.003459	−128.90	−28692	83.9	1574	1.544	.1627E−01	−.538

165	61.87	.003382	−126.31	−28271	84.4	1543	1.546	2629E−01	−.531
170	62.39	.003311	−123.78	−27847	85.0	1512	1.549	4108E−01	−.524
175	62.92	.003243	−121.31	−27421	85.4	1482	1.552	6222E−01	−.516
180	63.47	.003181	−118.89	−26993	85.9	1451	1.555	9272E−01	−.508
185	64.02	.003122	−116.53	−26562	86.5	1421	1.559	.1338	−.500
190	64.59	.003066	−114.22	−26128	87.0	1390	1.563	.1889	−.491
195	65.17	.003015	−111.95	−25692	87.5	1360	1.567	.2611	−.482
200	65.77	.002966	−109.73	−25253	88.1	1329	1.571	.3541	−.472
205	66.38	.002921	−107.55	−24811	88.7	1298	1.576	.4718	−.460
210	67.01	.002879	−105.40	−24366	89.4	1266	1.580	.6184	−.449
215	67.66	.002839	−103.29	−23917	90.1	1235	1.584	.7984	−.436
220	23707	.97205	−15.67	−4728	54.4	222.7	1.209	.9729	41.0
225	24295	.97403	−14.45	−4454	54.9	225.3	1.204	.9747	38.5
230	24881	.97582	−13.23	−4178	55.5	227.8	1.200	.9764	36.2
235	25464	.97746	−12.03	−3899	56.1	230.2	1.195	.9780	34.0
240	26046	.97895	−10.84	−3617	56.8	232.6	1.191	.9794	32.0
245	26625	.98031	−9.67	−3331	57.4	235.0	1.187	.9807	30.2
250	27203	.98156	−8.50	−3043	58.1	237.3	1.184	.9819	28.5
255	27780	.98270	−7.34	−2750	58.8	239.6	1.180	.9830	26.9
260	28355	.98375	−6.19	−2455	59.4	241.8	1.177	.9840	25.4
265	28929	.98472	−5.06	−2156	60.1	244.1	1.173	.9850	24.1
270	29501	.98562	−3.93	−1854	60.9	246.2	1.170	.9858	22.8
275	30073	.98644	−2.80	−1547	61.6	248.4	1.167	.9866	21.6
280	30643	.98721	−1.69	−1238	62.3	250.5	1.164	.9874	20.5
285	31213	.98792	−.58	−924	63.0	252.6	1.161	.9881	19.5

Table 2—continued

P/MPa				0.075					
T_σ/K				219.059					
Liq	V_σ 68.21	Z_σ .002808	S_σ -101.60	H_σ -23550	C_σ 90.7	w_σ 1209	γ_σ 1.588	$(f/P)_\sigma$ .9725	μ_σ -.424
Vap	23596	.97165	-15.91	-4779	54.3	222.2	1.210	.9725	41.5
$\dfrac{T}{K}$	$\dfrac{V}{cm^3\,mol^{-1}}$	Z	$\dfrac{S}{J\,K^{-1}\,mol^{-1}}$	$\dfrac{H}{J\,mol^{-1}}$	$\dfrac{C_P}{J\,K^{-1}\,mol^{-1}}$	$\dfrac{w}{m\,sec^{-1}}$	γ	(f/P)	$\dfrac{\mu}{K\,MPa^{-1}}$
290	31781	.98858	.53	-607	63.8	254.7	1.159	.9887	18.5
295	32349	.98919	1.62	-286	64.5	256.8	1.156	.9893	17.6
300	32917	.98976	2.71	38	65.3	258.8	1.154	.9899	16.8
305	33483	.99029	3.80	366	66.0	260.8	1.151	.9904	16.0
310	34049	.99078	4.88	699	66.8	262.8	1.149	.9909	15.3
315	34615	.99125	5.95	1035	67.6	264.8	1.147	.9913	14.6
320	35179	.99168	7.03	1374	68.3	266.7	1.144	.9917	13.9
325	35744	.99208	8.09	1718	69.1	268.7	1.142	.9921	13.3
330	36307	.99246	9.15	2066	69.9	270.6	1.140	.9925	12.7
335	36871	.99282	10.21	2417	70.7	272.5	1.138	.9929	12.2
340	37434	.99315	11.26	2772	71.4	274.4	1.136	.9932	11.6
345	37996	.99347	12.31	3131	72.2	276.3	1.134	.9935	11.1
350	38558	.99376	13.35	3494	73.0	278.1	1.133	.9938	10.7
355	39120	.99404	14.40	3861	73.8	280.0	1.131	.9941	10.2
360	39682	.99431	15.43	4232	74.5	281.8	1.129	.9943	9.83
365	40243	.99455	16.47	4606	75.3	283.6	1.128	.9946	9.44
370	40804	.99479	17.50	4985	76.1	285.4	1.126	.9948	9.07
380	41925	.99522	19.54	5753	77.6	289.0	1.123	.9952	8.39
390	43044	.99560	21.58	6537	79.2	292.5	1.120	.9956	7.77
400	44163	.99595	23.60	7337	80.7	295.9	1.117	.9960	7.21

410	45282	.99626	25.61	8151	82.2	299.4	1.115	.9963	6.70
420	46399	.99654	27.61	8980	83.7	302.8	1.112	.9966	6.24
430	47516	.99680	29.60	9825	85.2	306.1	1.110	.9968	5.83
440	48632	.99703	31.58	10684	86.7	309.4	1.108	.9970	5.45
450	49748	.99724	33.54	11558	88.1	312.7	1.106	.9972	5.10
475	52535	.99768	38.40	13806	91.7	320.7	1.101	.9977	4.35
500	55320	.99804	43.19	16142	95.2	328.5	1.097	.9980	3.74
525	58103	.99833	47.92	18564	98.6	336.1	1.093	.9983	3.25
550	60885	.99857	52.58	21069	101.8	343.6	1.090	.9986	2.84
575	63665	.99877	57.18	23654	105.0	350.9	1.087	.9988	2.49

Table 2—continued

P/MPa					0.100				
T_σ/K					225.171				
	V_σ	Z_σ	S_σ	H_σ	C_P	w_σ	γ_σ	$(f/P)_\sigma$	μ_σ
Liq	69.05	.003688	−99.09	−22992	91.7	1171	1.594	.9664	−.406
Vap	18068	.96511	−16.96	−4498	55.5	223.9	1.211	.9664	38.8
$\dfrac{T}{K}$	$\dfrac{V}{\text{cm}^3\,\text{mol}^{-1}}$	Z	$\dfrac{S}{\text{J K}^{-1}\,\text{mol}^{-1}}$	$\dfrac{H}{\text{J mol}^{-1}}$	$\dfrac{C_P}{\text{J K}^{-1}\,\text{mol}^{-1}}$	$\dfrac{w}{\text{m sec}^{-1}}$	γ	(f/P)	$\dfrac{\mu}{\text{K MPa}^{-1}}$
90	54.90	.007337	−174.46	−34245				.2067E−07	
95	55.32	.007003	−169.73	−33808				.1095E−06	
100	55.74	.006705	−165.70	−33416				.4838E−06	
105	56.18	.006435	−162.05	−33041				.1835E−05	
110	56.62	.006191	−158.60	−32671	74.2	1947	1.585	.6103E−05	−.631
115	57.07	.005969	−155.29	−32298	75.1	1905	1.583	.1808E−04	−.622
120	57.52	.005765	−152.07	−31920	76.3	1861	1.575	.4911E−04	−.610
125	57.98	.005579	−148.92	−31534	77.7	1819	1.566	.1215E−03	−.598
130	58.44	.005407	−145.85	−31143	78.9	1779	1.558	.2786E−03	−.587
135	58.91	.005249	−142.85	−30745	80.1	1741	1.552	.5968E−03	−.577
140	59.39	.005102	−139.92	−30343	81.1	1705	1.547	.1204E−02	−.568
145	59.87	.004966	−137.06	−29935	81.9	1671	1.544	.2300E−02	−.559
150	60.36	.004840	−134.27	−29524	82.7	1638	1.543	.4187E−02	−.552
155	60.85	.004722	−131.55	−29109	83.3	1606	1.543	.7297E−02	−.545
160	61.35	.004612	−128.90	−28691	83.9	1574	1.544	.1222E−01	−.538

165	61.87	.004510	−126.31	−28270	84.4	1543	1.546	.1975E−01	−.531
170	62.39	.004414	−123.78	−27846	84.9	1512	1.549	.3087E−01	−.524
175	62.92	.004324	−121.31	−27420	85.4	1482	1.552	.4680E−01	−.516
180	63.46	.004241	−118.90	−26992	85.9	1451	1.555	.6961E−01	−.509
185	64.02	.004162	−116.53	−26561	86.5	1421	1.559	.1005	−.500
190	64.59	.004088	−114.22	−26127	87.0	1390	1.563	.1418	−.491
195	65.17	.004020	−111.96	−25691	87.5	1360	1.567	.1960	−.482
200	65.77	.003955	−109.73	−25252	88.1	1329	1.571	.2658	−.472
205	66.38	.003894	−107.55	−24810	88.7	1298	1.576	.3542	−.461
210	67.01	.003838	−105.40	−24365	89.4	1267	1.580	.4643	−.449
215	67.66	.003785	−103.29	−23916	90.1	1235	1.584	.5994	−.436
220	68.33	.003736	−101.21	−23463	90.9	1204	1.589	.7631	−.422
225	69.03	.003690	−99.16	−23007	91.7	1172	1.593	.9587	−.406
230	18501	.96747	−15.78	−4229	56.1	226.4	1.206	.9686	36.5
235	18946	.96969	−14.57	−3947	56.6	228.9	1.201	.9706	34.3
240	19390	.97171	−13.37	−3663	57.2	231.4	1.197	.9725	32.3
245	19831	.97355	−12.18	−3375	57.8	233.8	1.192	.9743	30.4
250	20271	.97524	−11.01	−3084	58.5	236.2	1.188	.9759	28.7
255	20709	.97679	−9.84	−2790	59.1	238.5	1.184	.9773	27.1
260	21146	.97821	−8.69	−2493	59.8	240.8	1.181	.9787	25.6
265	21582	.97952	−7.54	−2192	60.5	243.1	1.177	.9799	24.2
270	22016	.98073	−6.41	−1888	61.2	245.3	1.174	.9811	23.0
275	22449	.98184	−5.28	−1581	61.9	247.5	1.170	.9822	21.8
280	22881	.98287	−4.16	−1270	62.6	249.7	1.167	.9832	20.6
285	23313	.98382	−3.04	−955	63.3	251.9	1.164	.9841	19.6

Table 2—continued

	V cm³ mol⁻¹	Z	S J K⁻¹ mol⁻¹	H J mol⁻¹	C_P J K⁻¹ mol⁻¹	w m sec⁻¹	γ	(f/P)	μ K MPa⁻¹
P/MPa					0.100				
T_σ/K					225.171				
Liq V_σ 69.05 / Vap 18068		Z_σ .003688 / .96511	S_σ −99.09 / −16.96	H_σ −22992 / −4498	C_P 91.7 / 55.5	w_σ 1171 / 223.9	γ_σ 1.594 / 1.211	$(f/P)_\sigma$ .9664 / .9664	μ_σ −.406 / 38.8
T/K									
290	23743	.98471	−1.94	−637	64.0	254.0	1.161	.9849	18.6
295	24172	.98553	−.84	−315	64.8	256.1	1.158	.9857	17.7
300	24601	.98630	.26	11	65.5	258.1	1.156	.9865	16.9
305	25029	.98701	1.35	340	66.2	260.2	1.153	.9872	16.1
310	25457	.98767	2.43	673	67.0	262.2	1.151	.9878	15.3
315	25884	.98829	3.51	1010	67.8	264.2	1.148	.9884	14.6
320	26310	.98887	4.58	1351	68.5	266.2	1.146	.9890	14.0
325	26736	.98942	5.65	1695	69.3	268.1	1.144	.9895	13.3
330	27161	.98993	6.71	2043	70.0	270.1	1.142	.9900	12.7
335	27586	.99040	7.77	2395	70.8	272.0	1.140	.9905	12.2
340	28010	.99085	8.83	2751	71.6	273.9	1.138	.9909	11.7
345	28434	.99127	9.88	3111	72.3	275.8	1.136	.9913	11.2
350	28858	.99167	10.92	3475	73.1	277.7	1.134	.9917	10.7
355	29281	.99204	11.97	3842	73.9	279.5	1.132	.9921	10.3
360	29704	.99240	13.01	4214	74.6	281.4	1.130	.9924	9.86
365	30127	.99273	14.04	4589	75.4	283.2	1.128	.9928	9.46
370	30549	.99304	15.07	4968	76.2	285.0	1.127	.9931	9.09
380	31393	.99361	17.12	5737	77.7	288.6	1.124	.9937	8.40
390	32236	.99413	19.16	6522	79.2	292.1	1.121	.9942	7.78
400	33078	.99459	21.19	7322	80.8	295.6	1.118	.9946	7.22

410	33919	.99501	23.20	8137	82.3	299.1	1.115	.9950	6.71
420	34759	.99538	25.20	8967	83.8	302.5	1.113	.9954	6.25
430	35599	.99572	27.19	9812	85.2	305.8	1.110	.9957	5.83
440	36438	.99603	29.17	10672	86.7	309.2	1.108	.9960	5.45
450	37277	.99631	31.13	11546	88.2	312.4	1.106	.9963	5.10
475	39371	.99691	35.99	13796	91.7	320.5	1.101	.9969	4.35
500	41463	.99739	40.79	16133	95.2	328.3	1.097	.9974	3.75
525	43553	.99778	45.52	18556	98.6	336.0	1.093	.9978	3.25
550	45642	.99809	50.18	21062	101.9	343.5	1.090	.9981	2.84
575	47729	.99835	54.78	23648	105.0	350.8	1.087	.9984	2.49

Table 2—continued

80

P/MPa	0.150
T_σ/K	234.470

	V_σ	Z_σ	S_σ	H_σ	C_P	w_σ	γ_σ	$(f/P)_\sigma$	μ_σ
Liq	70.40	.005417	−95.36	−22130	93.3	1111	1.603	.9556	−.374
Vap	12390	.95337	−18.36	−4077	57.6	225.9	1.215	.9556	35.3

$\dfrac{T}{K}$	$\dfrac{V}{cm^3\,mol^{-1}}$	Z	$\dfrac{S}{J\,K^{-1}\,mol^{-1}}$	$\dfrac{H}{J\,mol^{-1}}$	$\dfrac{C_P}{J\,K^{-1}\,mol^{-1}}$	$\dfrac{w}{m\,sec^{-1}}$	γ	(f/P)	$\dfrac{\mu}{K\,MPa^{-1}}$
90	54.90	.011005	−174.46	−34242				.1383E−07	
95	55.32	.010505	−169.73	−33806				.7324E−07	
100	55.74	.010057	−165.71	−33413				.3236E−06	
105	56.18	.009653	−162.05	−33039				.1228E−05	
110	56.62	.009286	−158.61	−32669	74.2	1948	1.585	.4084E−05	−.631
115	57.07	.008953	−155.29	−32296	75.1	1905	1.583	.1212E−04	−.622
120	57.52	.008648	−152.07	−31917	76.3	1861	1.575	.3283E−04	−.610
125	57.98	.008368	−148.93	−31532	77.7	1819	1.566	.8123E−04	−.598
130	58.44	.008110	−145.86	−31141	78.9	1779	1.558	.1862E−03	−.587
135	58.91	.007873	−142.86	−30743	80.1	1741	1.551	.3989E−03	−.577
140	59.39	.007653	−139.93	−30340	81.1	1706	1.547	.8045E−03	−.568
145	59.87	.007449	−137.07	−29933	81.9	1671	1.544	.1537E−02	−.559
150	60.35	.007259	−134.28	−29521	82.7	1638	1.543	.2798E−02	−.552
155	60.85	.007082	−131.56	−29106	83.3	1606	1.543	.4876E−02	−.545
160	61.35	.006918	−128.90	−28688	83.9	1574	1.544	.8169E−02	−.538

165	61.86	.006764	−126.31	−28268	84.4	1543	1.546	.1320E−01	−.531
170	62.39	.006621	−123.78	−27844	84.9	1513	1.549	.2065E−01	−.524
175	62.92	.006486	−121.32	−27418	85.4	1482	1.552	.3132E−01	−.516
180	63.46	.006360	−118.90	−26990	85.9	1452	1.555	.4651E−01	−.509
185	64.01	.006243	−116.54	−26559	86.4	1421	1.559	.6713E−01	−.500
190	64.58	.006132	−114.23	−26125	87.0	1391	1.563	.9474E−01	−.492
195	65.16	.006029	−111.96	−25689	87.5	1360	1.567	.1310	−.482
200	65.76	.005932	−109.74	−25250	88.1	1329	1.571	.1776	−.472
205	66.37	.005841	−107.56	−24808	88.7	1298	1.575	.2366	−.461
210	67.01	.005756	−105.41	−24363	89.4	1267	1.580	.3101	−.449
215	67.66	.005677	−103.30	−23914	90.1	1236	1.584	.4004	−.436
220	68.33	.005603	−101.22	−23462	90.8	1204	1.589	.5097	−.422
225	69.02	.005534	−99.17	−23005	91.7	1172	1.593	.6404	−.407
230	69.74	.005470	−97.14	−22545	92.5	1140	1.598	.7947	−.390
235	12423	.95372	−18.23	−4046	57.7	226.2	1.215	.9559	35.1
240	12729	.95687	−17.01	−3757	58.2	228.9	1.209	.9588	32.9
245	13033	.95973	−15.81	−3464	58.7	231.4	1.203	.9614	31.0
250	13335	.96235	−14.61	−3169	59.3	233.9	1.198	.9638	29.2
255	13636	.96474	−13.43	−2872	59.9	236.4	1.194	.9660	27.5
260	13935	.96693	−12.27	−2571	60.5	238.8	1.189	.9680	26.0
265	14232	.96894	−11.11	−2267	61.1	241.2	1.185	.9699	24.6
270	14529	.97079	−9.96	−1959	61.8	243.5	1.181	.9716	23.2
275	14824	.97250	−8.82	−1649	62.4	245.8	1.177	.9732	22.0
280	15118	.97407	−7.69	−1335	63.1	248.1	1.173	.9747	20.9
285	15411	.97553	−6.57	−1018	63.8	250.3	1.170	.9761	19.8

Table 2—continued

P/MPa									0.150
T_σ/K									234.470
Liq	V_σ 70.40	Z_σ .005417	S_σ −95.36	H_σ −22130	C_P 93.3	w_σ 1111	γ_σ 1.603	$(f/P)_\sigma$ .9556	μ_σ −.374
Vap	12390	.95337	−18.36	−4077	57.6	225.9	1.215	.9556	35.3
T / K	V cm³ mol⁻¹	Z	S J K⁻¹ mol⁻¹	H J mol⁻¹	C_P J K⁻¹ mol⁻¹	w m sec⁻¹	γ	(f/P)	μ K MPa⁻¹
290	15703	.97688	−5.45	−697	64.5	252.5	1.166	.9774	18.8
295	15994	.97814	−4.34	−373	65.2	254.6	1.163	.9786	17.9
300	16284	.97930	−3.24	−45	65.9	256.8	1.160	.9797	17.0
305	16574	.98039	−2.14	286	66.6	258.9	1.157	.9808	16.2
310	16863	.98140	−1.06	621	67.4	260.9	1.155	.9817	15.4
315	17152	.98234	.03	960	68.1	263.0	1.152	.9826	14.7
320	17440	.98322	1.11	1302	68.8	265.0	1.149	.9835	14.0
325	17727	.98405	2.18	1649	69.6	267.0	1.147	.9843	13.4
330	18014	.98482	3.25	1998	70.3	269.0	1.145	.9850	12.8
335	18300	.98554	4.31	2352	71.1	271.0	1.142	.9857	12.3
340	18586	.98622	5.37	2709	71.8	272.9	1.140	.9864	11.7
345	18872	.98686	6.43	3070	72.6	274.9	1.138	.9870	11.2
350	19157	.98746	7.48	3435	73.4	276.8	1.136	.9876	10.8
355	19442	.98802	8.52	3804	74.1	278.7	1.134	.9882	10.3
360	19726	.98855	9.56	4177	74.9	280.5	1.132	.9887	9.90
365	20010	.98905	10.60	4553	75.6	282.4	1.130	.9892	9.50
370	20294	.98953	11.64	4933	76.4	284.2	1.129	.9896	9.13
380	20861	.99040	13.69	5704	77.9	287.9	1.125	.9905	8.43
390	21426	.99117	15.74	6491	79.4	291.5	1.122	.9912	7.81
400	21991	.99187	17.77	7293	80.9	295.0	1.119	.9919	7.24

410	22555	.99250	19.78	8109	82.4	298.5	1.116	.9925	6.73
420	23119	.99306	21.79	8941	83.9	301.9	1.114	.9931	6.27
430	23681	.99358	23.78	9787	85.4	305.3	1.111	.9936	5.85
440	24243	.99404	25.76	10648	86.8	308.7	1.109	.9941	5.46
450	24805	.99446	27.72	11524	88.3	312.0	1.107	.9945	5.11
475	26207	.99536	32.59	13776	91.8	320.1	1.102	.9954	4.36
500	27606	.99608	37.39	16115	95.3	328.0	1.097	.9961	3.75
525	29003	.99666	42.12	18540	98.7	335.7	1.094	.9967	3.25
550	30399	.99714	46.79	21047	101.9	343.2	1.090	.9971	2.84
575	31793	.99753	51.39	23635	105.1	350.6	1.087	.9975	2.49

Table 2—continued

P/MPa						0.200				
T_σ/K						241.610				
Liq Vap	V_σ 71.51 9470	Z_σ .007119 .94282	S_σ −92.54 −19.29	H_σ −21457 −3759	C_P 94.7 59.4	w_σ 1065 227.1	γ_σ 1.610 1.220	$(f/P)_\sigma$ .9461 .9461	μ_σ −.347 32.9
$\dfrac{T}{\mathrm{K}}$	$\dfrac{V}{\mathrm{cm^3\,mol^{-1}}}$	Z	$\dfrac{S}{\mathrm{J\,K^{-1}\,mol^{-1}}}$	$\dfrac{H}{\mathrm{J\,mol^{-1}}}$	$\dfrac{C_P}{\mathrm{J\,K^{-1}\,mol^{-1}}}$	$\dfrac{w}{\mathrm{m\,sec^{-1}}}$	γ	(f/P)	$\dfrac{\mu}{\mathrm{K\,MPa^{-1}}}$
90	54.90	.014673	−174.46	−34240				.1041E−07	
95	55.32	.014006	−169.74	−33803				.5512E−07	
100	55.74	.013409	−165.71	−33411				.2435E−06	
105	56.18	.012870	−162.06	−33037				.9237E−06	
110	56.62	.012381	−158.61	−32666	74.2	1948	1.585	.3074E−05	−.631
115	57.07	.011937	−155.30	−32293	75.1	1905	1.582	.9125E−05	−.622
120	57.52	.011530	−152.08	−31915	76.3	1862	1.575	.2470E−04	−.610
125	57.98	.011157	−148.93	−31530	77.7	1820	1.566	.6109E−04	−.598
130	58.44	.010814	−145.86	−31138	78.9	1780	1.558	.1400E−03	−.587
135	58.91	.010497	−142.86	−30741	80.1	1742	1.551	.3000E−03	−.577
140	59.38	.010203	−139.93	−30338	81.0	1706	1.547	.6049E−03	−.568
145	59.86	.009931	−137.07	−29930	81.9	1671	1.544	.1156E−02	−.560
150	60.35	.009678	−134.28	−29519	82.6	1638	1.543	.2104E−02	−.552
155	60.85	.009443	−131.56	−29104	83.3	1606	1.543	.3666E−02	−.545
160	61.35	.009223	−128.91	−28686	83.9	1575	1.544	.6141E−02	−.538

165	61.86	.009019	−126.32	−28265	84.4	1544	1.546	.9926E−02	−.531
170	62.38	.008827	−123.79	−27842	84.9	1513	1.549	.1553E−01	−.524
175	62.91	.008648	−121.32	−27416	85.4	1483	1.552	.2357E−01	−.516
180	63.46	.008480	−118.91	−26987	85.9	1452	1.555	.3496E−01	−.509
185	64.01	.008323	−116.55	−26557	86.4	1422	1.559	.5045E−01	−.500
190	64.58	.008176	−114.23	−26123	87.0	1391	1.563	.7120E−01	−.492
195	65.16	.008038	−111.97	−25687	87.5	1360	1.567	.9841E−01	−.482
200	65.76	.007909	−109.74	−25248	88.1	1330	1.571	.1334	−.472
205	66.37	.007788	−107.56	−24806	88.7	1299	1.575	.1778	−.461
210	67.00	.007675	−105.42	−24361	89.4	1268	1.580	.2330	−.449
215	67.65	.007569	−103.30	−23912	90.1	1236	1.584	.3009	−.436
220	68.32	.007470	−101.22	−23460	90.8	1205	1.588	.3830	−.422
225	69.01	.007378	−99.17	−23003	91.6	1173	1.593	.4812	−.407
230	69.73	.007293	−97.15	−22543	92.5	1141	1.598	.5972	−.391
235	70.48	.007214	−95.15	−22078	93.4	1108	1.603	.7327	−.373
240	71.25	.007141	−93.18	−21609	94.4	1076	1.608	.8889	−.353
245	9630	.94546	−18.46	−3557	59.7	229.0	1.216	.9485	31.6
250	9864	.94906	−17.25	−3257	60.2	231.6	1.209	.9517	29.7
255	10096	.95235	−16.06	−2955	60.7	234.2	1.203	.9546	28.0
260	10326	.95535	−14.87	−2650	61.2	236.7	1.198	.9573	26.4
265	10555	.95811	−13.70	−2343	61.8	239.2	1.193	.9599	24.9
270	10783	.96064	−12.54	−2032	62.4	241.7	1.188	.9622	23.5
275	11009	.96296	−11.39	−1718	63.0	244.0	1.184	.9643	22.3
280	11234	.96511	−10.25	−1402	63.7	246.4	1.180	.9663	21.1
285	11458	.96709	−9.11	−1082	64.3	248.7	1.176	.9682	20.0

Table 2—continued

P/MPa									
T_σ/K									
	V_σ	Z_σ	S_σ	H_σ	C_P	w_σ	γ_σ	$(f/P)_\sigma$	μ_σ
Liq	71.51	.007119	-92.54	-21457	94.7	1065	1.610	.9461	-.347
Vap	9470	.94282	-19.29	-3759	59.4	227.1	1.220	.9461	32.9
$\dfrac{T}{\text{K}}$	$\dfrac{V}{\text{cm}^3\,\text{mol}^{-1}}$	Z	$\dfrac{S}{\text{J K}^{-1}\,\text{mol}^{-1}}$	$\dfrac{H}{\text{J mol}^{-1}}$	$\dfrac{C_P}{\text{J K}^{-1}\,\text{mol}^{-1}}$	$\dfrac{w}{\text{m sec}^{-1}}$	γ	(f/P)	$\dfrac{\mu}{\text{K MPa}^{-1}}$
290	11681	.96893	-7.99	-758	65.0	251.0	1.172	.9699	19.0
295	11903	.97063	-6.87	-432	65.7	253.2	1.168	.9715	18.0
300	12125	.97221	-5.76	-102	66.4	255.4	1.165	.9730	17.2
305	12346	.97368	-4.66	232	67.1	257.5	1.162	.9743	16.3
310	12566	.97504	-3.56	569	67.8	259.7	1.159	.9756	15.6
315	12785	.97632	-2.47	910	68.5	261.8	1.156	.9769	14.8
320	13004	.97751	-1.39	1254	69.2	263.9	1.153	.9780	14.1
325	13222	.97862	-.31	1602	69.9	265.9	1.150	.9791	13.5
330	13440	.97966	.76	1953	70.7	267.9	1.148	.9801	12.9
335	13657	.98063	1.83	2308	71.4	270.0	1.145	.9810	12.3
340	13874	.98155	2.89	2667	72.1	271.9	1.143	.9819	11.8
345	14090	.98241	3.95	3030	72.9	273.9	1.141	.9827	11.3
350	14306	.98321	5.00	3396	73.6	275.8	1.138	.9835	10.8
355	14521	.98397	6.05	3766	74.4	277.8	1.136	.9842	10.4
360	14737	.98469	7.10	4139	75.1	279.7	1.134	.9849	9.95
365	14952	.98536	8.14	4517	75.9	281.6	1.132	.9856	9.55
370	15166	.98600	9.18	4898	76.6	283.4	1.131	.9862	9.17
380	15594	.98716	11.24	5671	78.1	287.1	1.127	.9873	8.47
390	16022	.98820	13.29	6460	79.6	290.8	1.124	.9883	7.84
400	16448	.98914	15.32	7263	81.1	294.4	1.121	.9892	7.27

0.200

241.610

410	16874	.98998	17.34	8082	82.6	297.9	1.118	.9901	6.75
420	17298	.99074	19.35	8915	84.0	301.4	1.115	.9908	6.29
430	17723	.99142	21.34	9762	85.5	304.8	1.112	.9915	5.86
440	18146	.99204	23.33	10625	87.0	308.2	1.110	.9921	5.48
450	18569	.99261	25.30	11501	88.4	311.5	1.108	.9927	5.12
475	19624	.99381	30.17	13756	91.9	319.7	1.103	.9938	4.37
500	20677	.99477	34.98	16097	95.4	327.6	1.098	.9948	3.76
525	21728	.99555	39.71	18524	98.7	335.4	1.094	.9956	3.26
550	22777	.99619	44.38	21033	102.0	342.9	1.090	.9962	2.84
575	23825	.99671	48.98	23622	105.1	350.3	1.087	.9967	2.50

87

Table 2—*continued*

P/MPa					0.250				
T_σ/K					247.491				
Liq	V_σ 72.46	Z_σ .008804	S_σ -90.26	H_σ -20894	C_P 96.0	w_σ 1027	γ_σ 1.617	$(f/P)_\sigma$ .9376	μ_σ -.321
Vap	7680	.93308	-19.98	-3501	60.9	227.8	1.225	.9376	31.2
$\dfrac{T}{K}$	$\dfrac{V}{cm^3\,mol^{-1}}$	Z	$\dfrac{S}{J\,K^{-1}\,mol^{-1}}$	$\dfrac{H}{J\,mol^{-1}}$	$\dfrac{C_P}{J\,K^{-1}\,mol^{-1}}$	$\dfrac{w}{m\,sec^{-1}}$	γ	(f/P)	$\dfrac{\mu}{K\,MPa^{-1}}$
90	54.90	.018341	-174.47	-34238				.8362E-08	
95	55.31	.017507	-169.74	-33801				.4425E-07	
100	55.74	.016760	-165.72	-33409				.1955E-06	
105	56.18	.016087	-162.06	-33034				.7414E-06	
110	56.62	.015476	-158.62	-32664	74.2	1948	1.584	.2467E-05	-.632
115	57.06	.014920	-155.30	-32291	75.1	1905	1.582	.7327E-05	-.622
120	57.52	.014412	-152.08	-31913	76.3	1862	1.575	.1981E-04	-.610
125	57.98	.013946	-148.94	-31528	77.7	1820	1.566	.4901E-04	-.598
130	58.44	.013517	-145.87	-31136	78.9	1780	1.558	.1123E-03	-.587
135	58.91	.013120	-142.87	-30738	80.1	1742	1.551	.2406E-03	-.577
140	59.38	.012754	-139.94	-30336	81.0	1706	1.547	.4852E-03	-.568
145	59.86	.012413	-137.08	-29928	81.9	1672	1.544	.9269E-03	-.560
150	60.35	.012097	-134.29	-29517	82.6	1639	1.543	.1687E-02	-.552
155	60.84	.011803	-131.57	-29102	83.3	1606	1.543	.2940E-02	-.545
160	61.35	.011529	-128.91	-28684	83.9	1575	1.544	.4925E-02	-.538

165	61.86	.011273	−126.32	−28263	84.4	1544	1.546	.7960E−02	−.531
170	62.38	.011033	−123.80	−27840	84.9	1513	1.549	.1245E−01	−.524
175	62.91	.010809	−121.33	−27414	85.4	1483	1.552	.1890E−01	−.516
180	63.45	.010600	−118.91	−26985	85.9	1452	1.555	.2802E−01	−.509
185	64.01	.010403	−116.55	−26554	86.4	1422	1.559	.4045E−01	−.501
190	64.57	.010219	−114.24	−26121	87.0	1392	1.563	.5708E−01	−.492
195	65.16	.010047	−111.97	−25685	87.5	1361	1.567	.7889E−01	−.482
200	65.75	.009885	−109.75	−25246	88.1	1330	1.571	.1070	−.472
205	66.36	.009734	−107.57	−24804	88.7	1299	1.575	.1425	−.461
210	66.99	.009593	−105.42	−24359	89.4	1268	1.579	.1868	−.449
215	67.64	.009460	−103.31	−23910	90.1	1237	1.584	.2411	−.436
220	68.31	.009337	−101.23	−23458	90.8	1205	1.588	.3070	−.422
225	69.01	.009222	−99.18	−23002	91.6	1173	1.593	.3857	−.407
230	69.72	.009115	−97.16	−22541	92.5	1141	1.598	.4787	−.391
235	70.47	.009016	−95.16	−22077	93.4	1109	1.603	.5873	−.373
240	71.24	.008925	−93.18	−21607	94.4	1077	1.608	.7127	−.354
245	72.05	.008842	−91.23	−21133	95.4	1044	1.614	.8583	−.333
250	7777	.93537	−19.36	−3348	61.1	229.2	1.221	.9395	30.2
255	7968	.93961	−18.15	−3041	61.6	231.9	1.214	.9432	28.4
260	8158	.94348	−16.95	−2732	62.1	234.6	1.208	.9467	26.8
265	8346	.94701	−15.76	−2421	62.6	237.2	1.202	.9498	25.3
270	8533	.95026	−14.59	−2106	63.1	239.7	1.196	.9527	23.9
275	8718	.95323	−13.42	−1789	63.7	242.2	1.191	.9554	22.6
280	8902	.95597	−12.27	−1470	64.3	244.7	1.186	.9579	21.3
285	9085	.95850	−11.13	−1147	64.9	247.0	1.182	.9602	20.2

Table 2—*continued*

P/MPa					0.250				
T_σ/K					247.491				
Liq Vap	V_σ 72.46 7680	Z_σ .008804 .93308	S_σ -90.26 -19.98	H_σ -20894 -3501	C_P 96.0 60.9	w_σ 1027 227.8	γ_σ 1.617 1.225	$(f/P)_\sigma$ .9376 .9376	μ_σ -.321 31.2
$\dfrac{T}{K}$	$\dfrac{V}{cm^3\,mol^{-1}}$	Z	$\dfrac{S}{J\,K^{-1}\,mol^{-1}}$	$\dfrac{H}{J\,mol^{-1}}$	$\dfrac{C_P}{J\,K^{-1}\,mol^{-1}}$	$\dfrac{w}{m\,sec^{-1}}$	γ	(f/P)	$\dfrac{\mu}{K\,MPa^{-1}}$
290	9267	.96084	-9.99	-821	65.5	249.4	1.178	.9623	19.2
295	9448	.96300	-8.87	-491	66.2	251.7	1.174	.9643	18.2
300	9628	.96501	-7.75	-159	66.8	254.0	1.170	.9662	17.3
305	9807	.96687	-6.64	177	67.5	256.2	1.166	.9679	16.5
310	9986	.96861	-5.54	516	68.2	258.4	1.163	.9696	15.7
315	10164	.97022	-4.44	858	68.9	260.5	1.160	.9711	14.9
320	10341	.97173	-3.35	1205	69.6	262.7	1.157	.9725	14.2
325	10518	.97314	-2.27	1554	70.3	264.8	1.154	.9738	13.6
330	10695	.97445	-1.19	1907	71.0	266.9	1.151	.9751	13.0
335	10870	.97568	-.12	2264	71.7	268.9	1.148	.9762	12.4
340	11046	.97684	.95	2624	72.4	270.9	1.146	.9773	11.9
345	11220	.97792	2.01	2988	73.1	272.9	1.143	.9784	11.4
350	11395	.97894	3.07	3356	73.9	274.9	1.141	.9794	10.9
355	11569	.97989	4.12	3727	74.6	276.9	1.139	.9803	10.4
360	11743	.98079	5.17	4102	75.3	278.8	1.137	.9811	9.99
365	11916	.98164	6.22	4480	76.1	280.7	1.135	.9820	9.59
370	12089	.98244	7.26	4863	76.8	282.6	1.133	.9827	9.21
380	12434	.98391	9.32	5638	78.3	286.4	1.129	.9841	8.50
390	12779	.98522	11.38	6429	79.8	290.1	1.125	.9854	7.86
400	13122	.98639	13.42	7234	81.2	293.7	1.122	.9866	7.29

410	13464	.98745	15.44	8054	82.7	297.3	1.119	.9876	6.77
420	13806	.98840	17.45	8888	84.2	300.8	1.116	.9885	6.30
430	14147	.98926	19.45	9737	85.6	304.3	1.113	.9894	5.88
440	14488	.99004	21.43	10601	87.1	307.7	1.111	.9901	5.49
450	14827	.99075	23.41	11479	88.5	311.0	1.109	.9908	5.14
475	15675	.99226	28.29	13735	92.0	319.3	1.103	.9923	4.38
500	16520	.99346	33.10	16079	95.5	327.3	1.099	.9935	3.76
525	17363	.99444	37.83	18508	98.8	335.1	1.094	.9945	3.26
550	18204	.99523	42.51	21018	102.0	342.7	1.091	.9952	2.84
575	19044	.99589	47.11	23608	105.2	350.1	1.088	.9959	2.50

Table 2—continued

P/MPa					0.300				
T_σ/K					252.536				
Liq	V_σ 73.32	Z_σ .010476	S_σ -88.32	H_σ -20406	C_P 97.1	w_σ 994.7	γ_σ 1.623	$(f/P)_\sigma$ .9297	μ_σ -.298
Vap	6467	.92395	-20.52	-3284	62.4	228.1	1.230	.9297	29.8
$\dfrac{T}{K}$	$\dfrac{V}{\mathrm{cm^3\,mol^{-1}}}$	Z	$\dfrac{S}{\mathrm{J\,K^{-1}\,mol^{-1}}}$	$\dfrac{H}{\mathrm{J\,mol^{-1}}}$	$\dfrac{C_P}{\mathrm{J\,K^{-1}\,mol^{-1}}}$	$\dfrac{w}{\mathrm{m\,sec^{-1}}}$	γ	(f/P)	$\dfrac{\mu}{\mathrm{K\,MPa^{-1}}}$
90	54.90	.022009	-174.47	-34235				.6994E-08	
95	55.31	.021008	-169.75	-33798				.3701E-07	
100	55.74	.020112	-165.72	-33406				.1634E-06	
105	56.17	.019304	-162.07	-33032				.6198E-06	
110	56.62	.018571	-158.62	-32662	74.2	1949	1.584	.2063E-05	-.632
115	57.06	.017904	-155.31	-32289	75.1	1906	1.582	.6127E-05	-.622
120	57.52	.017294	-152.09	-31910	76.3	1862	1.575	.1656E-04	-.610
125	57.97	.016734	-148.94	-31525	77.7	1820	1.566	.4096E-04	-.598
130	58.44	.016219	-145.87	-31134	78.9	1780	1.558	.9386E-04	-.587
135	58.90	.015744	-142.87	-30736	80.1	1742	1.551	.2010E-03	-.577
140	59.38	.015304	-139.94	-30333	81.0	1706	1.547	.4053E-03	-.568
145	59.86	.014896	-137.08	-29926	81.9	1672	1.544	.7743E-03	-.560
150	60.35	.014516	-134.29	-29514	82.6	1639	1.543	.1409E-02	-.552
155	60.84	.014163	-131.57	-29100	83.3	1607	1.543	.2456E-02	-.545
160	61.34	.013834	-128.92	-28682	83.9	1575	1.544	.4114E-02	-.538

165	61.86	.013527	− 126.33	− 28261	84.4	1544	1.546	.6649E − 02	− .531
170	62.38	.013239	− 123.80	− 27837	84.9	1514	1.549	.1040E − 01	− .524
175	62.91	.012970	− 121.33	− 27412	85.4	1483	1.552	.1579E − 01	− .517
180	63.45	.012719	− 118.92	− 26983	85.9	1453	1.555	.2340E − 01	− .509
185	64.00	.012483	− 116.56	− 26552	86.4	1422	1.559	.3378E − 01	− .501
190	64.57	.012262	− 114.24	− 26119	87.0	1392	1.563	.4766E − 01	− .492
195	65.15	.012055	− 111.98	− 25683	87.5	1361	1.567	.6587E − 01	− .482
200	65.75	.011862	− 109.76	− 25244	88.1	1330	1.571	.8932E − 01	− .472
205	66.36	.011680	− 107.57	− 24802	88.7	1300	1.575	.1190	− .461
210	66.99	.011510	− 105.43	− 24357	89.4	1268	1.579	.1560	− .449
215	67.64	.011351	− 103.32	− 23908	90.1	1237	1.584	.2013	− .436
220	68.31	.011203	− 101.24	− 23456	90.8	1205	1.588	.2563	− .422
225	69.00	.011065	− 99.19	− 23000	91.6	1174	1.593	.3220	− .407
230	69.72	.010937	− 97.17	− 22540	92.5	1142	1.597	.3997	− .391
235	70.46	.010819	− 95.17	− 22075	93.4	1109	1.602	.4904	− .373
240	71.23	.010709	− 93.19	− 21606	94.4	1077	1.608	.5951	− .354
245	72.04	.010609	− 91.23	− 21131	95.4	1044	1.613	.7165	− .333
250	72.88	.010518	− 89.30	− 20651	96.5	1012	1.619	.8536	− .310
255	6548	.92648	− 19.91	− 3130	62.6	229.6	1.226	.9318	28.9
260	6711	.93127	− 18.69	− 2817	62.9	232.4	1.219	.9359	27.2
265	6872	.93564	− 17.49	− 2501	63.4	235.1	1.212	.9397	25.6
270	7031	.93963	− 16.30	− 2183	63.9	237.8	1.205	.9432	24.2
275	7189	.94329	− 15.13	− 1862	64.4	240.4	1.199	.9465	22.8
280	7346	.94666	− 13.96	− 1539	64.9	242.9	1.194	.9494	21.6
285	7502	.94976	− 12.81	− 1213	65.5	245.4	1.189	.9522	20.5

Table 2—continued

P/MPa									0.300	
T_σ/K									252.536	
Liq	V_σ 73.32	Z_σ .010476	S_σ -88.32	H_σ -20406	C_{P} 97.1	w_σ 994.7	γ_σ 1.623	$(f/P)_\sigma$ .9297	μ_σ -.298	
Vap	6467	.92395	-20.52	-3284	62.4	228.1	1.230	.9297	29.8	
$\dfrac{T}{K}$	$\dfrac{V}{cm^3\,mol^{-1}}$	Z	$\dfrac{S}{J\,K^{-1}\,mol^{-1}}$	$\dfrac{H}{J\,mol^{-1}}$	$\dfrac{C_P}{J\,K^{-1}\,mol^{-1}}$	$\dfrac{w}{m\,sec^{-1}}$	γ	(f/P)	$\dfrac{\mu}{K\,MPa^{-1}}$	
290	7656	.95262	-11.66	-884	66.1	247.8	1.184	.9548	19.4	
295	7810	.95526	-10.53	-552	66.7	250.2	1.179	.9572	18.4	
300	7963	.95771	-9.40	-217	67.3	252.5	1.175	.9594	17.5	
305	8115	.95998	-8.29	121	67.9	254.8	1.171	.9615	16.6	
310	8266	.96209	-7.18	462	68.6	257.1	1.167	.9635	15.8	
315	8416	.96405	-6.07	807	69.3	259.3	1.164	.9653	15.1	
320	8566	.96588	-4.98	1155	69.9	261.5	1.160	.9670	14.3	
325	8715	.96759	-3.89	1506	70.6	263.6	1.157	.9686	13.7	
330	8864	.96919	-2.80	1861	71.3	265.8	1.154	.9701	13.1	
335	9012	.97068	-1.73	2219	72.0	267.9	1.151	.9715	12.5	
340	9160	.97208	-.65	2581	72.7	269.9	1.149	.9728	11.9	
345	9307	.97339	.41	2946	73.4	272.0	1.146	.9741	11.4	
350	9454	.97463	1.47	3315	74.1	274.0	1.143	.9752	10.9	
355	9600	.97578	2.53	3688	74.9	276.0	1.141	.9763	10.5	
360	9746	.97687	3.58	4064	75.6	278.0	1.139	.9774	10.0	
365	9892	.97790	4.63	4444	76.3	279.9	1.137	.9783	9.63	
370	10038	.97887	5.67	4827	77.0	281.8	1.135	.9793	9.24	
380	10328	.98064	7.75	5605	78.5	285.7	1.131	.9810	8.53	
390	10616	.98222	9.81	6397	80.0	289.4	1.127	.9825	7.89	
400	10904	.98364	11.85	7204	81.4	293.1	1.123	.9839	7.31	

410	11192	.98491	13.88	8026	82.9	296.7	1.120	.9851	6.79
420	11478	.98606	15.89	8862	84.3	300.2	1.117	.9862	6.32
430	11763	.98710	17.89	9712	85.8	303.7	1.115	.9872	5.89
440	12048	.98804	19.88	10577	87.2	307.2	1.112	.9882	5.50
450	12333	.98889	21.86	11456	88.6	310.6	1.109	.9890	5.15
475	13042	.99070	26.74	13715	92.1	318.9	1.104	.9908	4.38
500	13748	.99215	31.55	16061	95.5	326.9	1.099	.9922	3.77
525	14453	.99332	36.30	18492	98.9	334.8	1.095	.9933	3.26
550	15156	.99428	40.97	21004	102.1	342.4	1.091	.9943	2.85
575	15857	.99507	45.58	23595	105.2	349.8	1.088	.9951	2.50

Table 2—continued

P/MPa	0.350								
T_σ/K	256.980								
Liq	V_σ 74.10	Z_σ .012139	S_σ −86.62	H_σ −19970	C_P 98.2	w_σ 965.8	γ_σ 1.628	$(f/P)_\sigma$ .9224	μ_σ −.275
Vap	5588	.91529	−20.96	−3096	63.7	228.3	1.236	.9224	28.7
$\dfrac{T}{\text{K}}$	$\dfrac{V}{\text{cm}^3\,\text{mol}^{-1}}$	Z	$\dfrac{S}{\text{J K}^{-1}\text{mol}^{-1}}$	$\dfrac{H}{\text{J mol}^{-1}}$	$\dfrac{C_P}{\text{J K}^{-1}\text{mol}^{-1}}$	$\dfrac{w}{\text{m sec}^{-1}}$	γ	(f/P)	$\dfrac{\mu}{\text{K MPa}^{-1}}$
90	54.90	.025676	−174.48	−34233				.6017E−08	
95	55.31	.024509	−169.75	−33796				.3183E−07	
100	55.74	.023463	−165.73	−33404				.1406E−06	
105	56.17	.022520	−162.07	−33029				.5330E−06	
110	56.61	.021666	−158.63	−32659	74.2	1949	1.584	.1774E−05	−.632
115	57.06	.020887	−155.31	−32286	75.1	1906	1.582	.5269E−05	−.622
120	57.51	.020176	−152.09	−31908	76.3	1863	1.575	.1424E−04	−.610
125	57.97	.019523	−148.95	−31523	77.7	1820	1.566	.3520E−04	−.598
130	58.43	.018922	−145.88	−31131	78.9	1780	1.558	.8067E−04	−.587
135	58.90	.018367	−142.88	−30734	80.1	1743	1.551	.1728E−03	−.577
140	59.38	.017854	−139.95	−30331	81.0	1707	1.547	.3483E−03	−.568
145	59.86	.017377	−137.09	−29924	81.9	1672	1.544	.6654E−03	−.560
150	60.34	.016935	−134.30	−29512	82.6	1639	1.543	.1211E−02	−.552
155	60.84	.016523	−131.58	−29097	83.3	1607	1.543	.2110E−02	−.545
160	61.34	.016139	−128.92	−28679	83.9	1576	1.544	.3534E−02	−.538

165	61.85	.015780	−126.33	−28259	84.4	1545	1.546	.5712E−02	−.531
170	62.37	.015445	−123.81	−27835	84.9	1514	1.549	.8936E−02	−.524
175	62.90	.015131	−121.34	−27409	85.4	1484	1.552	.1357E−01	−.517
180	63.45	.014838	−118.92	−26981	85.9	1453	1.555	.2010E−01	−.509
185	64.00	.014563	−116.56	−26550	86.4	1423	1.559	.2901E−01	−.501
190	64.57	.014305	−114.25	−26117	87.0	1392	1.563	.4094E−01	−.492
195	65.15	.014064	−111.98	−25681	87.5	1362	1.567	.5658E−01	−.482
200	65.74	.013838	−109.76	−25242	88.1	1331	1.571	.7671E−01	−.472
205	66.35	.013626	−107.58	−24800	88.7	1300	1.575	.1022	−.461
210	66.98	.013427	−105.43	−24355	89.4	1269	1.579	.1339	−.449
215	67.63	.013242	−103.32	−23906	90.1	1237	1.584	.1729	−.437
220	68.30	.013069	−101.25	−23454	90.8	1206	1.588	.2201	−.423
225	68.99	.012908	−99.20	−22998	91.6	1174	1.593	.2765	−.408
230	69.71	.012759	−97.17	−22538	92.5	1142	1.597	.3432	−.391
235	70.45	.012620	−95.17	−22073	93.4	1110	1.602	.4211	−.374
240	71.22	.012493	−93.20	−21604	94.4	1078	1.607	.5111	−.354
245	72.03	.012376	−91.24	−21130	95.4	1045	1.613	.6152	−.333
250	72.87	.012270	−89.30	−20650	96.5	1012	1.619	.7329	−.311
255	73.74	.012174	−87.38	−20164	97.7	979.0	1.626	.8655	−.286
260	5674	.91871	−20.22	−2903	63.9	230.1	1.231	.9252	27.6
265	5816	.92396	−18.99	−2583	64.3	232.9	1.222	.9296	26.0
270	5957	.92875	−17.79	−2261	64.7	235.7	1.215	.9337	24.5
275	6096	.93313	−16.60	−1936	65.1	238.4	1.208	.9375	23.1
280	6233	.93715	−15.42	−1610	65.6	241.1	1.202	.9410	21.9
285	6370	.94084	−14.26	−1281	66.1	243.7	1.196	.9442	20.7

Table 2—*continued*

P/MPa						0.350				
T_σ/K						256.980				
Liq	V_σ 74.10	Z_σ .012139	S_σ -86.62	H_σ -19970	C_P 98.2	w_σ 965.8	γ_σ 1.628	$(f/P)_\sigma$ .9224	μ_σ -.275	
Vap	5588	.91529	-20.96	-3096	63.7	228.3	1.236	.9224	28.7	
$\dfrac{T}{K}$	$\dfrac{V}{\mathrm{cm^3\,mol^{-1}}}$	Z	$\dfrac{S}{\mathrm{J\,K^{-1}\,mol^{-1}}}$	$\dfrac{H}{\mathrm{J\,mol^{-1}}}$	$\dfrac{C_P}{\mathrm{J\,K^{-1}\,mol^{-1}}}$	$\dfrac{w}{\mathrm{m\,sec^{-1}}}$	γ	(f/P)	$\dfrac{\mu}{\mathrm{K\,MPa^{-1}}}$	
290	6505	.94424	-13.10	-949	66.6	246.2	1.190	.9473	19.6	
295	6639	.94739	-11.96	-614	67.2	248.6	1.185	.9501	18.6	
300	6772	.95029	-10.82	-277	67.8	251.1	1.181	.9527	17.6	
305	6905	.95298	-9.70	64	68.4	253.4	1.176	.9551	16.8	
310	7036	.95548	-8.58	407	69.0	255.7	1.172	.9574	15.9	
315	7167	.95781	-7.47	754	69.7	258.0	1.168	.9595	15.2	
320	7297	.95997	-6.37	1104	70.3	260.3	1.164	.9615	14.5	
325	7427	.96199	-5.27	1457	71.0	262.5	1.161	.9634	13.8	
330	7556	.96388	-4.19	1814	71.7	264.6	1.158	.9651	13.2	
335	7685	.96564	-3.10	2174	72.3	266.8	1.155	.9667	12.6	
340	7813	.96729	-2.03	2537	73.0	268.9	1.152	.9683	12.0	
345	7940	.96883	-.95	2904	73.7	271.0	1.149	.9697	11.5	
350	8067	.97028	.11	3275	74.4	273.1	1.146	.9711	11.0	
355	8194	.97165	1.17	3649	75.1	275.1	1.144	.9724	10.5	
360	8320	.97293	2.23	4026	75.8	277.1	1.141	.9736	10.1	
365	8446	.97413	3.28	4407	76.6	279.1	1.139	.9747	9.67	
370	8572	.97527	4.33	4791	77.3	281.0	1.137	.9758	9.28	
380	8823	.97735	6.40	5571	78.7	284.9	1.132	.9778	8.56	
390	9072	.97921	8.47	6366	80.1	288.7	1.128	.9796	7.92	
400	9320	.98087	10.52	7174	81.6	292.4	1.125	.9812	7.34	

0.350 *MPa*

410	9568	.98237	12.55	7997	83.0	296.1	1.122	.9826	6.81
420	9815	.98371	14.57	8835	84.5	299.7	1.118	.9839	6.34
430	10061	.98493	16.57	9687	85.9	303.2	1.116	.9851	5.91
440	10306	.98603	18.56	10553	87.3	306.7	1.113	.9862	5.51
450	10551	.98702	20.54	11433	88.7	310.1	1.110	.9872	5.16
475	11161	.98915	25.43	13695	92.2	318.5	1.105	.9892	4.39
500	11769	.99084	30.25	16043	95.6	326.6	1.100	.9909	3.77
525	12374	.99221	34.99	18475	98.9	334.5	1.095	.9922	3.27
550	12978	.99333	39.67	20989	102.2	342.1	1.092	.9933	2.85
575	13581	.99425	44.28	23582	105.3	349.6	1.088	.9943	2.50

Table 2—continued

P/MPa	0.400								
T_σ/K	260.971								

Liq Vap	V_σ 74.84 4920	Z_σ .013796 .90701	S_σ −85.11 −21.33	H_σ −19575 −2930	C_P 99.2 65.0	w_σ 939.7 228.3	γ_σ 1.634 1.242	$(f/P)_\sigma$ .9155 .9155	μ_σ −.253 27.8

$\dfrac{T}{K}$	$\dfrac{V}{\text{cm}^3\,\text{mol}^{-1}}$	Z	$\dfrac{S}{\text{J K}^{-1}\,\text{mol}^{-1}}$	$\dfrac{H}{\text{J mol}^{-1}}$	$\dfrac{C_P}{\text{J K}^{-1}\,\text{mol}^{-1}}$	$\dfrac{w}{\text{m sec}^{-1}}$	γ	(f/P)	$\dfrac{\mu}{\text{K MPa}^{-1}}$
90	54.89	.029344	−174.48	−34231				.5285E−08	
95	55.31	.028010	−169.75	−33794				.2795E−07	
100	55.74	.026815	−165.73	−33401				.1234E−06	
105	56.17	.025737	−162.08	−33027				.4679E−06	
110	56.61	.024760	−158.63	−32657	74.2	1949	1.584	.1557E−05	−.632
115	57.06	.023871	−155.32	−32284	75.1	1906	1.582	.4626E−05	−.622
120	57.51	.023057	−152.10	−31906	76.3	1863	1.575	.1249E−04	−.610
125	57.97	.022311	−148.95	−31521	77.7	1821	1.566	.3089E−04	−.598
130	58.43	.021624	−145.88	−31129	78.9	1781	1.558	.7078E−04	−.587
135	58.90	.020990	−142.88	−30732	80.1	1743	1.551	.1516E−03	−.577
140	59.37	.020403	−139.95	−30329	81.0	1707	1.547	.3056E−03	−.568
145	59.85	.019859	−137.09	−29921	81.9	1673	1.544	.5837E−03	−.560
150	60.34	.019353	−134.30	−29510	82.6	1640	1.543	.1062E−02	−.552
155	60.84	.018883	−131.58	−29095	83.3	1607	1.543	.1851E−02	−.545
160	61.34	.018443	−128.93	−28677	83.9	1576	1.544	.3100E−02	−.538

165	61.85	.018034	−126.34	−28256	84.4	1545	1.546	.5010E−02	−.531
170	62.37	.017651	−123.81	−27833	84.9	1514	1.548	.7837E−02	−.524
175	62.90	.017292	−121.34	−27407	85.4	1484	1.552	.1190E−01	−.517
180	63.44	.016957	−118.93	−26979	85.9	1453	1.555	.1763E−01	−.509
185	64.00	.016642	−116.57	−26548	86.4	1423	1.559	.2544E−01	−.501
190	64.56	.016348	−114.26	−26114	86.9	1393	1.562	.3589E−01	−.492
195	65.14	.016072	−111.99	−25678	87.5	1362	1.567	.4960E−01	−.483
200	65.74	.015813	−109.77	−25239	88.1	1331	1.571	.6725E−01	−.472
205	66.35	.015571	−107.59	−24798	88.7	1300	1.575	.8959E−01	−.461
210	66.98	.015344	−105.44	−24353	89.3	1269	1.579	.1174	−.450
215	67.63	.015133	−103.33	−23904	90.0	1238	1.583	.1516	−.437
220	68.30	.014935	−101.25	−23452	90.8	1206	1.588	.1929	−.423
225	68.99	.014751	−99.20	−22996	91.6	1175	1.592	.2424	−.408
230	69.70	.014580	−97.18	−22536	92.5	1143	1.597	.3009	−.392
235	70.44	.014422	−95.18	−22071	93.4	1111	1.602	.3692	−.374
240	71.22	.014276	−93.21	−21602	94.3	1078	1.607	.4481	−.355
245	72.02	.014142	−91.25	−21128	95.4	1046	1.613	.5393	−.334
250	72.86	.014020	−89.31	−20648	96.5	1013	1.619	.6424	−.311
255	73.73	.013911	−87.39	−20163	97.7	979.6	1.625	.7587	−.286
260	74.65	.013814	−85.48	−19671	99.0	946.2	1.633	.8886	−.259
265	5023	.91195	−20.33	−2668	65.2	230.7	1.234	.9195	26.4
270	5150	.91759	−19.11	−2341	65.5	233.6	1.225	.9242	24.9
275	5274	.92273	−17.91	−2013	65.9	236.5	1.217	.9285	23.4
280	5398	.92743	−16.72	−1682	66.3	239.2	1.210	.9325	22.1
285	5520	.93175	−15.54	−1350	66.8	241.9	1.203	.9363	20.9

Table 2—*continued*

P/MPa										
					0.400					
T_σ/K					260.971					
Liq	V_σ 74.84	Z_σ .013796	S_σ −85.11	H_σ −19575	$C_{P\sigma}$ 99.2	w_σ 939.7	γ_σ 1.634	$(f/P)_\sigma$ .9155	μ_σ −.253	
Vap	4920	.90701	−21.33	−2930	65.0	228.3	1.242	.9155	27.8	
$\dfrac{T}{\text{K}}$	$\dfrac{V}{\text{cm}^3\,\text{mol}^{-1}}$	Z	$\dfrac{S}{\text{J K}^{-1}\,\text{mol}^{-1}}$	$\dfrac{H}{\text{J mol}^{-1}}$	$\dfrac{C_P}{\text{J K}^{-1}\,\text{mol}^{-1}}$	$\dfrac{w}{\text{m sec}^{-1}}$	γ	(f/P)	$\dfrac{\mu}{\text{K MPa}^{-1}}$	
290	5640	.93572	−14.37	−1015	67.3	244.5	1.197	.9397	19.8	
295	5760	.93938	−13.22	−677	67.8	247.1	1.191	.9429	18.8	
300	5879	.94276	−12.08	−337	68.3	249.6	1.186	.9459	17.8	
305	5997	.94589	−10.94	6	68.9	252.0	1.181	.9487	16.9	
310	6114	.94879	−9.82	352	69.5	254.4	1.177	.9513	16.1	
315	6230	.95149	−8.70	701	70.1	256.7	1.173	.9537	15.3	
320	6345	.95399	−7.59	1053	70.7	259.0	1.169	.9560	14.6	
325	6460	.95633	−6.49	1408	71.4	261.3	1.165	.9581	13.9	
330	6575	.95851	−5.40	1767	72.0	263.5	1.161	.9601	13.2	
335	6689	.96054	−4.31	2128	72.7	265.7	1.158	.9620	12.6	
340	6802	.96245	−3.23	2493	73.3	267.9	1.155	.9638	12.1	
345	6915	.96423	−2.15	2862	74.0	270.0	1.152	.9654	11.5	
350	7027	.96591	−1.08	3234	74.7	272.1	1.149	.9670	11.0	
355	7139	.96748	−.02	3609	75.4	274.2	1.146	.9685	10.6	
360	7251	.96895	1.04	3988	76.1	276.2	1.143	.9698	10.1	
365	7362	.97034	2.10	4370	76.8	278.2	1.141	.9711	9.72	
370	7473	.97165	3.15	4755	77.5	280.2	1.139	.9724	9.32	
380	7694	.97405	5.23	5538	78.9	284.2	1.134	.9746	8.60	
390	7913	.97618	7.30	6334	80.3	288.0	1.130	.9767	7.95	
400	8132	.97809	9.35	7144	81.8	291.8	1.126	.9785	7.36	

410	8350	.97981	11.39	7969	83.2	295.5	1.123	.9802	6.83
420	8567	.98135	13.41	8808	84.6	299.1	1.120	.9816	6.35
430	8784	.98275	15.42	9661	86.0	302.7	1.117	.9830	5.92
440	9000	.98401	17.41	10529	87.4	306.2	1.114	.9842	5.53
450	9215	.98516	19.39	11410	88.8	309.6	1.111	.9853	5.17
475	9751	.98759	24.29	13675	92.3	318.1	1.105	.9877	4.40
500	10284	.98953	29.11	16025	95.7	326.2	1.100	.9896	3.78
525	10815	.99110	33.86	18459	99.0	334.2	1.096	.9911	3.27
550	11345	.99238	38.54	20975	102.2	341.9	1.092	.9924	2.85
575	11873	.99343	43.15	23569	105.3	349.4	1.089	.9934	2.50

Table 2—continued

P/MPa									
T_σ/K				0.450					
				264.604					
	V_σ 75.53	Z_σ .015450	S_σ −83.74	H_σ −19212	C_P 100.2	w_σ 916.0	γ_σ 1.640	$(f/P)_\sigma$ .9090	μ_σ −.231
Liq	4395	.89905	−21.65	−2782	66.2	228.2	1.247	.9090	27.0
Vap									
$\dfrac{T}{K}$	$\dfrac{V}{\mathrm{cm^3\,mol^{-1}}}$	Z	$\dfrac{S}{\mathrm{J\,K^{-1}\,mol^{-1}}}$	$\dfrac{H}{\mathrm{J\,mol^{-1}}}$	$\dfrac{C_P}{\mathrm{J\,K^{-1}\,mol^{-1}}}$	$\dfrac{w}{\mathrm{m\,sec^{-1}}}$	γ	(f/P)	$\dfrac{\mu}{\mathrm{K\,MPa^{-1}}}$
90	54.89	.033011	−174.49	−34228				.4715E−08	
95	55.31	.031510	−169.76	−33791				.2493E−07	
100	55.74	.030166	−165.73	−33399				.1101E−06	
105	56.17	.028953	−162.08	−33025				.4172E−06	
110	56.61	.027854	−158.64	−32655	74.2	1949	1.584	.1388E−05	−.632
115	57.06	.026854	−155.32	−32282	75.1	1907	1.582	.4125E−05	−.622
120	57.51	.025939	−152.10	−31903	76.3	1863	1.575	.1114E−04	−.610
125	57.97	.025099	−148.96	−31518	77.7	1821	1.566	.2753E−04	−.598
130	58.43	.024326	−145.89	−31127	78.9	1781	1.558	.6308E−04	−.587
135	58.90	.023613	−142.89	−30729	80.1	1743	1.551	.1351E−03	−.577
140	59.37	.022953	−139.96	−30326	81.0	1707	1.547	.2723E−03	−.568
145	59.85	.022341	−137.10	−29919	81.9	1673	1.544	.5201E−03	−.560
150	60.34	.021772	−134.31	−29508	82.6	1640	1.543	.9465E−03	−.552
155	60.83	.021242	−131.59	−29093	83.3	1608	1.543	.1649E−02	−.545
160	61.34	.020748	−128.93	−28675	83.9	1576	1.544	.2762E−02	−.538

165	61.85	.020287	−126.34	−28254	84.4	1545	1.546	.4463E−02	−.531
170	62.37	.019856	−123.82	−27831	84.9	1515	1.548	.6982E−02	−.524
175	62.90	.019453	−121.35	−27405	85.4	1484	1.551	.1060E−01	−.517
180	63.44	.019075	−118.93	−26977	85.9	1454	1.555	.1570E−01	−.509
185	63.99	.018721	−116.57	−26546	86.4	1423	1.559	.2266E−01	−.501
190	64.56	.018390	−114.26	−26112	86.9	1393	1.562	.3197E−01	−.492
195	65.14	.018080	−112.00	−25676	87.5	1362	1.566	.4418E−01	−.483
200	65.73	.017789	−109.77	−25237	88.1	1332	1.570	.5990E−01	−.473
205	66.35	.017516	−107.59	−24796	88.7	1301	1.575	.7979E−01	−.462
210	66.97	.017261	−105.45	−24351	89.3	1270	1.579	.1046	−.450
215	67.62	.017023	−103.34	−23902	90.0	1238	1.583	.1350	−.437
220	68.29	.016800	−101.26	−23450	90.8	1207	1.588	.1718	−.423
225	68.98	.016593	−99.21	−22994	91.6	1175	1.592	.2159	−.408
230	69.70	.016401	−97.19	−22534	92.4	1143	1.597	.2679	−.392
235	70.44	.016222	−95.19	−22070	93.3	1111	1.602	.3288	−.374
240	71.21	.016058	−93.21	−21601	94.3	1079	1.607	.3991	−.355
245	72.01	.015908	−91.26	−21126	95.4	1046	1.612	.4802	−.334
250	72.85	.015771	−89.32	−20647	96.5	1013	1.618	.5721	−.311
255	73.72	.015648	−87.40	−20162	97.7	980.2	1.625	.6755	−.287
260	74.64	.015538	−85.49	−19670	98.9	946.9	1.632	.7912	−.259
265	4405	.89960	−21.55	−2755	66.2	228.4	1.246	.9094	26.9
270	4520	.90613	−20.31	−2424	66.4	231.5	1.236	.9147	25.2
275	4634	.91208	−19.09	−2091	66.7	234.4	1.227	.9196	23.8
280	4747	.91750	−17.89	−1757	67.1	237.3	1.219	.9241	22.4
285	4858	.92247	−16.69	−1420	67.5	240.1	1.211	.9283	21.2

Table 2—continued

P/MPa						0.450			
T_σ/K						264.604			
Liq	V_σ 75.53	Z_σ .015450	S_σ -83.74	H_σ -19212	C_P 100.2	w_σ 916.0	γ_σ 1.640	$(f/P)_\sigma$ .9090	μ_σ -.231
Vap	4395	.89905	-21.65	-2782	66.2	228.2	1.247	.9090	27.0
$\dfrac{T}{K}$	$\dfrac{V}{\text{cm}^3\,\text{mol}^{-1}}$	Z	$\dfrac{S}{\text{J K}^{-1}\text{mol}^{-1}}$	$\dfrac{H}{\text{J mol}^{-1}}$	$\dfrac{C_P}{\text{J K}^{-1}\text{mol}^{-1}}$	$\dfrac{w}{\text{m sec}^{-1}}$	γ	(f/P)	$\dfrac{\mu}{\text{K MPa}^{-1}}$
290	4967	.92704	-15.52	-1082	67.9	242.8	1.204	.9322	20.0
295	5076	.93124	-14.35	-741	68.4	245.5	1.198	.9358	19.0
300	5183	.93511	-13.20	-398	68.9	248.0	1.192	.9391	18.0
305	5290	.93869	-12.06	-53	69.4	250.5	1.187	.9423	17.1
310	5395	.94200	-10.92	296	70.0	253.0	1.182	.9452	16.2
315	5500	.94508	-9.80	647	70.5	255.4	1.177	.9479	15.4
320	5605	.94794	-8.68	1001	71.1	257.8	1.173	.9505	14.7
325	5708	.95060	-7.58	1358	71.7	260.1	1.169	.9529	14.0
330	5811	.95308	-6.48	1719	72.4	262.4	1.165	.9551	13.3
335	5914	.95540	-5.38	2082	73.0	264.6	1.161	.9573	12.7
340	6015	.95756	-4.30	2449	73.7	266.8	1.158	.9592	12.1
345	6117	.95959	-3.22	2819	74.3	269.0	1.155	.9611	11.6
350	6218	.96149	-2.14	3192	75.0	271.1	1.152	.9629	11.1
355	6318	.96327	-1.07	3569	75.7	273.3	1.149	.9645	10.6
360	6418	.96495	-.01	3949	76.4	275.3	1.146	.9661	10.2
365	6518	.96652	1.05	4332	77.0	277.4	1.143	.9675	9.76
370	6618	.96801	2.10	4719	77.7	279.4	1.141	.9689	9.36
380	6815	.97072	4.19	5503	79.1	283.4	1.136	.9715	8.63
390	7012	.97314	6.26	6302	80.5	287.3	1.132	.9738	7.97
400	7208	.97530	8.32	7114	81.9	291.1	1.128	.9758	7.38

410	7403	.97724	10.36	7941	83.4	294.9	1.124	.9777	6.85
420	7597	.97899	12.39	8781	84.8	298.5	1.121	.9794	6.37
430	7790	.98056	14.40	9636	86.2	302.1	1.118	.9809	5.93
440	7983	.98199	16.39	10504	87.6	305.7	1.115	.9822	5.54
450	8175	.98328	18.38	11387	89.0	309.2	1.112	.9835	5.18
475	8654	.98603	23.28	13655	92.4	317.7	1.106	.9862	4.41
500	9129	.98822	28.11	16007	95.8	325.9	1.101	.9883	3.78
525	9603	.98999	32.86	18443	99.1	333.9	1.096	.9900	3.27
550	10075	.99143	37.54	20960	102.3	341.6	1.092	.9915	2.85
575	10545	.99262	42.16	23556	105.4	349.1	1.089	.9926	2.51

Table 2—continued

P/MPa					0.500				
T_σ/K					267.947				
Liq	V_σ 76.19	Z_σ .017100	S_σ −82.49	H_σ −18874	C_P 101.1	w_σ 894.1	γ_σ 1.645	$(f/P)_\sigma$ .9028	μ_σ −.210
Vap	3972	.89136	−21.93	−2647	67.4	227.9	1.253	.9028	26.3
$\dfrac{T}{K}$	$\dfrac{V}{\text{cm}^3\,\text{mol}^{-1}}$	Z	$\dfrac{S}{\text{J K}^{-1}\,\text{mol}^{-1}}$	$\dfrac{H}{\text{J mol}^{-1}}$	$\dfrac{C_P}{\text{J K}^{-1}\,\text{mol}^{-1}}$	$\dfrac{w}{\text{m sec}^{-1}}$	γ	(f/P)	$\dfrac{\mu}{\text{K MPa}^{-1}}$
90	54.89	.036678	−174.49	−34226				.4259E−08	
95	55.31	.035011	−169.76	−33789				.2252E−07	
100	55.73	.033517	−165.74	−33397				.9938E−07	
105	56.17	.032169	−162.08	−33022				.3767E−06	
110	56.61	.030948	−158.64	−32652	74.2	1950	1.584	.1253E−05	−.632
115	57.06	.029836	−155.33	−32279	75.1	1907	1.582	.3724E−05	−.622
120	57.51	.028820	−152.10	−31901	76.3	1863	1.575	.1005E−04	−.610
125	57.97	.027887	−148.96	−31516	77.7	1821	1.566	.2485E−04	−.598
130	58.43	.027029	−145.89	−31124	78.9	1781	1.558	.5693E−04	−.587
135	58.90	.026236	−142.89	−30727	80.1	1743	1.551	.1219E−03	−.577
140	59.37	.025502	−139.96	−30324	81.0	1708	1.547	.2457E−03	−.568
145	59.85	.024822	−137.10	−29917	81.9	1673	1.544	.4692E−03	−.560
150	60.34	.024190	−134.31	−29505	82.6	1640	1.543	.8539E−03	−.552
155	60.83	.023601	−131.59	−29091	83.3	1608	1.543	.1488E−02	−.545
160	61.33	.023052	−128.94	−28673	83.9	1577	1.544	.2491E−02	−.538

0.500 MPa

165	61.84	.022540	−126.35	−28252	84.4	1546	1.546	.4026E−02	−.531
170	62.36	.022061	−123.82	−27829	84.9	1515	1.548	.6298E−02	−.524
175	62.89	.021613	−121.35	−27403	85.4	1485	1.551	.9565E−02	−.517
180	63.44	.021193	−118.94	−26974	85.9	1454	1.555	.1416E−01	−.509
185	63.99	.020800	−116.58	−26544	86.4	1424	1.558	.2044E−01	−.501
190	64.55	.020432	−114.27	−26110	86.9	1393	1.562	.2883E−01	−.492
195	65.13	.020087	−112.00	−25674	87.5	1363	1.566	.3984E−01	−.483
200	65.73	.019764	−109.78	−25235	88.1	1332	1.570	.5402E−01	−.473
205	66.34	.019461	−107.60	−24794	88.7	1301	1.575	.7195E−01	−.462
210	66.97	.019178	−105.45	−24349	89.3	1270	1.579	.9429E−01	−.450
215	67.62	.018913	−103.34	−23900	90.0	1239	1.583	.1217	−.437
220	68.28	.018665	−101.27	−23448	90.8	1207	1.587	.1549	−.423
225	68.97	.018435	−99.22	−22992	91.6	1176	1.592	.1946	−.408
230	69.69	.018221	−97.19	−22532	92.4	1144	1.597	.2416	−.392
235	70.43	.018023	−95.20	−22068	93.3	1112	1.601	.2964	−.374
240	71.20	.017840	−93.22	−21599	94.3	1079	1.607	.3598	−.355
245	72.00	.017673	−91.27	−21125	95.3	1047	1.612	.4330	−.335
250	72.84	.017521	−89.33	−20645	96.4	1014	1.618	.5158	−.312
255	73.71	.017384	−87.41	−20160	97.6	980.8	1.625	.6091	−.287
260	74.63	.017262	−85.50	−19669	98.9	947.5	1.632	.7133	−.260
265	75.60	.017155	−83.60	−19171	100.3	914.0	1.640	.8291	−.230
270	4015	.89435	−21.42	−2509	67.4	229.2	1.249	.9051	25.6
275	4121	.90115	−20.18	−2171	67.6	232.3	1.238	.9106	24.1
280	4225	.90734	−18.96	−1833	67.9	235.4	1.229	.9156	22.7
285	4327	.91300	−17.75	−1493	68.2	238.3	1.220	.9203	21.4

Table 2—*continued*

P/MPa	0.500								
T_σ/K	267.947								
	V_σ	Z_σ	S_σ	H_σ	C_P	w_σ	γ_σ	$(f/P)_\sigma$	μ_σ
Liq	76.19	.017100	−82.49	−18874	101.1	894.1	1.645	.9028	−.210
Vap	3972	.89136	−21.93	−2647	67.4	227.9	1.253	.9028	26.3
$\frac{T}{K}$	$\frac{V}{cm^3\,mol^{-1}}$	Z	$\frac{S}{J\,K^{-1}\,mol^{-1}}$	$\frac{H}{J\,mol^{-1}}$	$\frac{C_P}{J\,K^{-1}\,mol^{-1}}$	$\frac{w}{m\,sec^{-1}}$	γ	(f/P)	$\frac{\mu}{K\,MPa^{-1}}$
290	4428	.91818	−16.56	−1151	68.6	241.1	1.212	.9246	20.3
295	4527	.92294	−15.39	−807	69.0	243.8	1.205	.9286	19.2
300	4626	.92733	−14.22	−461	69.4	246.5	1.199	.9324	18.1
305	4724	.93138	−13.07	−112	69.9	249.1	1.193	.9359	17.2
310	4820	.93512	−11.93	239	70.5	251.6	1.187	.9391	16.3
315	4916	.93859	−10.80	592	71.0	254.1	1.182	.9422	15.5
320	5012	.94181	−9.68	949	71.6	256.5	1.177	.9450	14.8
325	5106	.94481	−8.56	1308	72.1	258.9	1.173	.9477	14.1
330	5200	.94760	−7.46	1670	72.7	261.2	1.169	.9502	13.4
335	5293	.95020	−6.36	2036	73.4	263.5	1.165	.9525	12.8
340	5386	.95264	−5.27	2404	74.0	265.8	1.161	.9547	12.2
345	5478	.95491	−4.18	2775	74.6	268.0	1.158	.9568	11.7
350	5570	.95704	−3.10	3150	75.3	270.2	1.154	.9587	11.2
355	5661	.95904	−2.03	3528	75.9	272.3	1.151	.9606	10.7
360	5752	.96092	−.96	3910	76.6	274.4	1.148	.9623	10.2
365	5843	.96268	.10	4295	77.3	276.5	1.146	.9639	9.81
370	5933	.96434	1.15	4683	78.0	278.6	1.143	.9655	9.40
380	6113	.96738	3.25	.5469	79.3	282.6	1.138	.9683	8.66
390	6291	.97008	5.33	6270	80.7	286.6	1.133	.9709	8.00
400	6469	.97250	7.39	7084	82.1	290.5	1.129	.9732	7.41

410	6645	.97467	9.44	7912	83.5	294.2	1.126	.9752	6.87
420	6821	.97662	11.47	8754	84.9	298.0	1.122	.9771	6.39
430	6996	.97837	13.48	9610	86.3	301.6	1.119	.9788	5.95
440	7170	.97996	15.48	10480	87.7	305.2	1.116	.9803	5.55
450	7344	.98141	17.47	11364	89.1	308.7	1.113	.9817	5.19
475	7776	.98446	22.38	13634	92.5	317.3	1.107	.9846	4.41
500	8205	.98690	27.21	15989	95.9	325.5	1.101	.9870	3.79
525	8633	.98887	31.96	18427	99.1	333.6	1.097	.9889	3.28
550	9059	.99048	36.65	20945	102.3	341.3	1.093	.9905	2.86
575	9483	.99180	41.27	23543	105.4	348.9	1.089	.9918	2.51

111

Table 2—*continued*

P/MPa				0.550				
T_σ/K				271.050				
	V_σ 76.82 / 3622	Z_σ .018749 / .88391	S_σ −81.33 / −22.18	H_σ −18558 / −2525	C_P 102.1 / 68.5	w_σ 873.7 / 227.6	γ_σ 1.650 / 1.259	$(f/P)_\sigma$ .8969 / .8969	μ_σ −.189 / 25.7
Liq / Vap									
$\dfrac{T}{K}$	$\dfrac{V}{\mathrm{cm^3\,mol^{-1}}}$	Z	$\dfrac{S}{\mathrm{J\,K^{-1}\,mol^{-1}}}$	$\dfrac{H}{\mathrm{J\,mol^{-1}}}$	$\dfrac{C_P}{\mathrm{J\,K^{-1}\,mol^{-1}}}$	$\dfrac{w}{\mathrm{m\,sec^{-1}}}$	γ	(f/P)	$\dfrac{\mu}{\mathrm{K\,MPa^{-1}}}$
90	54.89	.040345	−174.49	−34223				.3886E−08	
95	55.31	.038511	−169.77	−33787				.2054E−07	
100	55.73	.036867	−165.74	−33394				.9065E−07	
105	56.17	.035385	−162.09	−33020				.3436E−06	
110	56.61	.034042	−158.64	−32650	74.2	1950	1.584	.1143E−05	−.632
115	57.05	.032819	−155.33	−32277	75.1	1907	1.582	.3396E−05	−.622
120	57.51	.031701	−152.11	−31899	76.3	1864	1.575	.9164E−05	−.610
125	57.96	.030675	−148.97	−31514	77.7	1822	1.566	.2265E−04	−.598
130	58.43	.029730	−145.90	−31122	78.9	1782	1.558	.5189E−04	−.587
135	58.89	.028858	−142.90	−30725	80.0	1744	1.551	.1111E−03	−.577
140	59.37	.028052	−139.97	−30322	81.0	1708	1.547	.2239E−03	−.568
145	59.85	.027303	−137.11	−29914	81.9	1673	1.544	.4277E−03	−.560
150	60.33	.026608	−134.32	−29503	82.6	1640	1.543	.7781E−03	−.552
155	60.83	.025960	−131.60	−29088	83.3	1608	1.543	.1356E−02	−.545
160	61.33	.025356	−128.94	−28670	83.9	1577	1.544	.2270E−02	−.538

165	61.84	.024793	−126.35	−28250	84.4	1546	1.546	.3668E−02	−.531
170	62.36	.024266	−123.83	−27826	84.9	1515	1.548	.5739E−02	−.524
175	62.89	.023773	−121.36	−27400	85.4	1485	1.551	.8715E−02	−.517
180	63.43	.023311	−118.95	−26972	85.9	1455	1.555	.1290E−01	−.509
185	63.98	.022879	−116.58	−26541	86.4	1424	1.558	.1862E−01	−.501
190	64.55	.022474	−114.27	−26108	86.9	1394	1.562	.2626E−01	−.492
195	65.13	.022094	−112.01	−25672	87.5	1363	1.566	.3629E−01	−.483
200	65.72	.021739	−109.79	−25233	88.1	1332	1.570	.4920E−01	−.473
205	66.34	.021406	−107.60	−24791	88.7	1302	1.574	.6554E−01	−.462
210	66.96	.021094	−105.46	−24346	89.3	1270	1.579	.8589E−01	−.450
215	67.61	.020802	−103.35	−23898	90.0	1239	1.583	.1109	−.437
220	68.28	.020530	−101.27	−23446	90.8	1208	1.587	.1411	−.423
225	68.97	.020277	−99.22	−22990	91.6	1176	1.592	.1773	−.408
230	69.68	.020041	−97.20	−22530	92.4	1144	1.596	.2200	−.392
235	70.42	.019823	−95.20	−22066	93.3	1112	1.601	.2700	−.375
240	71.19	.019622	−93.23	−21597	94.3	1080	1.606	.3277	−.356
245	71.99	.019438	−91.28	−21123	95.3	1047	1.612	.3943	−.335
250	72.83	.019270	−89.34	−20644	96.4	1014	1.618	.4697	−.312
255	73.70	.019119	−87.42	−20159	97.6	981.4	1.624	.5546	−.288
260	74.62	.018985	−85.51	−19668	98.9	948.2	1.631	.6496	−.260
265	75.58	.018867	−83.61	−19170	100.3	914.7	1.639	.7550	−.230
270	76.60	.018768	−81.73	−18665	101.7	880.8	1.648	.8711	−.197
275	3700	.88993	−21.19	−2254	68.6	230.2	1.250	.9016	24.5
280	3797	.89694	−19.95	−1911	68.7	233.3	1.239	.9071	23.0
285	3892	.90332	−18.73	−1566	69.0	236.4	1.229	.9123	21.7

Table 2—*continued*

P/MPa						0.550			
T_σ/K						271.050			
	V_σ 76.82	Z_σ .018749	S_σ -81.33	H_σ -18558	C_P 102.1	w_σ 873.7	γ_σ 1.650	$(f/P)_\sigma$ .8969	μ_σ -.189
Liq	76.82	.018749	-81.33	-18558	102.1	873.7	1.650	.8969	-.189
Vap	3622	.88391	-22.18	-2525	68.5	227.6	1.259	.8969	25.7
$\dfrac{T}{\text{K}}$	$\dfrac{V}{\text{cm}^3\,\text{mol}^{-1}}$	Z	$\dfrac{S}{\text{J K}^{-1}\,\text{mol}^{-1}}$	$\dfrac{H}{\text{J mol}^{-1}}$	$\dfrac{C_P}{\text{J K}^{-1}\,\text{mol}^{-1}}$	$\dfrac{w}{\text{m sec}^{-1}}$	γ	(f/P)	$\dfrac{\mu}{\text{K MPa}^{-1}}$
290	3986	.90915	-17.53	-1221	69.3	239.3	1.221	.9170	20.5
295	4078	.91450	-16.34	-873	69.6	242.1	1.213	.9215	19.4
300	4170	.91942	-15.17	-524	70.0	244.9	1.205	.9256	18.3
305	4260	.92395	-14.01	-173	70.5	247.6	1.199	.9294	17.4
310	4350	.92814	-12.86	181	71.0	250.2	1.193	.9330	16.5
315	4438	.93202	-11.72	537	71.5	252.7	1.187	.9364	15.7
320	4526	.93561	-10.59	895	72.0	255.2	1.182	.9395	14.9
325	4613	.93895	-9.47	1257	72.6	257.7	1.177	.9424	14.2
330	4700	.94206	-8.35	1621	73.1	260.1	1.173	.9452	13.5
335	4785	.94496	-7.25	1988	73.7	262.4	1.168	.9478	12.9
340	4871	.94766	-6.15	2358	74.3	264.7	1.165	.9502	12.3
345	4956	.95019	-5.06	2732	75.0	267.0	1.161	.9525	11.7
350	5040	.95256	-3.98	3108	75.6	269.2	1.157	.9546	11.2
355	5124	.95478	-2.90	3488	76.2	271.4	1.154	.9566	10.7
360	5207	.95686	-1.83	3870	76.9	273.5	1.151	.9586	10.3
365	5290	.95882	-.77	4257	77.5	275.7	1.148	.9603	9.85
370	5373	.96066	.29	4646	78.2	277.8	1.145	.9620	9.44
380	5538	.96402	2.40	5435	79.6	281.9	1.140	.9652	8.70
390	5701	.96701	4.48	6237	80.9	285.9	1.135	.9680	8.03
400	5863	.96969	6.55	7053	82.3	289.8	1.131	.9705	7.43

410	6025	.97208	8.59	7883	83.7	293.6	1.127	.9727	6.89
420	6186	.97424	10.63	8727	85.1	297.4	1.123	.9748	6.40
430	6345	.97618	12.65	9584	86.4	301.1	1.120	.9766	5.96
440	6505	.97793	14.65	10456	87.8	304.7	1.117	.9783	5.56
450	6663	.97952	16.64	11341	89.2	308.2	1.114	.9798	5.20
475	7058	.98290	21.55	13614	92.6	316.9	1.108	.9831	4.42
500	7450	.98559	26.39	15971	96.0	325.2	1.102	.9857	3.79
525	7839	.98776	31.15	18411	99.2	333.3	1.097	.9878	3.28
550	8227	.98953	35.84	20931	102.4	341.1	1.093	.9896	2.86
575	8614	.99099	40.46	23529	105.5	348.7	1.090	.9910	2.51

Table 2—continued

P/MPa					0.600				
T_{cr}/K					273.951				

	V_σ 77.43	Z_σ .020397	S_σ −80.25	H_σ −18260	$C_{P\sigma}$ 103.0	w_σ 854.6	γ_σ 1.656	$(f/P)_\sigma$ .8912	μ_σ −.169
Liq									
Vap	3328	.87665	−22.40	−2412	69.6	227.3	1.266	.8912	25.2

$\dfrac{T}{K}$	$\dfrac{V}{cm^3\,mol^{-1}}$	Z	$\dfrac{S}{J\,K^{-1}\,mol^{-1}}$	$\dfrac{H}{J\,mol^{-1}}$	$\dfrac{C_P}{J\,K^{-1}\,mol^{-1}}$	$\dfrac{w}{m\,sec^{-1}}$	γ	(f/P)	$\dfrac{\mu}{K\,MPa^{-1}}$
90	54.89	.044012	−174.50	−34221				.3575E−08	
95	55.30	.042011	−169.77	−33784				.1890E−07	
100	55.73	.040218	−165.75	−33392				.8338E−07	
105	56.17	.038601	−162.09	−33018				.3160E−06	
110	56.61	.037136	−158.65	−32648	74.2	1950	1.584	.1051E−05	−.632
115	57.05	.035802	−155.33	−32275	75.1	1907	1.582	.3123E−05	−.622
120	57.50	.034582	−152.11	−31896	76.3	1864	1.574	.8424E−05	−.610
125	57.96	.033462	−148.97	−31511	77.7	1822	1.566	.2082E−04	−.598
130	58.42	.032432	−145.90	−31120	78.9	1782	1.557	.4770E−04	−.587
135	58.89	.031481	−142.90	−30722	80.0	1744	1.551	.1021E−03	−.577
140	59.37	.030601	−139.97	−30320	81.0	1708	1.547	.2058E−03	−.568
145	59.85	.029784	−137.11	−29912	81.9	1674	1.544	.3930E−03	−.560
150	60.33	.029025	−134.32	−29501	82.6	1641	1.543	.7150E−03	−.552
155	60.83	.028319	−131.60	−29086	83.3	1609	1.543	.1245E−02	−.545
160	61.33	.027660	−128.95	−28668	83.9	1577	1.544	.2086E−02	−.538

165	61.84	.027045	−126.36	−28247	84.4	1546	1.546	.3370E−02	−.531
170	62.36	.026470	−123.83	−27824	84.9	1516	1.548	.5272E−02	−.524
175	62.89	.025933	−121.36	−27398	85.4	1485	1.551	.8007E−02	−.517
180	63.43	.025429	−118.95	−26970	85.9	1455	1.555	.1185E−01	−.509
185	63.98	.024958	−116.59	−26539	86.4	1424	1.558	.1710E−01	−.501
190	64.55	.024516	−114.28	−26106	86.9	1394	1.562	.2412E−01	−.492
195	65.13	.024101	−112.01	−25670	87.5	1364	1.566	.3334E−01	−.483
200	65.72	.023713	−109.79	−25231	88.0	1333	1.570	.4519E−01	−.473
205	66.33	.023350	−107.61	−24789	88.7	1302	1.574	.6019E−01	−.462
210	66.96	.023010	−105.47	−24344	89.3	1271	1.578	.7888E−01	−.450
215	67.60	.022691	−103.36	−23896	90.0	1240	1.583	.1018	−.437
220	68.27	.022395	−101.28	−23444	90.8	1208	1.587	.1296	−.424
225	68.96	.022118	−99.23	−22989	91.6	1177	1.592	.1628	−.409
230	69.67	.021861	−97.21	−22529	92.4	1145	1.596	.2021	−.393
235	70.41	.021623	−95.21	−22064	93.3	1113	1.601	.2479	−.375
240	71.18	.021404	−93.24	−21595	94.3	1080	1.606	.3010	−.356
245	71.98	.021202	−91.28	−21122	95.3	1048	1.612	.3621	−.335
250	72.82	.021019	−89.35	−20642	96.4	1015	1.617	.4313	−.313
255	73.69	.020854	−87.43	−20157	97.6	982.0	1.624	.5093	−.288
260	74.61	.020707	−85.52	−19666	98.9	948.8	1.631	.5965	−.261
265	75.57	.020579	−83.62	−19169	100.2	915.3	1.639	.6933	−.231
270	76.59	.020470	−81.74	−18664	101.7	881.5	1.648	.7999	−.198
275	3347	.87840	−22.14	−2339	69.6	228.0	1.263	.8925	24.8
280	3439	.88627	−20.88	−1991	69.7	231.3	1.250	.8986	23.3
285	3528	.89341	−19.65	−1642	69.8	234.4	1.239	.9042	22.0

Table 2—continued

P/MPa							0.600		
T_σ/K							273.951		
Liq	V_σ 77.43	Z_σ .020397	S_σ −80.25	H_σ −18260	C_P 103.0	w_σ 854.6	γ_σ 1.656	$(f/P)_b$ .8912	μ_σ −.169
Vap	3328	.87665	−22.40	−2412	69.6	227.3	1.266	.8912	25.2
$\dfrac{T}{K}$	$\dfrac{V}{cm^3\,mol^{-1}}$	Z	$\dfrac{S}{J\,K^{-1}\,mol^{-1}}$	$\dfrac{H}{J\,mol^{-1}}$	$\dfrac{C_P}{J\,K^{-1}\,mol^{-1}}$	$\dfrac{w}{m\,sec^{-1}}$	γ	(f/P)	$\dfrac{\mu}{K\,MPa^{-1}}$
290	3616	.89993	−18.43	−1293	70.1	237.5	1.229	.9094	20.7
295	3703	.90590	−17.23	−942	70.3	240.4	1.221	.9143	19.6
300	3789	.91137	−16.05	−589	70.7	243.3	1.213	.9188	18.5
305	3873	.91641	−14.87	−235	71.1	246.0	1.205	.9230	17.5
310	3957	.92106	−13.72	122	71.5	248.7	1.199	.9269	16.6
315	4039	.92535	−12.57	480	72.0	251.4	1.193	.9306	15.8
320	4121	.92933	−11.43	841	72.5	253.9	1.187	.9340	15.0
325	4202	.93303	−10.30	1205	73.0	256.4	1.182	.9372	14.3
330	4282	.93646	−9.18	1571	73.5	258.9	1.177	.9402	13.6
335	4362	.93966	−8.07	1941	74.1	261.3	1.172	.9430	13.0
340	4441	.94264	−6.97	2313	74.7	263.6	1.168	.9457	12.4
345	4520	.94543	−5.88	2687	75.3	265.9	1.164	.9482	11.8
350	4598	.94804	−4.79	3065	75.9	268.2	1.160	.9505	11.3
355	4676	.95048	−3.71	3447	76.5	270.4	1.157	.9527	10.8
360	4753	.95277	−2.63	3831	77.2	272.6	1.154	.9548	10.3
365	4830	.95492	−1.56	4218	77.8	274.8	1.150	.9568	9.89
370	4906	.95694	−.50	4609	78.5	276.9	1.147	.9586	9.48
380	5058	.96064	1.61	5400	79.8	281.1	1.142	.9620	8.73
390	5209	.96393	3.70	6205	81.1	285.2	1.137	.9651	8.06
400	5359	.96686	5.77	7023	82.5	289.1	1.132	.9678	7.45

410	5508	.96949	7.82	7854	83.8	293.0	1.128	.9703	6.91
420	5656	.97185	9.86	8700	85.2	296.8	1.125	.9725	6.42
430	5804	.97398	11.88	9559	86.6	300.5	1.121	.9745	5.98
440	5950	.97590	13.89	10431	88.0	304.2	1.118	.9764	5.58
450	6096	.97764	15.88	11318	89.3	307.8	1.115	.9780	5.21
475	6459	.98133	20.80	13593	92.7	316.5	1.108	.9816	4.43
500	6820	.98427	25.64	15953	96.0	324.8	1.103	.9844	3.80
525	7178	.98665	30.40	18394	99.3	332.9	1.098	.9867	3.28
550	7534	.98858	35.10	20916	102.5	340.8	1.094	.9886	2.86
575	7890	.99017	39.72	23516	105.5	348.4	1.090	.9902	2.51

Table 2—continued

P/MPa					0.650				
T_σ/K					276.678				
	V_σ cm³ mol⁻¹	Z_σ	S_σ	H_σ	C_P	w_σ	γ_σ	$(f/P)_\sigma$	μ_σ K MPa⁻¹
Liq	78.02	.022046	−79.24	−17977	103.8	836.6	1.661	.8857	−.148
Vap	3077	.86958	−22.61	−2309	70.7	226.8	1.272	.8857	24.7
$\frac{T}{K}$	$\frac{V}{cm^3\,mol^{-1}}$	Z	$\frac{S}{J\,K^{-1}\,mol^{-1}}$	$\frac{H}{J\,mol^{-1}}$	$\frac{C_P}{J\,K^{-1}\,mol^{-1}}$	$\frac{w}{m\,sec^{-1}}$	γ	(f/P)	$\frac{\mu}{K\,MPa^{-1}}$
90	54.89	.047678	−174.50	−34219				.3313E−08	
95	55.30	.045511	−169.78	−33782				.1750E−07	
100	55.73	.043568	−165.75	−33390				.7722E−07	
105	56.16	.041817	−162.10	−33015				.2926E−06	
110	56.60	.040229	−158.65	−32645	74.2	1951	1.584	.9733E−06	−.632
115	57.05	.038784	−155.34	−32272	75.1	1908	1.582	.2891E−05	−.622
120	57.50	.037462	−152.12	−31894	76.3	1864	1.574	.7799E−05	−.610
125	57.96	.036250	−148.98	−31509	77.7	1822	1.566	.1928E−04	−.598
130	58.42	.035133	−145.90	−31117	78.9	1782	1.557	.4415E−04	−.587
135	58.89	.034103	−142.90	−30720	80.0	1744	1.551	.9450E−04	−.577
140	59.36	.033149	−139.98	−30317	81.0	1708	1.547	.1904E−03	−.568
145	59.84	.032265	−137.12	−29910	81.9	1674	1.544	.3637E−03	−.560
150	60.33	.031443	−134.33	−29499	82.6	1641	1.543	.6616E−03	−.552
155	60.82	.030677	−131.61	−29084	83.3	1609	1.543	.1152E−02	−.545
160	61.32	.029964	−128.95	−28666	83.9	1577	1.544	.1930E−02	−.538

165	61.83	.029298	−126.36	−28245	84.4	1547	1.546	.3118E−02	−.531
170	62.35	.028675	−123.84	−27822	84.9	1516	1.548	.4878E−02	−.524
175	62.88	.028092	−121.37	−27396	85.4	1486	1.551	.7407E−02	−.517
180	63.42	.027547	−118.96	−26968	85.9	1455	1.554	.1096E−01	−.509
185	63.98	.027036	−116.60	−26537	86.4	1425	1.558	.1582E−01	−.501
190	64.54	.026557	−114.28	−26104	86.9	1394	1.562	.2231E−01	−.492
195	65.12	.026108	−112.02	−25668	87.5	1364	1.566	.3083E−01	−.483
200	65.72	.025688	−109.80	−25229	88.0	1333	1.570	.4180E−01	−.473
205	66.33	.025294	−107.62	−24787	88.7	1302	1.574	.5567E−01	−.462
210	66.95	.024925	−105.47	−24342	89.3	1271	1.578	.7295E−01	−.450
215	67.60	.024580	−103.36	−23894	90.0	1240	1.583	.9416E−01	−.438
220	68.27	.024259	−101.29	−23442	90.7	1209	1.587	.1198	−.424
225	68.95	.023959	−99.24	−22987	91.5	1177	1.591	.1506	−.409
230	69.67	.023680	−97.22	−22527	92.4	1145	1.596	.1869	−.393
235	70.41	.023422	−95.22	−22063	93.3	1113	1.601	.2293	−.375
240	71.17	.023184	−93.25	−21594	94.3	1081	1.606	.2783	−.356
245	71.97	.022966	−91.29	−21120	95.3	1048	1.611	.3348	−.336
250	72.81	.022768	−89.36	−20641	96.4	1016	1.617	.3988	−.313
255	73.68	.022589	−87.44	−20156	97.6	982.7	1.624	.4709	−.289
260	74.59	.022429	−85.53	−19665	98.8	949.5	1.631	.5516	−.261
265	75.56	.022290	−83.63	−19167	100.2	916.0	1.639	.6410	−.232
270	76.57	.022171	−81.75	−18663	101.7	882.3	1.647	.7396	−.198
275	77.65	.022074	−79.87	−18151	103.3	848.2	1.658	.8474	−.161
280	3135	.87532	−21.76	−2074	70.7	229.1	1.263	.8901	23.7
285	3220	.88327	−20.51	−1720	70.7	232.4	1.250	.8962	22.3

121

Table 2—continued

P/MPa										
T_σ/K										
	0.650									
	276.678									
Liq	V_σ 78.02	Z_σ 022046	S_σ −79.24	H_σ −17977	$C_{P\sigma}$ 103.8	w_σ 836.6	γ_σ 1.661	$(f/P)_\sigma$ .8857	μ_σ −.148	
Vap	3077	.86958	−22.61	−2309	70.7	226.8	1.272	.8857	24.7	
$\dfrac{T}{K}$	$\dfrac{V}{cm^3\,mol^{-1}}$	Z	$\dfrac{S}{J\,K^{-1}\,mol^{-1}}$	$\dfrac{H}{J\,mol^{-1}}$	$\dfrac{C_P}{J\,K^{-1}\,mol^{-1}}$	$\dfrac{w}{m\,sec^{-1}}$	γ	(f/P)	$\dfrac{\mu}{K\,MPa^{-1}}$
290	3303	.89051	−19.28	−1366	70.9	235.6	1.239	.9019	21.0
295	3385	.89712	−18.07	−1011	71.1	238.7	1.229	.9071	19.8
300	3466	.90318	−16.87	−655	71.4	241.6	1.220	.9120	18.7
305	3545	.90874	−15.69	−298	71.7	244.5	1.212	.9166	17.7
310	3624	.91386	−14.52	62	72.1	247.3	1.205	.9208	16.8
315	3701	.91859	−13.36	423	72.5	250.0	1.198	.9248	15.9
320	3778	.92297	−12.22	787	72.9	252.6	1.192	.9285	15.1
325	3854	.92703	−11.08	1153	73.4	255.2	1.186	.9320	14.4
330	3929	.93080	−9.96	1521	74.0	257.7	1.181	.9352	13.7
335	4004	.93430	−8.84	1892	74.5	260.1	1.176	.9383	13.0
340	4078	.93757	−7.73	2266	75.1	262.5	1.172	.9412	12.4
345	4151	.94062	−6.63	2643	75.6	264.9	1.167	.9439	11.9
350	4224	.94348	−5.54	3022	76.2	267.2	1.163	.9464	11.3
355	4296	.94615	−4.46	3405	76.8	269.5	1.160	.9488	10.8
360	4368	.94865	−3.38	3791	77.4	271.7	1.156	.9510	10.4
365	4440	.95100	−2.31	4180	78.1	273.9	1.153	.9532	9.94
370	4511	.95321	−1.24	4571	78.7	276.1	1.150	.9552	9.53
380	4653	.95724	.88	5365	80.0	280.3	1.144	.9589	8.77
390	4793	.96083	2.97	6172	81.3	284.5	1.139	.9622	8.09
400	4932	.96402	5.05	6992	82.7	288.5	1.134	.9651	7.48

410	5071	.96688	7.11	7825	84.0	292.4	1.130	.9678	6.93
420	5208	.96945	9.15	8672	85.4	296.2	1.126	.9702	6.44
430	5345	.97177	11.17	9533	86.7	300.0	1.122	.9724	5.99
440	5481	.97386	13.18	10407	88.1	303.7	1.119	.9744	5.59
450	5617	.97575	15.18	11294	89.4	307.3	1.116	.9762	5.22
475	5953	.97976	20.10	13573	92.8	316.1	1.109	.9801	4.43
500	6287	.98296	24.95	15934	96.1	324.5	1.103	.9831	3.80
525	6618	.98553	29.72	18378	99.4	332.6	1.098	.9856	3.29
550	6948	.98763	34.41	20902	102.5	340.5	1.094	.9877	2.86
575	7277	.98936	39.04	23503	105.6	348.2	1.090	.9894	2.51

Table 2—continued

P/MPa									
				0.700					
T_σ/K				**279.254**					
	V_σ	Z_σ	S_σ	H_σ	C_P	w_σ	γ_σ	$(f/P)_\sigma$	μ_σ
Liq	78.60	.023697	−78.28	−17707	104.7	819.6	1.667	.8805	−.127
Vap	2861	.86266	−22.79	−2212	71.8	226.4	1.278	.8805	24.3
$\dfrac{T}{K}$	$\dfrac{V}{\text{cm}^3\,\text{mol}^{-1}}$	Z	$\dfrac{S}{\text{J K}^{-1}\,\text{mol}^{-1}}$	$\dfrac{H}{\text{J mol}^{-1}}$	$\dfrac{C_P}{\text{J K}^{-1}\,\text{mol}^{-1}}$	$\dfrac{w}{\text{m sec}^{-1}}$	γ	(f/P)	$\dfrac{\mu}{\text{K MPa}^{-1}}$
90	54.89	.051344	−174.51	−34216				.3087E−08	
95	55.30	.049010	−169.78	−33780				.1631E−07	
100	55.73	.046918	−165.76	−33387				.7195E−07	
105	56.16	.045032	−162.10	−33013				.2726E−06	
110	56.60	.043323	−158.66	−32643	74.2	1951	1.584	.9066E−06	−.632
115	57.05	.041766	−155.34	−32270	75.1	1908	1.582	.2693E−05	−.622
120	57.50	.040343	−152.12	−31892	76.3	1865	1.574	.7263E−05	−.610
125	57.96	.039037	−148.98	−31507	77.7	1822	1.565	.1795E−04	−.598
130	58.42	.037835	−145.91	−31115	78.9	1782	1.557	.4111E−04	−.587
135	58.89	.036725	−142.91	−30718	80.0	1745	1.551	.8798E−04	−.577
140	59.36	.035698	−139.98	−30315	81.0	1709	1.547	.1773E−03	−.568
145	59.84	.034745	−137.12	−29908	81.9	1674	1.544	.3385E−03	−.560
150	60.33	.033860	−134.33	−29496	82.6	1641	1.542	.6159E−03	−.552
155	60.82	.033036	−131.61	−29081	83.3	1609	1.543	.1073E−02	−.545
160	61.32	.032267	−128.96	−28664	83.9	1578	1.544	.1796E−02	−.538

165	61.83	.031550	− 126.37	− 28243	84.4	1547	1.545	.2902E − 02	− .531
170	62.35	.030879	− 123.84	− 27820	84.9	1516	1.548	.4539E − 02	− .524
175	62.88	.030252	− 121.37	− 27394	85.4	1486	1.551	.6893E − 02	− .517
180	63.42	.029664	− 118.96	− 26966	85.9	1456	1.554	.1020E − 01	− .509
185	63.97	.029114	− 116.60	− 26535	86.4	1425	1.558	.1472E − 01	− .501
190	64.54	.028598	− 114.29	− 26102	86.9	1395	1.562	.2076E − 01	− .493
195	65.12	.028115	− 112.03	− 25666	87.5	1364	1.566	.2869E − 01	− .483
200	65.71	.027662	− 109.80	− 25227	88.0	1334	1.570	.3889E − 01	− .473
205	66.32	.027237	− 107.62	− 24785	88.6	1303	1.574	.5180E − 01	− .462
210	66.95	.026840	− 105.48	− 24340	89.3	1272	1.578	.6787E − 01	− .450
215	67.59	.026469	− 103.37	− 23892	90.0	1240	1.582	.8760E − 01	− .438
220	68.26	.026122	− 101.29	− 23440	90.7	1209	1.587	.1115	− .424
225	68.95	.025799	− 99.24	− 22985	91.5	1177	1.591	.1401	− .409
230	69.66	.025499	− 97.22	− 22525	92.4	1146	1.596	.1738	− .393
235	70.40	.025221	− 95.23	− 22061	93.3	1114	1.601	.2133	− .376
240	71.17	.024965	− 93.25	− 21592	94.2	1081	1.606	.2589	− .357
245	71.96	.024730	− 91.30	− 21118	95.3	1049	1.611	.3114	− .336
250	72.80	.024516	− 89.36	− 20639	96.4	1016	1.617	.3710	− .314
255	73.67	.024323	− 87.44	− 20154	97.5	983.3	1.623	.4381	− .289
260	74.58	.024151	− 85.54	− 19664	98.8	950.1	1.630	.5131	− .262
265	75.54	.024000	− 83.64	− 19166	100.2	916.7	1.638	.5963	− .232
270	76.56	.023872	− 81.76	− 18662	101.6	883.0	1.647	.6879	− .199
275	77.63	.023767	− 79.88	− 18150	103.2	848.9	1.657	.7882	− .162
280	2874	.86406	− 22.60	− 2159	71.8	226.9	1.276	.8815	24.0
285	2955	.87288	− 21.33	− 1800	71.7	230.4	1.262	.8881	22.6

Table 2—continued

P/MPa										0.700
T_σ/K										279.254

	V_σ 78.60 / 2861 $\frac{V}{cm^3\,mol^{-1}}$	Z_σ .023697 / .86266 Z	S_σ -78.28 / -22.79 $\frac{S}{J\,K^{-1}\,mol^{-1}}$	H_σ -17707 / -2212 $\frac{H}{J\,mol^{-1}}$	$C_{P\sigma}$ 104.7 / 71.8 $\frac{C_P}{J\,K^{-1}\,mol^{-1}}$	w_σ 819.6 / 226.4 $\frac{w}{m\,sec^{-1}}$	γ_σ 1.667 / 1.278 γ	$(f/P)_\sigma$ .8805 / .8805 (f/P)	μ_σ -.127 / 24.3 $\frac{\mu}{K\,MPa^{-1}}$
Liq / Vap	78.60 / 2861	.023697 / .86266	-78.28 / -22.79	-17707 / -2212	104.7 / 71.8	819.6 / 226.4	1.667 / 1.278	.8805 / .8805	-.127 / 24.3
$\frac{T}{K}$									
290	3034	.88088	-20.09	-1441	71.7	233.7	1.249	.8943	21.2
295	3112	.88817	-18.86	-1082	71.9	236.9	1.238	.8999	20.0
300	3189	.89483	-17.65	-723	72.1	239.9	1.228	.9052	18.9
305	3264	.90094	-16.46	-362	72.3	242.9	1.219	.9101	17.9
310	3338	.90656	-15.28	1	72.7	245.8	1.211	.9147	16.9
315	3411	.91174	-14.11	365	73.0	248.5	1.204	.9190	16.1
320	3484	.91652	-12.96	731	73.4	251.3	1.197	.9230	15.2
325	3555	.92096	-11.82	1100	73.9	253.9	1.191	.9267	14.5
330	3626	.92507	-10.68	1470	74.4	256.5	1.186	.9303	13.8
335	3696	.92889	-9.56	1843	74.9	259.0	1.180	.9335	13.1
340	3766	.93245	-8.45	2219	75.4	261.4	1.175	.9366	12.5
345	3835	.93578	-7.34	2598	76.0	263.9	1.171	.9395	11.9
350	3903	.93888	-6.25	2979	76.5	266.2	1.167	.9423	11.4
355	3971	.94178	-5.16	3363	77.1	268.5	1.163	.9449	10.9
360	4039	.94450	-4.07	3750	77.7	270.8	1.159	.9473	10.4
365	4106	.94706	-3.00	4141	78.3	273.0	1.156	.9496	9.99
370	4173	.94945	-1.93	4534	79.0	275.3	1.152	.9517	9.57
380	4305	.95383	.20	5330	80.2	279.6	1.146	.9557	8.80
390	4436	.95771	2.30	6139	81.5	283.7	1.141	.9593	8.12
400	4567	.96117	4.38	6961	82.9	287.8	1.136	.9625	7.50

410	4696	.96427	6.44	7796	84.2	291.8	1.131	.9653	6.95
420	4824	.96705	8.49	8645	85.5	295.7	1.127	.9679	6.46
430	4952	.96955	10.51	9507	86.9	299.4	1.123	.9703	6.01
440	5079	.97181	12.53	10382	88.2	303.2	1.120	.9724	5.60
450	5205	.97386	14.52	11271	89.6	306.8	1.117	.9744	5.23
475	5519	.97819	19.46	13552	92.9	315.7	1.110	.9785	4.44
500	5830	.98164	24.31	15916	96.2	324.2	1.104	.9819	3.81
525	6139	.98442	29.08	18362	99.4	332.3	1.099	.9845	3.29
550	6446	.98668	33.78	20887	102.6	340.3	1.094	.9868	2.87
575	6751	.98854	38.40	23490	105.6	348.0	1.091	.9886	2.51

Table 2—continued

P/MPa					0.750				
T_σ/K					281.698				
	V_σ	Z_σ	S_σ	H_σ	C_P	w_σ	γ_σ	$(f/P)_\sigma$	μ_σ
Liq	79.16	.025349	−77.38	−17450	105.5	803.5	1.672	.8754	−.107
Vap	2673	.85589	−22.97	−2122	72.9	225.9	1.285	.8754	23.9
$\dfrac{T}{\text{K}}$	$\dfrac{V}{\text{cm}^3\,\text{mol}^{-1}}$	Z	$\dfrac{S}{\text{J K}^{-1}\,\text{mol}^{-1}}$	$\dfrac{H}{\text{J mol}^{-1}}$	$\dfrac{C_P}{\text{J K}^{-1}\,\text{mol}^{-1}}$	$\dfrac{w}{\text{m sec}^{-1}}$	γ	(f/P)	$\dfrac{\mu}{\text{K MPa}^{-1}}$
90	54.89	.055011	−174.51	−34214				.2892E−08	
95	55.30	.052510	−169.78	−33777				.1528E−07	
100	55.73	.050269	−165.76	−33385				.6738E−07	
105	56.16	.048248	−162.11	−33011				.2552E−06	
110	56.60	.046416	−158.66	−32641	74.2	1951	1.584	.8488E−06	−.632
115	57.05	.044748	−155.35	−32268	75.1	1908	1.582	.2521E−05	−.622
120	57.50	.043223	−152.13	−31889	76.3	1865	1.574	.6798E−05	−.610
125	57.96	.041824	−148.98	−31504	77.7	1823	1.565	.1680E−04	−.598
130	58.42	.040536	−145.91	−31113	78.9	1783	1.557	.3847E−04	−.587
135	58.89	.039347	−142.91	−30715	80.0	1745	1.551	.8233E−04	−.577
140	59.36	.038246	−139.98	−30313	81.0	1709	1.546	.1659E−03	−.568
145	59.84	.037226	−137.13	−29905	81.9	1675	1.544	.3167E−03	−.560
150	60.32	.036277	−134.34	−29494	82.6	1642	1.542	.5762E−03	−.552
155	60.82	.035394	−131.62	−29079	83.3	1609	1.542	.1003E−02	−.545
160	61.32	.034571	−128.96	−28661	83.9	1578	1.544	.1680E−02	−.538

165	61.83	.033802	−126.37	−28241	84.4	1547	1.545	.2715E−02	−.531
170	62.35	.033083	−123.85	−27817	84.9	1517	1.548	.4246E−02	−.524
175	62.88	.032411	−121.38	−27392	85.4	1486	1.551	.6448E−02	−.517
180	63.42	.031781	−118.97	−26963	85.9	1456	1.554	.9541E−02	−.510
185	63.97	.031191	−116.61	−26533	86.4	1426	1.558	.1377E−01	−.501
190	64.53	.030639	−114.30	−26099	86.9	1395	1.562	.1942E−01	−.493
195	65.11	.030121	−112.03	−25664	87.5	1365	1.566	.2683E−01	−.483
200	65.71	.029635	−109.81	−25225	88.0	1334	1.570	.3637E−01	−.473
205	66.32	.029181	−107.63	−24783	88.6	1303	1.574	.4844E−01	−.462
210	66.94	.028755	−105.49	−24338	89.3	1272	1.578	.6347E−01	−.451
215	67.59	.028357	−103.38	−23890	90.0	1241	1.582	.8192E−01	−.438
220	68.25	.027986	−101.30	−23438	90.7	1210	1.587	.1043	−.424
225	68.94	.027640	−99.25	−22983	91.5	1178	1.591	.1310	−.409
230	69.65	.027318	−97.23	−22523	92.4	1146	1.596	.1625	−.393
235	70.39	.027020	−95.24	−22059	93.3	1114	1.600	.1994	−.376
240	71.16	.026745	−93.26	−21590	94.2	1082	1.605	.2421	−.357
245	71.96	.026493	−91.31	−21117	95.3	1049	1.611	.2912	−.336
250	72.79	.026263	−89.37	−20638	96.4	1017	1.617	.3469	−.314
255	73.66	.026056	−87.45	−20153	97.5	983.9	1.623	.4096	−.290
260	74.57	.025872	−85.55	−19662	98.8	950.8	1.630	.4797	−.263
265	75.53	.025710	−83.65	−19165	100.1	917.4	1.638	.5575	−.233
270	76.54	.025572	−81.77	−18661	101.6	883.7	1.646	.6432	−.200
275	77.61	.025459	−79.89	−18149	103.2	849.7	1.656	.7369	−.163
280	78.76	.025372	−78.01	−17629	104.9	815.3	1.668	.8389	−.122
285	2724	.86221	−22.12	−1882	72.8	228.2	1.275	.8801	22.9

Table 2—continued

P/MPa				0.750					
Tσ/K				281.698					
	V_σ 79.16	Z_σ .025349	S_σ -77.38	H_σ -17450	C_P 105.5	w_σ 803.5	γ_σ 1.672	$(f/P)_\sigma$ .8754	μ_σ -.107
Liq	2673	.85589	-22.97	-2122	72.9	225.9	1.285	.8754	23.9
Vap									
$\frac{T}{K}$	$\frac{V}{cm^3\,mol^{-1}}$	Z	$\frac{S}{J\,K^{-1}\,mol^{-1}}$	$\frac{H}{J\,mol^{-1}}$	$\frac{C_P}{J\,K^{-1}\,mol^{-1}}$	$\frac{w}{m\,sec^{-1}}$	γ	(f/P)	$\frac{\mu}{K\,MPa^{-1}}$
290	2800	.87102	-20.85	-1519	72.7	231.7	1.260	.8866	21.5
295	2875	.87903	-19.61	-1155	72.7	235.0	1.248	.8927	20.3
300	2948	.88633	-18.39	-792	72.8	238.2	1.237	.8984	19.1
305	3019	.89300	-17.18	-427	73.0	241.3	1.227	.9037	18.1
310	3090	.89914	-16.00	-61	73.3	244.2	1.218	.9086	17.1
315	3160	.90478	-14.82	306	73.6	247.1	1.210	.9132	16.2
320	3228	.90999	-13.66	675	74.0	249.9	1.203	.9175	15.4
325	3296	.91481	-12.51	1046	74.4	252.6	1.196	.9215	14.6
330	3363	.91928	-11.37	1419	74.8	255.2	1.190	.9253	13.9
335	3429	.92342	-10.24	1794	75.3	257.8	1.185	.9288	13.2
340	3495	.92729	-9.12	2172	75.8	260.3	1.179	.9321	12.6
345	3560	.93088	-8.01	2552	76.3	262.8	1.175	.9352	12.0
350	3625	.93424	-6.91	2935	76.9	265.2	1.170	.9382	11.5
355	3689	.93738	-5.81	3321	77.5	267.6	1.166	.9409	11.0
360	3753	.94032	-4.73	3710	78.0	269.9	1.162	.9435	10.5
365	3816	.94308	-3.65	4101	78.6	272.2	1.158	.9460	10.0
370	3879	.94567	-2.57	4496	79.2	274.4	1.155	.9483	9.61
380	4004	.95039	-.44	5294	80.5	278.8	1.148	.9526	8.83
390	4127	.95458	1.66	6106	81.8	283.0	1.143	.9564	8.14
400	4249	.95831	3.75	6930	83.0	287.1	1.137	.9598	7.53

410	4371	.96165	5.82	7767	84.4	291.1	1.133	.9629	6.97
420	4491	.96464	7.87	8617	85.7	295.1	1.128	.9657	6.47
430	4611	.96733	9.90	9480	87.0	298.9	1.125	.9682	6.02
440	4730	.96976	11.91	10357	88.4	302.7	1.121	.9705	5.61
450	4849	.97196	13.91	11248	89.7	306.3	1.118	.9726	5.24
475	5143	.97662	18.85	13531	93.0	315.3	1.111	.9770	4.45
500	5434	.98032	23.71	15898	96.3	323.8	1.104	.9806	3.81
525	5723	.98331	28.48	18345	99.5	332.0	1.099	.9834	3.29
550	6010	.98573	33.18	20872	102.6	340.0	1.095	.9858	2.87
575	6296	.98773	37.81	23476	105.7	347.8	1.091	.9878	2.51

Table 2—continued

P/MPa									0.800
Tσ/K									284.025
	V_σ 79.71 2507	Z_σ .027003 .84925	S_σ -76.52 -23.13	H_σ -17203 -2039	C_P 106.4 74.0	w_σ 788.1 225.3	γ_σ 1.678 1.292	$(f/P)_\sigma$ .8705 .8705	μ_σ -.086 23.5
Liq Vap									
$\dfrac{T}{K}$	$\dfrac{V}{cm^3\,mol^{-1}}$	Z	$\dfrac{S}{J\,K^{-1}\,mol^{-1}}$	$\dfrac{H}{J\,mol^{-1}}$	$\dfrac{C_P}{J\,K^{-1}\,mol^{-1}}$	$\dfrac{w}{m\,sec^{-1}}$	γ	(f/P)	$\dfrac{\mu}{K\,MPa^{-1}}$
90	54.88	.058677	-174.51	-34212				.2721E-08	
95	55.30	.056009	-169.79	-33775				.1437E-07	
100	55.73	.053618	-165.76	-33383				.6338E-07	
105	56.16	.051463	-162.11	-33008				.2401E-06	
110	56.60	.049509	-158.67	-32638	74.1	1951	1.584	.7982E-06	-.632
115	57.05	.047730	-155.35	-32265	75.1	1909	1.582	.2371E-05	-.622
120	57.50	.046103	-152.13	-31887	76.3	1865	1.574	.6391E-05	-.610
125	57.95	.044611	-148.99	-31502	77.7	1823	1.565	.1579E-04	-.598
130	58.42	.043237	-145.92	-31110	78.9	1783	1.557	.3616E-04	-.587
135	58.88	.041968	-142.92	-30713	80.0	1745	1.551	.7739E-04	-.577
140	59.36	.040795	-139.99	-30310	81.0	1709	1.546	.1559E-03	-.568
145	59.84	.039706	-137.13	-29903	81.9	1675	1.544	.2977E-03	-.560
150	60.32	.038694	-134.34	-29492	82.6	1642	1.542	.5415E-03	-.552
155	60.81	.037752	-131.62	-29077	83.3	1610	1.542	.9430E-03	-.545
160	61.32	.036874	-128.97	-28659	83.9	1578	1.543	.1579E-02	-.538

165	61.83	.036054	− 126.38	− 28238	84.4	1548	1.545	.2551E − 02	− .531
170	62.34	.035287	− 123.85	− 27815	84.9	1517	1.548	.3990E − 02	− .524
175	62.87	.034570	− 121.38	− 27389	85.4	1487	1.551	.6058E − 02	− .517
180	63.41	.033898	− 118.97	− 26961	85.9	1456	1.554	.8964E − 02	− .510
185	63.97	.033269	− 116.61	− 26531	86.4	1426	1.558	.1293E − 01	− .501
190	64.53	.032679	− 114.30	− 26097	86.9	1396	1.562	.1824E − 01	− .493
195	65.11	.032127	− 112.04	− 25661	87.4	1365	1.566	.2520E − 01	− .483
200	65.70	.031609	− 109.82	− 25223	88.0	1334	1.570	.3416E − 01	− .473
205	66.31	.031124	− 107.64	− 24781	88.6	1304	1.574	.4550E − 01	− .462
210	66.94	.030670	− 105.49	− 24336	89.3	1273	1.578	.5962E − 01	− .451
215	67.58	.030245	− 103.38	− 23888	90.0	1241	1.582	.7694E − 01	− .438
220	68.25	.029849	− 101.31	− 23437	90.7	1210	1.586	.9792E − 01	− .424
225	68.94	.029480	− 99.26	− 22981	91.5	1178	1.591	.1230	− .410
230	69.65	.029136	− 97.24	− 22521	92.4	1147	1.595	.1527	− .394
235	70.38	.028818	− 95.24	− 22057	93.3	1115	1.600	.1873	− .376
240	71.15	.028525	− 93.27	− 21589	94.2	1082	1.605	.2274	− .357
245	71.95	.028256	− 91.32	− 21115	95.2	1050	1.610	.2735	− .337
250	72.78	.028011	− 89.38	− 20636	96.3	1017	1.616	.3258	− .314
255	73.65	.027789	− 87.46	− 20152	97.5	984.5	1.623	.3846	− .290
260	74.56	.027592	− 85.56	− 19661	98.8	951.4	1.629	.4505	− .263
265	75.52	.027420	− 83.66	− 19164	100.1	918.0	1.637	.5235	− .234
270	76.53	.027272	− 81.78	− 18660	101.6	884.4	1.646	.6040	− .201
275	77.60	.027151	− 79.90	− 18148	103.1	850.4	1.656	.6921	− .164
280	78.74	.027057	− 78.03	− 17628	104.9	816.1	1.667	.7878	− .123
285	2521.	.85125	− 22.88	− 1967	73.9	226.0	1.289	.8720	23.2

Table 2—continued

P/MPa					0.800					
T_σ/K					284.025					
	V_σ	Z_σ	S_σ	H_σ	C_P	w_σ	γ_σ	$(f/P)_\sigma$	μ_σ	
Liq	79.71	.027003	−76.52	−17203	106.4	788.1	1.678	.8705	−.086	
Vap	2507	.84925	−23.13	−2039	74.0	225.3	1.292	.8705	23.5	
$\dfrac{T}{K}$	$\dfrac{V}{cm^3\,mol^{-1}}$	Z	$\dfrac{S}{J\,K^{-1}\,mol^{-1}}$	$\dfrac{H}{J\,mol^{-1}}$	$\dfrac{C_P}{J\,K^{-1}\,mol^{-1}}$	$\dfrac{w}{m\,sec^{-1}}$	γ	(f/P)	$\dfrac{\mu}{K\,MPa^{-1}}$	
290	2595	.86092	−21.59	−1598	73.7	229.7	1.272	.8790	21.8	
295	2666	.86968	−20.33	−1230	73.6	233.1	1.258	.8855	20.5	
300	2736	.87765	−19.10	−862	73.6	236.4	1.246	.8916	19.3	
305	2805	.88492	−17.88	−494	73.7	239.6	1.235	.8972	18.3	
310	2873	.89159	−16.68	−125	73.9	242.7	1.226	.9025	17.3	
315	2939	.89772	−15.49	246	74.2	245.6	1.217	.9074	16.3	
320	3004	.90337	−14.32	617	74.5	248.5	1.209	.9120	15.5	
325	3069	.90858	−13.17	991	74.9	251.3	1.202	.9163	14.7	
330	3133	.91341	−12.02	1366	75.3	254.0	1.195	.9203	14.0	
335	3196	.91790	−10.88	1744	75.7	256.6	1.189	.9241	13.3	
340	3258	.92206	−9.76	2124	76.2	259.2	1.183	.9276	12.7	
345	3320	.92595	−8.64	2506	76.7	261.7	1.178	.9309	12.1	
350	3381	.92957	−7.54	2891	77.2	264.2	1.173	.9341	11.5	
355	3442	.93295	−6.44	3278	77.8	266.6	1.169	.9370	11.0	
360	3502	.93611	−5.34	3669	78.3	268.9	1.165	.9398	10.5	
365	3562	.93908	−4.26	4062	78.9	271.3	1.161	.9424	10.1	
370	3622	.94186	−3.18	4458	79.5	273.5	1.157	.9449	9.65	
380	3740	.94694	−1.05	5259	80.7	278.0	1.150	.9494	8.87	
390	3856	.95143	1.07	6072	82.0	282.3	1.144	.9535	8.17	
400	3972	.95544	3.16	6898	83.2	286.5	1.139	.9571	7.55	

410	4086	.95901	5.23	7737	84.5	290.5	1.134	.9604	6.99
420	4200	.96222	7.28	8589	85.8	294.5	1.130	.9634	6.49
430	4313	.96511	9.32	9454	87.2	298.4	1.126	.9661	6.04
440	4425	.96771	11.34	10332	88.5	302.2	1.122	.9685	5.62
450	4537	.97006	13.34	11224	89.8	305.9	1.119	.9707	5.25
475	4813	.97505	18.29	13511	93.1	314.8	1.111	.9755	4.45
500	5087	.97901	23.14	15879	96.4	323.5	1.105	.9793	3.82
525	5359	.98219	27.92	18329	99.6	331.7	1.100	.9824	3.30
550	5629	.98479	32.63	20857	102.7	339.8	1.095	.9849	2.87
575	5898	.98692	37.26	23463	105.7	347.5	1.091	.9870	2.52

Table 2—continued

P/MPa					0.850					
T_σ/K					286.248					
	V_σ 80.25 2360	Z_σ .028661 .84272	S_σ −75.70 −23.28	H_σ −16965 −1960	$C_{P\sigma}$ 107.2 75.0	w_σ 773.4 224.7	γ_σ 1.683 1.299	$(f/P)_\sigma$ .8658 .8658		μ_σ −.065 23.2
Liq Vap										
$\dfrac{T}{K}$	$\dfrac{V}{\text{cm}^3\,\text{mol}^{-1}}$	Z	$\dfrac{S}{\text{J K}^{-1}\,\text{mol}^{-1}}$	$\dfrac{H}{\text{J mol}^{-1}}$	$\dfrac{C_P}{\text{J K}^{-1}\,\text{mol}^{-1}}$	$\dfrac{w}{\text{m sec}^{-1}}$	γ	(f/P)		$\dfrac{\mu}{\text{K MPa}^{-1}}$
90	54.88	.062343	−174.52	−34209				.2571E−08		
95	55.30	.059508	−169.79	−33772				.1357E−07		
100	55.72	.056968	−165.77	−33380				.5985E−07		
105	56.16	.054678	−162.12	−33006				.2267E−06		
110	56.60	.052602	−158.67	−32636	74.1	1952	1.584	.7536E−06		−.632
115	57.04	.050712	−155.36	−32263	75.1	1909	1.582	.2238E−05		−.622
120	57.50	.048983	−152.14	−31885	76.3	1865	1.574	.6033E−05		−.610
125	57.95	.047397	−148.99	−31500	77.7	1823	1.565	.1491E−04		−.598
130	58.41	.045938	−145.92	−31108	78.9	1783	1.557	.3413E−04		−.587
135	58.88	.044590	−142.92	−30711	80.0	1745	1.551	.7303E−04		−.577
140	59.35	.043343	−139.99	−30308	81.0	1710	1.546	.1471E−03		−.568
145	59.83	.042186	−137.14	−29901	81.9	1675	1.544	.2809E−03		−.560
150	60.32	.041111	−134.35	−29489	82.6	1642	1.542	.5109E−03		−.552
155	60.81	.040110	−131.63	−29075	83.3	1610	1.542	.8896E−03		−.545
160	61.31	.039176	−128.97	−28657	83.9	1579	1.543	.1489E−02		−.538

165	61.82	.038305	−126.39	−28236	84.4	1548	1.545	.2406E−02	−.532
170	62.34	.037490	−123.86	−27813	84.9	1517	1.548	.3763E−02	−.525
175	62.87	.036728	−121.39	−27387	85.4	1487	1.551	.5715E−02	−.517
180	63.41	.036015	−118.98	−26959	85.9	1457	1.554	.8454E−02	−.510
185	63.96	.035346	−116.62	−26528	86.4	1426	1.558	.1220E−01	−.502
190	64.53	.034720	−114.31	−26095	86.9	1396	1.562	.1720E−01	−.493
195	65.10	.034133	−112.04	−25659	87.4	1365	1.565	.2377E−01	−.484
200	65.70	.033582	−109.82	−25221	88.0	1335	1.569	.3222E−01	−.473
205	66.31	.033067	−107.64	−24779	88.6	1304	1.574	.4291E−01	−.463
210	66.93	.032584	−105.50	−24334	89.3	1273	1.578	.5622E−01	−.451
215	67.58	.032133	−103.39	−23886	90.0	1242	1.582	.7255E−01	−.438
220	68.24	.031712	−101.31	−23435	90.7	1210	1.586	.9234E−01	−.425
225	68.93	.031319	−99.27	−22979	91.5	1179	1.591	.1160	−.410
230	69.64	.030954	−97.25	−22520	92.3	1147	1.595	.1439	−.394
235	70.38	.030616	−95.25	−22056	93.2	1115	1.600	.1766	−.376
240	71.14	.030304	−93.28	−21587	94.2	1083	1.605	.2144	−.358
245	71.94	.030018	−91.32	−21114	95.2	1051	1.610	.2578	−.337
250	72.77	.029757	−89.39	−20635	96.3	1018	1.616	.3071	−.315
255	73.64	.029522	−87.47	−20150	97.5	985.1	1.622	.3626	−.291
260	74.55	.029312	−85.57	−19660	98.7	952.0	1.629	.4247	−.264
265	75.50	.029128	−83.67	−19163	100.1	918.7	1.637	.4936	−.234
270	76.51	.028971	−81.79	−18659	101.5	885.1	1.645	.5694	−.202
275	77.58	.028842	−79.91	−18147	103.1	851.2	1.655	.6525	−.165
280	78.72	.028742	−78.04	−17627	104.8	816.9	1.667	.7427	−.124
285	79.93	.028673	−76.17	−17099	106.7	782.1	1.680	.8402	−.078

Table 2—continued

P/MPa										0.850
T_σ/K										286.248
Liq	V_σ 80.25	Z_σ .028661	S_σ -75.70	H_σ -16965	C_P 107.2	w_σ 773.4	γ_σ 1.683	$(f/P)_\sigma$ .8658	μ_σ -.065	
Vap	2360	.84272	-23.28	-1960	75.0	224.7	1.299	.8658	23.2	
$\dfrac{T}{K}$	$\dfrac{V}{cm^3\,mol^{-1}}$	Z	$\dfrac{S}{J\,K^{-1}\,mol^{-1}}$	$\dfrac{H}{J\,mol^{-1}}$	$\dfrac{C_P}{J\,K^{-1}\,mol^{-1}}$	$\dfrac{w}{m\,sec^{-1}}$	γ	(f/P)	$\dfrac{\mu}{K\,MPa^{-1}}$	
290	2413	.85056	-22.30	-1679	74.7	227.6	1.286	.8714	22.1	
295	2482	.86012	-21.03	-1306	74.5	231.2	1.270	.8783	20.8	
300	2549	.86879	-19.78	-934	74.4	234.6	1.256	.8848	19.6	
305	2616	.87669	-18.55	-562	74.5	237.9	1.244	.8908	18.4	
310	2680	.88392	-17.33	-189	74.6	241.1	1.233	.8964	17.4	
315	2744	.89055	-16.14	185	74.8	244.1	1.224	.9016	16.5	
320	2807	.89665	-14.96	559	75.1	247.1	1.215	.9065	15.6	
325	2868	.90228	-13.79	935	75.4	249.9	1.207	.9110	14.8	
330	2929	.90748	-12.64	1313	75.8	252.7	1.200	.9153	14.1	
335	2989	.91231	-11.50	1693	76.2	255.4	1.194	.9193	13.4	
340	3049	.91679	-10.37	2075	76.6	258.1	1.188	.9231	12.8	
345	3108	.92096	-9.24	2459	77.1	260.6	1.182	.9266	12.2	
350	3166	.92485	-8.13	2846	77.6	263.1	1.177	.9299	11.6	
355	3224	.92848	-7.03	3235	78.1	265.6	1.172	.9331	11.1	
360	3281	.93187	-5.93	3627	78.6	268.0	1.168	.9360	10.6	
365	3338	.93505	-4.84	4022	79.2	270.4	1.164	.9388	10.1	
370	3395	.93803	-3.76	4419	79.8	272.7	1.160	.9414	9.69	
380	3507	.94346	-1.62	5223	81.0	277.2	1.153	.9463	8.90	
390	3617	.94827	.50	6039	82.2	281.5	1.146	.9506	8.20	
400	3727	.95255.	2.60	6867	83.4	285.8	1.141	.9545	7.57	

0.850 *MPa*

410	3835	.95637	4.67	7708	84.7	289.9	1.136	.9580	7.01
420	3943	.95980	6.73	8561	86.0	293.9	1.131	.9611	6.51
430	4050	.96287	8.77	9428	87.3	297.8	1.127	.9640	6.05
440	4156	.96565	10.79	10308	88.6	301.6	1.123	.9666	5.64
450	4262	.96816	12.80	11200	89.9	305.4	1.120	.9689	5.26
475	4523	.97347	17.75	13490	93.2	314.4	1.112	.9740	4.46
500	4782	.97769	22.61	15861	96.5	323.1	1.106	.9780	3.82
525	5038	.98108	27.40	18312	99.6	331.4	1.100	.9813	3.30
550	5293	.98384	32.11	20843	102.8	339.5	1.096	.9839	2.87
575	5546	.98611	36.74	23450	105.8	347.3	1.092	.9861	2.52

Table 2—continued

					0.900				
P/MPa									
T_σ/K					288.377				
	V_σ 80.78	Z_σ .030322	S_σ −74.91	H_σ −16736	C_P 108.1	w_σ 759.2	γ_σ 1.689	$(f/P)_\sigma$ .8612	μ_σ −.044
Liq Vap	2228	.83630	−23.42	−1887	76.1	224.1	1.306	.8612	22.9
$\dfrac{T}{K}$	$\dfrac{V}{\mathrm{cm^3\,mol^{-1}}}$	Z	$\dfrac{S}{\mathrm{J\,K^{-1}\,mol^{-1}}}$	$\dfrac{H}{\mathrm{J\,mol^{-1}}}$	$\dfrac{C_P}{\mathrm{J\,K^{-1}\,mol^{-1}}}$	$\dfrac{w}{\mathrm{m\,sec^{-1}}}$	γ	(f/P)	$\dfrac{\mu}{\mathrm{K\,MPa^{-1}}}$
90	54.88	.066008	−174.52	−34207				.2437E−08	
95	55.30	.063007	−169.80	−33770				.1287E−07	
100	55.72	.060318	−165.77	−33378				.5671E−07	
105	56.16	.057893	−162.12	−33004				.2148E−06	
110	56.60	.055695	−158.68	−32633	74.1	1952	1.584	.7140E−06	−.632
115	57.04	.053693	−155.36	−32261	75.1	1909	1.582	.2120E−05	−.622
120	57.49	.051863	−152.14	−31882	76.3	1866	1.574	.5714E−05	−.611
125	57.95	.050184	−149.00	−31497	77.7	1824	1.565	.1412E−04	−.598
130	58.41	.048638	−145.93	−31106	78.9	1784	1.557	.3232E−04	−.587
135	58.88	.047211	−142.93	−30708	80.0	1746	1.551	.6915E−04	−.577
140	59.35	.045891	−140.00	−30306	81.0	1710	1.546	.1393E−03	−.568
145	59.83	.044666	−137.14	−29898	81.9	1676	1.544	.2659E−03	−.560
150	60.32	.043527	−134.35	−29487	82.6	1642	1.542	.4837E−03	−.552
155	60.81	.042467	−131.63	−29072	83.3	1610	1.542	.8422E−03	−.545
160	61.31	.041479	−128.98	−28655	83.9	1579	1.543	.1410E−02	−.538

165	61.82	.040556	−126.39	−28234	84.4	1548	1.545	.2278E−02	−.532
170	62.34	.039694	−123.86	−27811	84.9	1518	1.548	.3562E−02	−.525
175	62.87	.038887	−121.40	−27385	85.4	1487	1.551	.5409E−02	−.517
180	63.41	.038131	−118.98	−26957	85.9	1457	1.554	.8002E−02	−.510
185	63.96	.037423	−116.62	−26526	86.4	1427	1.558	.1154E−01	−.502
190	64.52	.036760	−114.31	−26093	86.9	1396	1.561	.1628E−01	−.493
195	65.10	.036138	−112.05	−25657	87.4	1366	1.565	.2249E−01	−.484
200	65.69	.035555	−109.83	−25219	88.0	1335	1.569	.3049E−01	−.474
205	66.30	.035009	−107.65	−24777	88.6	1304	1.573	.4060E−01	−.463
210	66.93	.034498	−105.50	−24332	89.3	1273	1.578	.5320E−01	−.451
215	67.57	.034020	−103.40	−23884	90.0	1242	1.582	.6865E−01	−.438
220	68.24	.033574	−101.32	−23433	90.7	1211	1.586	.8737E−01	−.425
225	68.92	.033158	−99.27	−22977	91.5	1179	1.590	.1097	−.410
230	69.63	.032772	−97.25	−22518	92.3	1148	1.595	.1362	−.394
235	70.37	.032413	−95.26	−22054	93.2	1116	1.600	.1671	−.377
240	71.13	.032083	−93.28	−21585	94.2	1083	1.605	.2029	−.358
245	71.93	.031780	−91.33	−21112	95.2	1051	1.610	.2440	−.338
250	72.76	.031503	−89.40	−20633	96.3	1018	1.616	.2906	−.315
255	73.63	.031254	−87.48	−20149	97.5	985.7	1.622	.3431	−.291
260	74.53	.031031	−85.58	−19658	98.7	952.7	1.629	.4018	−.264
265	75.49	.030836	−83.68	−19162	100.1	919.4	1.636	.4670	−.235
270	76.50	.030669	−81.80	−18658	101.5	885.8	1.645	.5387	−.202
275	77.57	.030532	−79.92	−18146	103.1	851.9	1.655	.6173	−.166
280	78.70	.030425	−78.05	−17627	104.8	817.7	1.666	.7026	−.125
285	79.91	.030352	−76.18	−17098	106.7	783.0	1.679	.7949	−.079

141

Table 2—continued

P/MPa					0.900					
T_σ/K					288.377					
	V_σ	Z_σ	S_σ	H_σ	C_P	w_σ	γ_σ	$(f/P)_\sigma$	μ_σ	
Liq	80.78	.030322	−74.91	−16736	108.1	759.2	1.689	.8612	−.044	
Vap	2228	.83630	−23.42	−1887	76.1	224.1	1.306	.8612	22.9	
$\dfrac{T}{K}$	$\dfrac{V}{\text{cm}^3\,\text{mol}^{-1}}$	Z	$\dfrac{S}{\text{J K}^{-1}\,\text{mol}^{-1}}$	$\dfrac{H}{\text{J mol}^{-1}}$	$\dfrac{C_P}{\text{J K}^{-1}\,\text{mol}^{-1}}$	$\dfrac{w}{\text{m sec}^{-1}}$	γ	(f/P)	$\dfrac{\mu}{\text{K MPa}^{-1}}$	
290	2250	.83992	−23.00	−1763	75.9	225.4	1.300	.8637	22.4	
295	2317	.85033	−21.70	−1385	75.5	229.2	1.282	.8711	21.1	
300	2383	.85974	−20.43	−1008	75.3	232.8	1.267	.8779	19.8	
305	2447	.86830	−19.19	−631	75.3	236.2	1.253	.8843	18.6	
310	2509	.87611	−17.96	−255	75.3	239.5	1.241	.8903	17.6	
315	2570	.88326	−16.76	122	75.5	242.6	1.231	.8958	16.6	
320	2631	.88983	−15.57	500	75.7	245.6	1.221	.9010	15.8	
325	2690	.89589	−14.39	879	75.9	248.6	1.213	.9058	14.9	
330	2748	.90148	−13.23	1260	76.3	251.4	1.205	.9103	14.2	
335	2806	.90666	−12.08	1642	76.6	254.2	1.198	.9146	13.5	
340	2863	.91146	−10.94	2026	77.0	256.9	1.192	.9186	12.8	
345	2919	.91593	−9.82	2412	77.5	259.5	1.186	.9223	12.2	
350	2975	.92009	−8.70	2801	78.0	262.1	1.181	.9258	11.7	
355	3030	.92397	−7.59	3192	78.4	264.6	1.176	.9291	11.1	
360	3085	.92760	−6.49	3585	79.0	267.1	1.171	.9323	10.6	
365	3139	.93099	−5.40	3981	79.5	269.5	1.167	.9352	10.2	
370	3193	.93418	−4.31	4380	80.1	271.8	1.162	.9380	9.74	
380	3300	.93997	−2.16	5187	81.2	276.4	1.155	.9431	8.94	
390	3405	.94510	−.03	6005	82.4	280.8	1.148	.9477	8.23	
400	3509	.94965	2.07	6835	83.6	285.1	1.143	.9518	7.60	

410	3612	.95372	4.15	7678	84.9	289.3	1.137	.9555	7.03
420	3715	.95736	6.21	8533	86.2	293.3	1.133	.9588	6.52
430	3816	.96063	8.25	9401	87.5	297.3	1.128	.9619	6.06
440	3917	.96359	10.28	10283	88.8	301.1	1.124	.9646	5.65
450	4017	.96625	12.29	11177	90.1	304.9	1.121	.9671	5.27
475	4265	.97189	17.24	13469	93.3	314.0	1.113	.9724	4.47
500	4510	.97637	22.11	15843	96.5	322.8	1.106	.9767	3.83
525	4753	.97997	26.90	18296	99.7	331.1	1.101	.9802	3.30
550	4994	.98290	31.61	20828	102.8	339.2	1.096	.9830	2.88
575	5234	.98530	36.25	23436	105.8	347.1	1.092	.9853	2.52

Table 2—continued

P/MPa									
	0.950								

T_σ/K									
	290.421								

	V_σ	Z_σ	S_σ	H_σ	$C_{P,\sigma}$	w_σ	γ_σ	$(f/P)_\sigma$	μ_σ
Liq	81.30	.031987	−74.16	−16514	108.9	745.6	1.695	.8567	−.023
Vap	2110	.82997	−23.56	−1817	77.1	223.5	1.314	.8567	22.6

$\dfrac{T}{K}$	$\dfrac{V}{\text{cm}^3\,\text{mol}^{-1}}$	Z	$\dfrac{S}{\text{J K}^{-1}\,\text{mol}^{-1}}$	$\dfrac{H}{\text{J mol}^{-1}}$	$\dfrac{C_P}{\text{J K}^{-1}\,\text{mol}^{-1}}$	$\dfrac{w}{\text{m sec}^{-1}}$	γ	(f/P)	$\dfrac{\mu}{\text{K MPa}^{-1}}$
90	54.88	.069674	−174.53	−34204				.2317E−08	
95	55.30	.066506	−169.80	−33768				.1223E−07	
100	55.72	.063667	−165.78	−33376				.5391E−07	
105	56.15	.061107	−162.12	−33001				.2041E−06	
110	56.60	.058787	−158.68	−32631	74.1	1952	1.584	.6785E−06	−.632
115	57.04	.056674	−155.37	−32258	75.1	1909	1.582	.2015E−05	−.622
120	57.49	.054743	−152.15	−31880	76.3	1866	1.574	.5429E−05	−.611
125	57.95	.052970	−149.00	−31495	77.7	1824	1.565	.1341E−04	−.598
130	58.41	.051338	−145.93	−31104	78.9	1784	1.557	.3070E−04	−.587
135	58.88	.049832	−142.93	−30706	80.0	1746	1.551	.6568E−04	−.577
140	59.35	.048438	−140.00	−30303	81.0	1710	1.546	.1323E−03	−.568
145	59.83	.047145	−137.15	−29896	81.9	1676	1.543	.2526E−03	−.560
150	60.31	.045944	−134.36	−29485	82.6	1643	1.542	.4593E−03	−.552
155	60.81	.044825	−131.64	−29070	83.3	1611	1.542	.7997E−03	−.545
160	61.31	.043782	−128.98	−28652	83.8	1579	1.543	.1339E−02	−.538

165	61.82	.042807	−126.40	−28232	84.4	1548	1.545	.2163E−02	−.532
170	62.34	.041897	−123.87	−27808	84.9	1518	1.548	.3382E−02	−.525
175	62.86	.041045	−121.40	−27383	85.4	1488	1.551	.5136E−02	−.517
180	63.40	.040247	−118.99	−26955	85.9	1457	1.554	.7597E−02	−.510
185	63.95	.039500	−116.63	−26524	86.4	1427	1.558	.1096E−01	−.502
190	64.52	.038800	−114.32	−26091	86.9	1397	1.561	.1546E−01	−.493
195	65.10	.038143	−112.06	−25655	87.4	1366	1.565	.2135E−01	−.484
200	65.69	.037528	−109.83	−25217	88.0	1336	1.569	.2894E−01	−.474
205	66.30	.036952	−107.65	−24775	88.6	1305	1.573	.3854E−01	−.463
210	66.92	.036412	−105.51	−24330	89.3	1274	1.577	.5049E−01	−.451
215	67.57	.035907	−103.40	−23882	90.0	1243	1.582	.6516E−01	−.439
220	68.23	.035436	−101.33	−23431	90.7	1211	1.586	.8293E−01	−.425
225	68.92	.034997	−99.28	−22975	91.5	1180	1.590	.1042	−.410
230	69.63	.034589	−97.26	−22516	92.3	1148	1.595	.1293	−.394
235	70.36	.034211	−95.27	−22052	93.2	1116	1.599	.1586	−.377
240	71.12	.033861	−93.29	−21584	94.2	1084	1.604	.1925	−.358
245	71.92	.033541	−91.34	−21110	95.2	1052	1.610	.2315	−.338
250	72.75	.033249	−89.41	−20632	96.3	1019	1.615	.2758	−.316
255	73.61	.032985	−87.49	−20147	97.4	986.3	1.621	.3256	−.291
260	74.52	.032750	−85.58	−19657	98.7	953.3	1.628	.3813	−.265
265	75.48	.032544	−83.69	−19160	100.0	920.1	1.636	.4431	−.236
270	76.48	.032367	−81.81	−18657	101.5	886.5	1.644	.5112	−.203
275	77.55	.032221	−79.93	−18145	103.0	852.7	1.654	.5858	−.167
280	78.68	.032108	−78.06	−17626	104.7	818.5	1.665	.6668	−.126
285	79.89	.032029	−76.19	−17098	106.6	783.8	1.678	.7543	−.080

Table 2—*continued*

P/MPa					0.950				
T_σ/K					290.421				
Liq	V_σ 81.30	Z_σ .031987	S_σ -74.16	H_σ -16514	C_P 108.9	w_σ 745.6	γ_σ 1.695	$(f/P)_\sigma$ .8567	μ_σ -.023
Vap	2110	.82997	-23.56	-1817	77.1	223.5	1.314	.8567	22.6
T/K	V cm³mol⁻¹	Z	S JK⁻¹mol⁻¹	H Jmol⁻¹	C_P JK⁻¹mol⁻¹	w m sec⁻¹	γ	(f/P)	μ K MPa⁻¹
290	81.19	.031989	-74.32	-16559	108.7	748.6	1.693	.8485	-.028
295	2169	.84029	-22.35	-1465	76.6	227.1	1.295	.8638	21.3
300	2233	.85049	-21.07	-1083	76.3	230.8	1.278	.8711	20.0
305	2295	.85974	-19.81	-702	76.1	234.4	1.263	.8778	18.9
310	2355	.86816	-18.57	-322	76.1	237.8	1.250	.8841	17.8
315	2415	.87586	-17.36	59	76.1	241.1	1.239	.8900	16.8
320	2473	.88292	-16.16	440	76.3	244.2	1.228	.8955	15.9
325	2530	.88941	-14.97	822	76.5	247.2	1.219	.9006	15.1
330	2586	.89540	-13.80	1205	76.8	250.1	1.211	.9054	14.3
335	2641	.90094	-12.64	1590	77.1	253.0	1.203	.9098	13.6
340	2696	.90608	-11.50	1976	77.5	255.7	1.196	.9140	12.9
345	2750	.91085	-10.37	2365	77.9	258.4	1.190	.9180	12.3
350	2804	.91529	-9.24	2755	78.3	261.0	1.184	.9217	11.7
355	2857	.91943	-8.13	3148	78.8	263.6	1.179	.9252	11.2
360	2909	.92329	-7.02	3543	79.3	266.1	1.174	.9285	10.7
365	2961	.92691	-5.92	3941	79.8	268.5	1.169	.9316	10.2
370	3012	.93029	-4.83	4341	80.3	270.9	1.165	.9346	9.78
380	3114	.93646	-2.68	5150	81.5	275.6	1.157	.9400	8.97
390	3215	.94190	-.55	5971	82.6	280.1	1.151	.9448	8.26
400	3314	.94674	1.56	6803	83.8	284.4	1.144	.9491	7.62

410	3413	.95106	3.65	7648	85.1	288.6	1.139	.9530	7.05
420	3510	.95492	5.71	8505	86.3	292.7	1.134	.9566	6.54
430	3607	.95839	7.76	9375	87.6	296.7	1.130	.9598	6.08
440	3703	.96152	9.79	10257	88.9	300.6	1.125	.9627	5.66
450	3798	.96434	11.80	11153	90.2	304.4	1.122	.9653	5.28
475	4034	.97031	16.76	13448	93.4	313.6	1.114	.9709	4.48
500	4267	.97505	21.64	15824	96.6	322.4	1.107	.9754	3.83
525	4498	.97886	26.43	18279	99.8	330.8	1.101	.9791	3.31
550	4727	.98195	31.14	20813	102.9	339.0	1.096	.9821	2.88
575	4954	.98449	35.78	23423	105.9	346.8	1.092	.9845	2.52

Table 2—continued

P/MPa					1.0				
T_σ/K					292.387				
	V_σ	Z_σ	S_σ	H_σ	C_P	w_σ	γ_σ	$(f/P)_\sigma$	μ_σ
Liq	81.82	.033657	−73.44	−16299	109.7	732.6	1.701	.8523	−.002
Vap	2002	.82373	−23.69	−1752	78.2	222.9	1.321	.8523	22.4
$\dfrac{T}{K}$	$\dfrac{V}{\mathrm{cm^3\,mol^{-1}}}$	Z	$\dfrac{S}{\mathrm{J\,K^{-1}\,mol^{-1}}}$	$\dfrac{H}{\mathrm{J\,mol^{-1}}}$	$\dfrac{C_P}{\mathrm{J\,K^{-1}\,mol^{-1}}}$	$\dfrac{w}{\mathrm{m\,sec^{-1}}}$	γ	(f/P)	$\dfrac{\mu}{\mathrm{K\,MPa^{-1}}}$
90	54.88	.073340	−174.53	−34202				.2209E−08	
95	55.29	.070005	−169.81	−33765				.1166E−07	
100	55.72	.067016	−165.78	−33373				.5139E−07	
105	56.15	.064322	−162.13	−32999				.1945E−06	
110	56.59	.061879	−158.68	−32629	74.1	1953	1.584	.6466E−06	−.632
115	57.04	.059655	−155.37	−32256	75.1	1910	1.582	.1920E−05	−.622
120	57.49	.057622	−152.15	−31878	76.3	1866	1.574	.5172E−05	−.611
125	57.95	.055756	−149.01	−31493	77.7	1824	1.565	.1278E−04	−.599
130	58.41	.054039	−145.94	−31101	78.9	1784	1.557	.2924E−04	−.587
135	58.88	.052453	−142.94	−30704	80.0	1746	1.551	.6256E−04	−.577
140	59.35	.050986	−140.01	−30301	81.0	1710	1.546	.1260E−03	−.568
145	59.83	.049625	−137.15	−29894	81.9	1676	1.543	.2405E−03	−.560
150	60.31	.048360	−134.36	−29483	82.6	1643	1.542	.4374E−03	−.552
155	60.80	.047182	−131.64	−29068	83.3	1611	1.542	.7615E−03	−.545
160	61.30	.046084	−128.99	−28650	83.8	1580	1.543	.1275E−02	−.539

165	61.81	.045058	−126.40	−28229	84.4	1549	1.545	.2059E−02	−.532
170	62.33	.044100	−123.87	−27806	84.9	1518	1.548	.3220E−02	−.525
175	62.86	.043203	−121.41	−27381	85.4	1488	1.551	.4890E−02	−.517
180	63.40	.042363	−118.99	−26952	85.9	1458	1.554	.7226E−02	−.510
185	63.95	.041577	−116.63	−26522	86.4	1427	1.557	.1043E−01	−.502
190	64.51	.040839	−114.32	−26089	86.9	1397	1.561	.1471E−01	−.493
195	65.09	.040148	−112.06	−25653	87.4	1367	1.565	.2033E−01	−.484
200	65.68	.039500	−109.84	−25214	88.0	1336	1.569	.2755E−01	−.474
205	66.29	.038894	−107.66	−24773	88.6	1305	1.573	.3668E−01	−.463
210	66.92	.038325	−105.52	−24328	89.3	1274	1.577	.4806E−01	−.451
215	67.56	.037794	−103.41	−23880	89.9	1243	1.581	.6202E−01	−.439
220	68.22	.037298	−101.33	−23429	90.7	1212	1.586	.7893E−01	−.425
225	68.91	.036836	−99.29	−22974	91.5	1180	1.590	.9914E−01	−.410
230	69.62	.036406	−97.27	−22514	92.3	1149	1.594	.1230	−.395
235	70.35	.036007	−95.27	−22050	93.2	1117	1.599	.1509	−.377
240	71.12	.035639	−93.30	−21582	94.2	1084	1.604	.1832	−.359
245	71.91	.035302	−91.35	−21109	95.2	1052	1.609	.2203	−.338
250	72.74	.034994	−89.42	−20630	96.3	1020	1.615	.2624	−.316
255	73.60	.034716	−87.50	−20146	97.4	986.9	1.621	.3099	−.292
260	74.51	.034468	−85.59	−19656	98.7	953.9	1.628	.3629	−.265
265	75.46	.034251	−83.70	−19159	100.0	920.7	1.635	.4217	−.236
270	76.47	.034064	−81.82	−18656	101.4	887.2	1.644	.4865	−.204
275	77.53	.033910	−79.94	−18145	103.0	853.4	1.654	.5574	−.168
280	78.66	.033790	−78.07	−17625	104.7	819.3	1.665	.6345	−.127
285	79.87	.033706	−76.20	−17097	106.6	784.7	1.677	.7178	−.082

Table 2—*continued*

	1.0								
	T_σ/K 292.387								
Liq Vap	V_σ 81.82 2002	Z_σ .033657 .82373	S_σ −73.44 −23.69	H_σ −16299 −1752	C_p 109.7 78.2	w_σ 732.6 222.9	γ_σ 1.701 1.321	$(f/P)_\sigma$ .8523 .8523	μ_σ −.002 22.4
T K	V cm³ mol⁻¹	Z	S J K⁻¹ mol⁻¹	H J mol⁻¹	C_p J K⁻¹ mol⁻¹	w m sec⁻¹	γ	(f/P)	μ K MPa⁻¹
290	81.17	.033663	−74.33	−16559	108.6	749.5	1.693	.8074	−.029
295	2036	.82998	−22.99	−1548	77.8	225.0	1.310	.8566	21.6
300	2098	.84103	−21.69	−1160	77.3	228.9	1.290	.8642	20.3
305	2158	.85101	−20.41	−775	77.0	232.6	1.274	.8714	19.1
310	2217	.86007	−19.16	−390	76.9	236.1	1.259	.8780	18.0
315	2274	.86833	−17.93	−6	76.9	239.5	1.247	.8842	17.0
320	2330	.87590	−16.72	379	76.9	242.7	1.235	.8899	16.0
325	2386	.88285	−15.53	764	77.1	245.8	1.226	.8953	15.2
330	2440	.88925	−14.35	1150	77.3	248.8	1.217	.9004	14.4
335	2493	.89516	−13.18	1537	77.6	251.7	1.208	.9051	13.7
340	2546	.90064	−12.03	1926	77.9	254.6	1.201	.9095	13.0
345	2598	.90572	−10.89	2316	78.3	257.3	1.194	.9137	12.4
350	2649	.91044	−9.76	2709	78.7	260.0	1.188	.9176	11.8
355	2700	.91484	−8.64	3104	79.2	262.6	1.183	.9213	11.3
360	2751	.91895	−7.53	3501	79.6	265.1	1.177	.9248	10.7
365	2800	.92279	−6.43	3900	80.1	267.6	1.172	.9280	10.3
370	2850	.92639	−5.34	4302	80.6	270.0	1.168	.9311	9.82
380	2948	.93292	−3.17	5114	81.7	274.8	1.160	.9368	9.01
390	3044	.93869	−1.03	5937	82.9	279.3	1.153	.9419	8.29
400	3139	.94382	1.08	6771	84.1	283.7	1.146	.9465	7.65

410	3233	.94838	3.17	7618	85.3	288.0	1.141	.9506	7.07
420	3326	.95247	5.24	8477	86.5	292.1	1.135	.9543	6.56
430	3418	.95614	7.29	9348	87.8	296.2	1.131	.9577	6.09
440	3510	.95944	9.32	10232	89.0	300.1	1.127	.9607	5.67
450	3601	.96243	11.34	11129	90.3	304.0	1.123	.9635	5.29
475	3826	.96873	16.31	13427	93.5	313.2	1.114	.9694	4.48
500	4048	.97373	21.18	15806	96.7	322.1	1.108	.9741	3.83
525	4268	.97774	25.98	18263	99.9	330.5	1.102	.9780	3.31
550	4486	.98101	30.70	20798	102.9	338.7	1.097	.9811	2.88
575	4703	.98368	35.34	23410	106.0	346.6	1.093	.9837	2.52

Table 2—continued

P/MPa					1.5				
T_σ/K					308.930				
Liq / Vap	V_σ 86.79 / 1310	Z_σ .050684 / .76489	S_σ −67.34 / −24.76	H_σ −14424 / −1268	C_P 118.3 / 89.2	w_σ 621.5 / 215.5	γ_σ 1.768 / 1.410	$(f/P)_\sigma$ .8138 / .8138	μ_σ .230 / 20.6
$\dfrac{T}{K}$	$\dfrac{V}{\text{cm}^3\,\text{mol}^{-1}}$	Z	$\dfrac{S}{\text{J K}^{-1}\text{mol}^{-1}}$	$\dfrac{H}{\text{J mol}^{-1}}$	$\dfrac{C_P}{\text{J K}^{-1}\text{mol}^{-1}}$	$\dfrac{w}{\text{m sec}^{-1}}$	γ	(f/P)	$\dfrac{\mu}{\text{K MPa}^{-1}}$
90	54.87	.10999	−174.57	−34178				.1528E−08	
95	55.28	.10498	−169.85	−33742				.8050E−08	
100	55.71	.10050	−165.82	−33350				.3542E−07	
105	56.14	.096457	−162.17	−32975				.1339E−06	
110	56.58	.092793	−158.73	−32605	74.1	1955	1.584	.4446E−06	−.632
115	57.02	.089457	−155.42	−32233	75.0	1913	1.582	.1319E−05	−.623
120	57.47	.086406	−152.20	−31854	76.3	1869	1.574	.3546E−05	−.611
125	57.93	.083607	−149.05	−31469	77.6	1827	1.565	.8758E−05	−.599
130	58.39	.081030	−145.98	−31078	78.9	1787	1.557	.2003E−04	−.588
135	58.85	.078652	−142.98	−30681	80.0	1749	1.550	.4282E−04	−.577
140	59.33	.076450	−140.06	−30278	81.0	1713	1.546	.8618E−04	−.568
145	59.80	.074408	−137.20	−29871	81.9	1679	1.543	.1644E−03	−.560
150	60.29	.072510	−134.41	−29460	82.6	1646	1.542	.2987E−03	−.553
155	60.78	.070743	−131.69	−29045	83.2	1614	1.542	.5198E−03	−.546
160	61.28	.069095	−129.04	−28627	83.8	1583	1.543	.8697E−03	−.539

165	61.78	.067556	−126.45	−28207	84.4	1552	1.544	.1404E−02	−.532
170	62.30	.066117	−123.93	−27784	84.9	1521	1.547	.2195E−02	−.525
175	62.83	.064771	−121.46	−27358	85.3	1491	1.550	.3331E−02	−.518
180	63.37	.063510	−119.05	−26931	85.8	1461	1.553	.4922E−02	−.511
185	63.91	.062329	−116.69	−26500	86.3	1431	1.557	.7101E−02	−.503
190	64.48	.061221	−114.38	−26067	86.8	1401	1.560	.1001E−01	−.494
195	65.05	.060183	−112.12	−25632	87.4	1370	1.564	.1382E−01	−.485
200	65.64	.059210	−109.90	−25194	87.9	1340	1.568	.1873E−01	−.475
205	66.24	.058298	−107.72	−24752	88.5	1309	1.572	.2494E−01	−.464
210	66.87	.057444	−105.58	−24308	89.2	1278	1.576	.3266E−01	−.453
215	67.51	.056645	−103.47	−23861	89.9	1247	1.580	.4214E−01	−.441
220	68.16	.055898	−101.40	−23410	90.6	1216	1.584	.5361E−01	−.427
225	68.85	.055202	−99.36	−22955	91.4	1185	1.588	.6732E−01	−.413
230	69.55	.054554	−97.34	−22496	92.2	1153	1.592	.8352E−01	−.397
235	70.28	.053953	−95.35	−22033	93.1	1122	1.597	.1025	−.380
240	71.04	.053398	−93.38	−21565	94.0	1090	1.602	.1244	−.362
245	71.82	.052888	−91.43	−21093	95.0	1058	1.607	.1495	−.342
250	72.64	.052422	−89.50	−20615	96.1	1025	1.612	.1780	−.320
255	73.50	.051999	−87.59	−20132	97.2	992.9	1.618	.2102	−.297
260	74.39	.051621	−85.69	−19643	98.4	960.3	1.624	.2461	−.271
265	75.34	.051288	−83.80	−19147	99.7	927.4	1.631	.2860	−.243
270	76.33	.051000	−81.92	−18645	101.1	894.2	1.639	.3299	−.211
275	77.37	.050760	−80.05	−18136	102.6	860.8	1.648	.3780	−.176
280	78.48	.050569	−78.19	−17618	104.3	827.1	1.658	.4302	−.138
285	79.67	.050431	−76.33	−17093	106.1	793.0	1.670	.4867	−.094

Table 2—*continued*

P/MPa		1.5							
T_σ/K		308.930							
Liq	V_σ 86.79	Z_σ .050684	S_σ -67.34	H_σ -14424	C_P 118.3	w_σ 621.5	γ_σ 1.768	$(f/P)_\sigma$ .8138	μ_σ .230
Vap	1310	.76489	-24.76	-1268	89.2	215.5	1.410	.8138	20.6
$\dfrac{T}{K}$	$\dfrac{V}{\mathrm{cm^3\,mol^{-1}}}$	Z	$\dfrac{S}{\mathrm{J\,K^{-1}\,mol^{-1}}}$	$\dfrac{H}{\mathrm{J\,mol^{-1}}}$	$\dfrac{C_P}{\mathrm{J\,K^{-1}\,mol^{-1}}}$	$\dfrac{w}{\mathrm{m\,sec^{-1}}}$	γ	(f/P)	$\dfrac{\mu}{\mathrm{K\,MPa^{-1}}}$
290	80.93	.050350	-74.47	-16557	108.1	758.5	1.684	.5473	-.044
295	82.30	.050331	-72.60	-16012	110.3	723.4	1.701	.6122	.013
300	83.78	.050382	-70.73	-15454	112.8	687.7	1.720	.6811	.079
305	85.40	.050514	-68.84	-14883	115.7	651.1	1.745	.7539	.158
310	1321	.76856	-24.45	-1173	88.7	216.6	1.401	.8159	20.3
315	1370	.78446	-23.04	-734	87.0	221.6	1.365	.8255	18.9
320	1416	.79859	-21.69	-302	85.7	226.1	1.337	.8344	17.7
325	1461	.81128	-20.36	124	84.8	230.4	1.313	.8426	16.6
330	1505	.82273	-19.07	546	84.2	234.4	1.293	.8503	15.6
335	1547	.83313	-17.81	966	83.7	238.2	1.275	.8575	14.8
340	1588	.84263	-16.57	1384	83.5	241.9	1.260	.8643	13.9
345	1628	.85133	-15.36	1801	83.3	245.3	1.247	.8706	13.2
350	1667	.85933	-14.16	2217	83.3	248.7	1.236	.8765	12.5
355	1705	.86672	-12.98	2634	83.4	251.9	1.226	.8821	11.9
360	1743	.87356	-11.81	3051	83.5	255.0	1.216	.8873	11.3
365	1780	.87990	-10.65	3469	83.7	258.0	1.208	.8923	10.8
370	1817	.88579	-9.51	3889	84.0	260.9	1.201	.8969	10.3
380	1888	.89642	-7.27	4732	84.6	266.4	1.187	.9055	9.38
390	1958	.90571	-5.06	5582	85.4	271.7	1.176	.9131	8.59
400	2026	.91388	-2.88	6440	86.3	276.7	1.167	.9200	7.89

1.5 MPa

410	2093	.92112	−.74	7308	87.3	281.5	1.158	.9262	7.27
420	2159	.92756	1.38	8186	88.3	286.2	1.151	.9317	6.72
430	2224	.93330	3.47	9075	89.4	290.6	1.145	.9368	6.23
440	2289	.93846	5.53	9975	90.5	295.0	1.139	.9413	5.79
450	2352	.94310	7.58	10886	91.7	299.2	1.134	.9455	5.39
475	2509	.95284	12.62	13215	94.6	309.2	1.123	.9543	4.55
500	2662	.96051	17.55	15618	97.6	318.6	1.114	.9614	3.88
525	2813	.96664	22.38	18096	100.6	327.6	1.107	.9672	3.34
550	2962	.97161	27.13	20649	103.6	336.2	1.101	.9719	2.90
575	3110	.97567	31.80	23275	106.5	344.4	1.096	.9758	2.53

155

Table 2—continued

P/MPa									2.0		
T_σ/K									321.816		
Liq	V_σ 91.74	Z_σ .068570	S_σ −62.53	H_σ −12862	$C_{P\sigma}$ 128.2	w_σ 533.0	γ_σ 1.859	$(f/P)_\sigma$ .7818	μ_σ .516		
Vap	949.7	.70984	−25.67	−998	102.2	207.2	1.532	.7818	19.6		
$\dfrac{T}{K}$	$\dfrac{V}{cm^3\,mol^{-1}}$	Z	$\dfrac{S}{J\,K^{-1}\,mol^{-1}}$	$\dfrac{H}{J\,mol^{-1}}$	$\dfrac{C_P}{J\,K^{-1}\,mol^{-1}}$	$\dfrac{w}{m\,sec^{-1}}$	γ	(f/P)	$\dfrac{\mu}{K\,MPa^{-1}}$		
90	54.86	.14661	−174.61	−34155				.1189E−08			
95	55.27	.13994	−169.89	−33718				.6253E−08			
100	55.69	.13397	−165.87	−33326				.2747E−07			
105	56.12	.12858	−162.22	−32952				.1037E−06			
110	56.56	.12369	−158.77	−32582	74.1	1958	1.583	.3440E−06	−.632		
115	57.01	.11924	−155.46	−32209	75.0	1915	1.581	.1019E−05	−.623		
120	57.46	.11517	−152.24	−31831	76.3	1872	1.573	.2738E−05	−.611		
125	57.91	.11144	−149.10	−31446	77.6	1830	1.565	.6754E−05	−.599		
130	58.37	.10800	−146.03	−31055	78.9	1790	1.556	.1543E−04	−.588		
135	58.83	.10483	−143.03	−30658	80.0	1752	1.550	.3297E−04	−.578		
140	59.30	.10190	−140.10	−30255	81.0	1716	1.545	.6630E−04	−.569		
145	59.78	.099172	−137.25	−29848	81.8	1682	1.543	.1264E−03	−.561		
150	60.26	.096640	−134.46	−29437	82.6	1649	1.541	.2295E−03	−.553		
155	60.75	.094283	−131.74	−29022	83.2	1617	1.541	.3992E−03	−.546		
160	61.25	.092085	−129.09	−28605	83.8	1586	1.542	.6675E−03	−.539		

165	61.76	.090032	−126.50	−28185	84.3	1555	1.544	.1077E−02	−.533
170	62.27	.088113	−123.98	−27762	84.8	1525	1.546	.1683E−02	−.526
175	62.80	.086316	−121.51	−27336	85.3	1494	1.549	.2553E−02	−.519
180	63.33	.084634	−119.10	−26909	85.8	1464	1.552	.3771E−02	−.511
185	63.88	.083057	−116.75	−26479	86.3	1434	1.556	.5438E−02	−.503
190	64.44	.081579	−114.44	−26046	86.8	1404	1.559	.7663E−02	−.495
195	65.01	.080193	−112.18	−25611	87.3	1374	1.563	.1058E−01	−.486
200	65.59	.078893	−109.96	−25173	87.9	1344	1.567	.1433E−01	−.476
205	66.20	.077675	−107.78	−24732	88.5	1313	1.571	.1907E−01	−.466
210	66.81	.076534	−105.64	−24288	89.1	1282	1.574	.2497E−01	−.454
215	67.45	.075465	−103.54	−23841	89.8	1252	1.578	.3220E−01	−.442
220	68.11	.074467	−101.47	−23390	90.5	1221	1.582	.4096E−01	−.429
225	68.78	.073535	−99.42	−22936	91.3	1189	1.586	.5143E−01	−.415
230	69.48	.072667	−97.41	−22478	92.1	1158	1.590	.6379E−01	−.399
235	70.20	.071862	−95.42	−22015	92.9	1127	1.595	.7824E−01	−.383
240	70.95	.071117	−93.45	−21548	93.9	1095	1.599	.9495E−01	−.365
245	71.73	.070431	−91.51	−21076	94.8	1063	1.604	.1141	−.345
250	72.55	.069804	−89.58	−20599	95.9	1031	1.609	.1359	−.324
255	73.39	.069234	−87.67	−20117	97.0	998.9	1.614	.1604	−.301
260	74.28	.068722	−85.78	−19629	98.2	966.5	1.620	.1878	−.276
265	75.21	.068269	−83.90	−19135	99.5	933.9	1.627	.2182	−.249
270	76.19	.067875	−82.02	−18634	100.8	901.1	1.635	.2517	−.218
275	77.22	.067543	−80.16	−18126	102.3	868.1	1.643	.2883	−.185
280	78.31	.067275	−78.30	−17611	103.9	834.8	1.652	.3281	−.147
285	79.47	.067075	−76.45	−17087	105.6	801.2	1.663	.3712	−.106

Table 2—continued

P/MPa					2.0				
T_σ/K					321.816				
Liq	V_σ 91.74	Z_σ .068570	S_σ -62.53	H_σ -12862	C_P 128.2	w_σ 533.0	γ_σ 1.859	$(f/P)_\sigma$ .7818	μ_σ .516
Vap	949.7	.70984	-25.67	-998	102.2	207.2	1.532	.7818	19.6
$\frac{T}{K}$	$\frac{V}{cm^3\,mol^{-1}}$	Z	$\frac{S}{J\,K^{-1}\,mol^{-1}}$	$\frac{H}{J\,mol^{-1}}$	$\frac{C_P}{J\,K^{-1}\,mol^{-1}}$	$\frac{w}{m\,sec^{-1}}$	γ	(f/P)	$\frac{\mu}{K\,MPa^{-1}}$
290	80.71	.066947	-74.60	-16555	107.5	767.2	1.676	.4175	-.058
295	82.04	.066899	-72.74	-16012	109.6	732.8	1.691	.4669	-.004
300	83.48	.066938	-70.88	-15458	112.0	697.8	1.709	.5195	.058
305	85.05	.067078	-69.01	-14891	114.7	662.0	1.731	.5750	.132
310	86.78	.067335	-67.12	-14310	117.8	625.4	1.758	.6334	.219
315	88.70	.067732	-65.20	-13712	121.6	587.5	1.792	.6946	.325
320	90.87	.068307	-63.25	-13093	126.2	547.9	1.838	.7581	.459
325	978.1	.72394	-24.68	-678	99.4	211.3	1.487	.7890	18.6
330	1020	.74361	-23.18	-189	96.1	217.1	1.432	.7997	17.3
335	1060	.76087	-21.76	285	93.8	222.3	1.391	.8095	16.1
340	1097	.77619	-20.38	749	92.1	227.2	1.357	.8187	15.1
345	1133	.78993	-19.05	1206	90.8	231.7	1.330	.8273	14.2
350	1167	.80234	-17.75	1658	89.9	236.0	1.307	.8354	13.4
355	1201	.81362	-16.48	2106	89.2	240.0	1.288	.8429	12.6
360	1233	.82392	-15.23	2551	88.8	243.8	1.271	.8499	12.0
365	1265	.83338	-14.01	2994	88.5	247.5	1.257	.8566	11.3
370	1295	.84208	-12.81	3436	88.3	251.0	1.244	.8628	10.8
380	1355	.85758	-10.45	4318	88.3	257.5	1.223	.8743	9.77
390	1412	.87096	-8.16	5202	88.5	263.7	1.206	.8845	8.90
400	1468	.88261	-5.91	6089	89.0	269.4	1.191	.8937	8.14

410	1522	.89284	− 3.71	6982	89.6	274.9	1.180	.9019	7.48
420	1575	.90186	− 1.54	7882	90.4	280.1	1.169	.9093	6.89
430	1626	.90988	.60	8791	91.3	285.0	1.160	.9160	6.37
440	1677	.91703	2.71	9708	92.2	289.8	1.153	.9221	5.90
450	1727	.92344	4.79	10635	93.2	294.4	1.146	.9276	5.48
475	1850	.93681	9.90	12997	95.8	305.2	1.132	.9394	4.61
500	1969	.94728	14.88	15427	98.6	315.3	1.121	.9488	3.92
525	2086	.95560	19.76	17927	101.4	324.7	1.113	.9565	3.36
550	2200	.96231	24.55	20498	104.3	333.7	1.106	.9627	2.91
575	2313	.96778	29.24	23140	107.1	342.3	1.100	.9679	2.54

Table 2—continued

P/MPa					2.5				
T_σ/K					332.513				
	V_σ 97.01 725.5	Z_σ .087727 .65610	S_σ −58.43 −26.58	H_σ −11474 −884	C_P 141.2 119.3	w_σ 456.9 198.3	γ_σ 1.991 1.709	$(f/P)_\sigma$ .7541 .7541	μ_σ .894 19.0
$\dfrac{T}{K}$	$\dfrac{V}{\text{cm}^3\,\text{mol}^{-1}}$	Z	$\dfrac{S}{\text{J K}^{-1}\text{mol}^{-1}}$	$\dfrac{H}{\text{J mol}^{-1}}$	$\dfrac{C_P}{\text{J K}^{-1}\text{mol}^{-1}}$	$\dfrac{w}{\text{m sec}^{-1}}$	γ	(f/P)	$\dfrac{\mu}{\text{K MPa}^{-1}}$
90	54.84	.18323	−174.65	−34131				.9865E−09	
95	55.26	.17489	−169.93	−33694				.5180E−08	
100	55.68	.16742	−165.91	−33302				.2273E−07	
105	56.11	.16068	−162.26	−32928				.8569E−07	
110	56.55	.15457	−158.82	−32559	74.1	1961	1.583	.2838E−06	−.632
115	56.99	.14901	−155.51	−32186	75.0	1918	1.581	.8401E−06	−.623
120	57.44	.14392	−152.29	−31808	76.3	1875	1.573	.2255E−05	−.611
125	57.89	.13926	−149.14	−31423	77.6	1832	1.564	.5556E−05	−.599
130	58.35	.13496	−146.08	−31032	78.9	1792	1.556	.1268E−04	−.588
135	58.81	.13099	−143.08	−30634	80.0	1755	1.550	.2707E−04	−.578
140	59.28	.12732	−140.15	−30232	81.0	1719	1.545	.5441E−04	−.569
145	59.76	.12392	−137.29	−29825	81.8	1685	1.542	.1036E−03	−.561
150	60.24	.12075	−134.51	−29414	82.6	1652	1.541	.1881E−03	−.554
155	60.73	.11780	−131.79	−29000	83.2	1620	1.541	.3269E−03	−.547
160	61.22	.11505	−129.14	−28582	83.8	1589	1.541	.5464E−03	−.540

165	61.73	.11249	−126.55	−28162	84.3	1558	1.543	.8813E−03	−.533
170	62.24	.11009	−124.03	−27739	84.8	1528	1.545	.1376E−02	−.526
175	62.76	.10784	−121.57	−27314	85.3	1498	1.548	.2087E−02	−.519
180	63.30	.10573	−119.16	−26887	85.7	1468	1.551	.3082E−02	−.512
185	63.84	.10376	−116.80	−26457	86.2	1438	1.555	.4441E−02	−.504
190	64.40	.10191	−114.49	−26024	86.7	1408	1.558	.6257E−02	−.496
195	64.97	.10018	−112.24	−25589	87.3	1378	1.562	.8634E−02	−.487
200	65.55	.098550	−110.02	−25152	87.8	1347	1.566	.1169E−01	−.477
205	66.15	.097024	−107.84	−24711	88.4	1317	1.569	.1555E−01	−.467
210	66.76	.095595	−105.71	−24268	89.0	1287	1.573	.2036E−01	−.456
215	67.40	.094256	−103.60	−23821	89.7	1256	1.577	.2625E−01	−.444
220	68.05	.093004	−101.53	−23371	90.4	1225	1.581	.3339E−01	−.431
225	68.72	.091835	−99.49	−22917	91.2	1194	1.584	.4191E−01	−.417
230	69.41	.090746	−97.48	−22459	92.0	1163	1.588	.5197E−01	−.402
235	70.13	.089734	−95.49	−21997	92.8	1132	1.592	.6373E−01	−.386
240	70.88	.088797	−93.53	−21531	93.7	1100	1.597	.7733E−01	−.368
245	71.65	.087933	−91.59	−21060	94.7	1068	1.601	.9292E−01	−.349
250	72.45	.087141	−89.67	−20584	95.7	1037	1.606	.1106	−.328
255	73.29	.086421	−87.76	−20102	96.8	1005	1.611	.1306	−.306
260	74.17	.085772	−85.87	−19615	98.0	972.6	1.617	.1528	−.281
265	75.08	.085195	−83.99	−19123	99.2	940.4	1.623	.1775	−.255
270	76.05	.084691	−82.12	−18623	100.6	907.9	1.630	.2048	−.225
275	77.06	.084262	−80.26	−18117	102.0	875.3	1.638	.2346	−.193
280	78.14	.083911	−78.41	−17603	103.5	842.4	1.647	.2670	−.157
285	79.28	.083641	−76.57	−17081	105.2	809.2	1.657	.3020	−.117

Table 2—continued

| | | | | | | | | | | | **2.5** | | | | | | | | | | |

	$\dfrac{V}{\text{cm}^3\,\text{mol}^{-1}}$	Z	$\dfrac{S}{\text{J K}^{-1}\text{mol}^{-1}}$	$\dfrac{H}{\text{J mol}^{-1}}$	$\dfrac{C_P}{\text{J K}^{-1}\text{mol}^{-1}}$	$\dfrac{w}{\text{m sec}^{-1}}$	γ	(f/P)	$\dfrac{\mu}{\text{K MPa}^{-1}}$
P/MPa									
T_c/K				**332.513**					
Liq $\;V_\sigma$ 97.01	Z_σ .087727	S_σ −58.43	H_σ −11474	$C_{P\sigma}$ 141.2	w_σ 456.9	γ_σ 1.991	$(f/P)_\sigma$ .7541	μ_σ .894
Vap $\;$ 725.5	.65610	−26.58	−884	119.3	198.3	1.709	.7541	19.0
T/K									
290	80.49	.083459	−74.72	−16551	107.0	775.8	1.669	.3396	−.071
295	81.79	.083371	−72.88	−16011	109.0	741.9	1.682	.3798	−.020
300	83.20	.083387	−71.03	−15461	111.2	707.6	1.699	.4226	.039
305	84.72	.083520	−69.17	−14898	113.8	672.6	1.718	.4678	.107
310	86.38	.083789	−67.29	−14322	116.7	636.9	1.742	.5153	.188
315	88.23	.084217	−65.40	−13731	120.1	600.2	1.772	.5651	.285
320	90.29	.084841	−63.48	−13120	124.2	562.1	1.811	.6169	.405
325	92.65	.085718	−61.51	−12487	129.5	522.2	1.864	.6706	.558
330	95.41	.086937	−59.49	−11823	136.5	479.7	1.939	.7258	.762
335	747.2	.67067	−25.71	−593	114.7	202.3	1.642	.7604	18.1
340	787.2	.69621	−24.07	−37	108.0	209.4	1.545	.7725	16.7
345	823.8	.71799	−22.52	491	103.5	215.8	1.476	.7836	15.5
350	857.8	.73697	−21.06	1000	100.3	221.4	1.425	.7939	14.4
355	889.9	.75375	−19.65	1496	98.0	226.7	1.385	.8035	13.5
360	920.4	.76875	−18.29	1981	96.3	231.5	1.354	.8125	12.7
365	949.6	.78227	−16.98	2460	95.1	236.0	1.327	.8209	12.0
370	977.7	.79455	−15.69	2932	94.1	240.3	1.305	.8288	11.3
380	1031	.81603	−13.20	3867	92.9	248.1	1.270	.8432	10.2
390	1082	.83425	−10.79	4792	92.3	255.3	1.243	.8561	9.22
400	1131	.84990	−8.46	5715	92.2	261.9	1.222	.8675	8.40

410	1177	.86349	− 6.18	6637	92.4	268.1	1.205	.8778	7.68
420	1223	.87539	− 3.95	7563	92.8	273.9	1.191	.8871	7.05
430	1267	.88588	− 1.76	8494	93.4	279.4	1.179	.8955	6.50
440	1310	.89519	.40	9431	94.0	284.6	1.169	.9031	6.01
450	1352	.90349	2.52	10375	94.8	289.6	1.160	.9100	5.57
475	1454	.92070	7.70	12773	97.1	301.2	1.142	.9247	4.66
500	1553	.93406	12.75	15232	99.6	312.0	1.129	.9364	3.95
525	1649	.94464	17.67	17755	102.3	322.0	1.119	.9459	3.38
550	1743	.95313	22.49	20345	105.0	331.4	1.111	.9537	2.93
575	1836	.96003	27.22	23003	107.7	340.3	1.104	.9602	2.55

Table 2—continued

P/MPa									
T_σ/K									
	3.0								
	341.716								
	V_σ	Z_σ	S_σ	H_σ	C_P	w_σ	γ_σ	$(f/P)_\sigma$	μ_σ
Liq	103.0	.10877	−54.75	−10181	160.6	388.0	2.206	.7294	1.43
Vap	569.9	.60173	−27.61	−908	144.8	189.0	1.990	.7294	18.5
$\dfrac{T}{K}$	$\dfrac{V}{\text{cm}^3\,\text{mol}^{-1}}$	Z	$\dfrac{S}{\text{J K}^{-1}\,\text{mol}^{-1}}$	$\dfrac{H}{\text{J mol}^{-1}}$	$\dfrac{C_P}{\text{J K}^{-1}\,\text{mol}^{-1}}$	$\dfrac{w}{\text{m sec}^{-1}}$	γ	(f/P)	$\dfrac{\mu}{\text{K MPa}^{-1}}$
90	54.83	.21983	−174.69	−34107				.8527E − 09	
95	55.24	.20982	−169.97	−33671				.4470E − 08	
100	55.66	.20085	−165.95	−33279				.1958E − 07	
105	56.09	.19276	−162.30	−32905				.7374E − 07	
110	56.53	.18543	−158.86	−32535	74.1	1964	1.583	.2439E − 06	−.632
115	56.97	.17876	−155.55	−32163	75.0	1921	1.580	.7213E − 06	−.623
120	57.42	.17265	−152.33	−31784	76.3	1877	1.573	.1934E − 05	−.611
125	57.87	.16705	−149.19	−31400	77.6	1835	1.564	.4761E − 05	−.599
130	58.33	.16190	−146.12	−31008	78.9	1795	1.556	.1086E − 04	−.588
135	58.79	.15714	−143.12	−30611	80.0	1757	1.549	.2316E − 04	−.578
140	59.26	.15273	−140.20	−30209	81.0	1722	1.544	.4651E − 04	−.569
145	59.73	.14864	−137.34	−29802	81.8	1688	1.542	.8853E − 04	−.561
150	60.21	.14484	−134.56	−29391	82.5	1655	1.540	.1606E − 03	−.554
155	60.70	.14131	−131.84	−28977	83.2	1623	1.540	.2790E − 03	−.547
160	61.20	.13800	−129.19	−28560	83.7	1592	1.541	.4659E − 03	−.540

165	61.70	.13492	−126.60	−28140	84.3	1561	1.542	.7511E−03	−.534
170	62.21	.13204	−124.08	−27717	84.8	1531	1.545	.1173E−02	−.527
175	62.73	.12934	−121.62	−27292	85.2	1501	1.547	.1777E−02	−.520
180	63.26	.12681	−119.21	−26865	85.7	1471	1.550	.2623E−02	−.513
185	63.80	.12444	−116.86	−26435	86.2	1441	1.554	.3779E−02	−.505
190	64.36	.12222	−114.55	−26003	86.7	1411	1.557	.5321E−02	−.497
195	64.93	.12014	−112.29	−25568	87.2	1381	1.561	.7341E−02	−.488
200	65.51	.11818	−110.08	−25131	87.8	1351	1.564	.9936E−02	−.478
205	66.10	.11635	−107.90	−24691	88.3	1321	1.568	.1322E−01	−.468
210	66.71	.11463	−105.77	−24247	89.0	1291	1.572	.1729E−01	−.457
215	67.34	.11302	−103.67	−23801	89.6	1260	1.575	.2229E−01	−.445
220	67.99	.11151	−101.60	−23351	90.3	1229	1.579	.2834E−01	−.433
225	68.66	.11010	−99.56	−22898	91.1	1198	1.583	.3557E−01	−.419
230	69.35	.10879	−97.55	−22441	91.9	1168	1.586	.4410E−01	−.404
235	70.06	.10757	−95.57	−21979	92.7	1136	1.590	.5407E−01	−.388
240	70.80	.10644	−93.61	−21514	93.6	1105	1.594	.6559E−01	−.371
245	71.56	.10539	−91.67	−21043	94.5	1074	1.599	.7880E−01	−.352
250	72.36	.10444	−89.75	−20568	95.6	1042	1.603	.9380E−01	−.332
255	73.19	.10356	−87.84	−20088	96.6	1010	1.608	.1107	−.310
260	74.06	.10277	−85.96	−19602	97.8	978.7	1.613	.1296	−.286
265	74.96	.10207	−84.08	−19110	99.0	946.7	1.619	.1505	−.260
270	75.91	.10145	−82.22	−18612	100.3	914.6	1.626	.1736	−.232
275	76.92	.10092	−80.37	−18107	101.7	882.3	1.633	.1988	−.200
280	77.97	.10048	−78.52	−17595	103.2	849.8	1.641	.2262	−.166
285	79.09	.10013	−76.68	−17075	104.8	817.1	1.651	.2559	−.127

Table 2—continued

P/MPa					3.0				
T_σ/K					341.716				

| Liq | V_σ 103.0 | Z_σ .10877 | S_σ −54.75 | H_σ −10181 | C_P 160.6 | w_σ 388.0 | γ_σ 2.206 | $(f/P)_\sigma$ .7294 | μ_σ 1.43 |
| Vap | 569.9 | .60173 | −27.61 | −908 | 144.8 | 189.0 | 1.990 | .7294 | 18.5 |

T K	V cm³ mol⁻¹	Z	S J K⁻¹ mol⁻¹	H J mol⁻¹	C_P J K⁻¹ mol⁻¹	w m sec⁻¹	γ	(f/P)	μ K MPa⁻¹
290	80.28	.099889	−74.85	−16547	106.5	784.1	1.661	.2878	−.084
295	81.56	.099753	−73.01	−16010	108.4	750.8	1.674	.3218	−.035
300	82.92	.099736	−71.17	−15462	110.6	717.1	1.689	.3581	.021
305	84.40	.099850	−69.32	−14904	112.9	682.9	1.707	.3964	.085
310	86.01	.10011	−67.47	−14332	115.6	648.0	1.728	.4367	.160
315	87.78	.10055	−65.59	−13747	118.8	612.3	1.754	.4789	.249
320	89.76	.10121	−63.69	−13144	122.5	575.6	1.788	.5229	.358
325	91.98	.10212	−61.76	−12520	127.1	537.3	1.832	.5684	.494
330	94.55	.10338	−59.78	−11871	133.0	497.1	1.893	.6154	.669
335	97.61	.10513	−57.72	−11186	141.3	453.9	1.982	.6635	.909
340	101.4	.10764	−55.54	−10451	154.2	406.0	2.129	.7125	1.26
345	599.1	.62661	−26.29	−455	132.2	195.5	1.821	.7386	17.3
350	638.1	.65779	−24.48	174	120.5	203.9	1.662	.7516	15.9
355	672.6	.68361	−22.83	757	113.3	211.1	1.561	.7636	14.6
360	704.1	.70570	−21.28	1311	108.4	217.5	1.490	.7747	13.6
365	733.4	.72498	−19.81	1844	104.9	223.3	1.437	.7850	12.7
370	760.9	.74207	−18.40	2362	102.4	228.6	1.395	.7947	12.0
380	812.2	.77121	−15.71	3368	99.1	238.1	1.335	.8122	10.6
390	859.6	.79530	−13.17	4348	97.2	246.5	1.292	.8277	9.56
400	904.2	.81562	−10.72	5314	96.1	254.1	1.260	.8416	8.65

410	946.6	.83304	− 8.36	6272	95.6	261.1	1.236	.8540	7.88
420	987.2	.84813	− 6.05	7228	95.5	267.6	1.216	.8651	7.21
430	1026	.86133	− 3.80	8184	95.7	273.7	1.200	.8752	6.62
440	1065	.87296	− 1.60	9143	96.1	279.4	1.187	.8842	6.11
450	1102	.88328	.57	10106	96.7	284.8	1.176	.8925	5.65
475	1191	.90453	5.84	12545	98.5	297.4	1.154	.9101	4.71
500	1276	.92090	10.94	15034	100.7	308.8	1.138	.9241	3.98
525	1359	.93379	15.92	17581	103.1	319.3	1.125	.9355	3.40
550	1439	.94409	20.77	20191	105.7	329.1	1.116	.9449	2.93
575	1518	.95244	25.53	22865	108.3	338.4	1.108	.9526	2.55

Table 2—continued

P/MPa				**3.5**					
T_σ/K				**349.816**					
	V_σ	Z_σ	S_σ	H_σ	C_P	w_σ	γ_σ	$(f/P)_\sigma$	μ_σ
Liq	110.4	.13284	−51.25	−8918	195.7	322.7	2.617	.7070	2.27
Vap	452.2	.54422	−28.86	−1087	189.8	179.2	2.509	.7070	18.1
$\dfrac{T}{K}$	$\dfrac{V}{cm^3\,mol^{-1}}$	Z	$\dfrac{S}{J\,K^{-1}\,mol^{-1}}$	$\dfrac{H}{J\,mol^{-1}}$	$\dfrac{C_P}{J\,K^{-1}\,mol^{-1}}$	$\dfrac{w}{m\,sec^{-1}}$	γ	(f/P)	$\dfrac{\mu}{K\,MPa^{-1}}$
90	54.82	.25641	−174.73	−34083				.7582E−09	
95	55.23	.24473	−170.01	−33647				.3968E−08	
100	55.65	.23427	−165.99	−33255				.1736E−07	
105	56.08	.22483	−162.35	−32881				.6527E−07	
110	56.52	.21628	−158.91	−32512	74.1	1967	1.582	.2157E−06	−.633
115	56.96	.20849	−155.59	−32139	75.0	1924	1.580	.6370E−06	−.623
120	57.40	.20137	−152.38	−31761	76.3	1880	1.572	.1706E−05	−.611
125	57.85	.19483	−149.24	−31376	77.6	1838	1.563	.4196E−05	−.599
130	58.31	.18882	−146.17	−30985	78.8	1798	1.555	.9563E−05	−.588
135	58.77	.18326	−143.17	−30588	80.0	1760	1.549	.2038E−04	−.578
140	59.24	.17812	−140.24	−30186	80.9	1724	1.544	.4090E−04	−.569
145	59.71	.17335	−137.39	−29779	81.8	1690	1.541	.7779E−04	−.561
150	60.19	.16892	−134.60	−29368	82.5	1658	1.540	.1410E−03	−.554
155	60.68	.16479	−131.89	−28954	83.1	1626	1.539	.2448E−03	−.547
160	61.17	.16093	−129.24	−28537	83.7	1595	1.540	.4087E−03	−.541

165	61.67	.15734	−126.66	−28117	84.2	1564	1.542	.6585E − 03	−.534
170	62.18	.15397	−124.13	−27695	84.7	1534	1.544	.1027E − 02	−.527
175	62.70	.15082	−121.67	−27270	85.2	1504	1.547	.1557E − 02	−.521
180	63.23	.14787	−119.26	−26843	85.7	1474	1.550	.2296E − 02	−.513
185	63.77	.14510	−116.91	−26413	86.1	1445	1.553	.3307E − 02	−.506
190	64.32	.14251	−114.61	−25981	86.6	1415	1.556	.4655E − 02	−.498
195	64.88	.14007	−112.35	−25547	87.1	1385	1.560	.6419E − 02	−.489
200	65.46	.13779	−110.14	−25110	87.7	1355	1.563	.8686E − 02	−.480
205	66.06	.13564	−107.96	−24670	88.3	1325	1.567	.1155E − 01	−.470
210	66.66	.13363	−105.83	−24227	88.9	1295	1.570	.1511E − 01	−.459
215	67.29	.13175	−103.73	−23781	89.5	1264	1.574	.1947E − 01	−.447
220	67.93	.12999	−101.66	−23332	90.2	1234	1.577	.2475E − 01	−.434
225	68.60	.12834	−99.63	−22879	91.0	1203	1.581	.3105E − 01	−.421
230	69.28	.12680	−97.62	−22422	91.7	1172	1.584	.3849E − 01	−.406
235	69.99	.12537	−95.64	−21961	92.6	1141	1.588	.4718E − 01	−.391
240	70.72	.12404	−93.68	−21496	93.5	1110	1.592	.5723E − 01	−.374
245	71.48	.12282	−91.74	−21026	94.4	1079	1.596	.6874E − 01	−.356
250	72.27	.12169	−89.83	−20552	95.4	1048	1.600	.8181E − 01	−.336
255	73.09	.12066	−87.93	−20072	96.4	1016	1.605	.9653E − 01	−.314
260	73.95	.11973	−86.04	−19588	97.6	984.6	1.610	.1130	−.291
265	74.84	.11889	−84.18	−19097	98.7	952.9	1.616	.1312	−.266
270	75.78	.11815	−82.32	−18600	100.0	921.1	1.622	.1513	−.238
275	76.77	.11752	−80.47	−18096	101.4	889.2	1.628	.1733	−.208
280	77.81	.11698	−78.63	−17586	102.8	857.1	1.636	.1972	−.174
285	78.91	.11655	−76.80	−17068	104.4	824.8	1.645	.2230	−.137

Table 2—continued

	V σ / Z σ etc.								

P/MPa						3.5			
T_σ/K						349.816			
Liq	V_σ 110.4	Z_σ .13284	S_σ −51.25	H_σ −8918	C_P 195.7	w_σ 322.7	γ_σ 2.617	$(f/P)_\sigma$ .7070	μ_σ 2.27
Vap	452.2	.54422	−28.86	−1087	189.8	179.2	2.509	.7070	18.1
$\dfrac{T}{K}$	$\dfrac{V}{cm^3\,mol^{-1}}$	Z	$\dfrac{S}{J\,K^{-1}\,mol^{-1}}$	$\dfrac{H}{J\,mol^{-1}}$	$\dfrac{C_P}{J\,K^{-1}\,mol^{-1}}$	$\dfrac{w}{m\,sec}$	γ	(f/P)	$\dfrac{\mu}{K\,MPa^{-1}}$
290	80.08	.11624	−74.97	−16542	106.1	792.2	1.655	.2508	−.096
295	81.32	.11605	−73.14	−16007	107.9	759.5	1.666	.2805	−.049
300	82.66	.11599	−71.31	−15463	109.9	726.3	1.680	.3120	.004
305	84.10	.11607	−69.48	−14908	112.2	692.8	1.696	.3454	.064
310	85.66	.11632	−67.63	−14341	114.7	658.7	1.715	.3806	.134
315	87.37	.11676	−65.77	−13760	117.6	624.0	1.738	.4174	.217
320	89.26	.11742	−63.90	−13164	121.0	588.4	1.768	.4558	.316
325	91.37	.11835	−61.99	−12550	125.0	551.6	1.805	.4956	.438
330	93.78	.11963	−60.04	−11912	130.2	513.2	1.855	.5366	.591
335	96.59	.12137	−58.04	−11245	136.9	472.6	1.926	.5787	.794
340	99.99	.12380	−55.94	−10538	146.7	428.7	2.033	.6217	1.08
345	104.4	.12733	−53.70	−9769	162.7	379.5	2.218	.6651	1.51
350	454.3	.54638	−28.76	−1052	187.3	179.7	2.478	.7076	18.1
355	501.1	.59418	−26.42	−228	148.9	191.7	1.986	.7227	16.2
360	538.5	.62969	−24.47	470	131.8	201.0	1.762	.7363	14.8
365	570.9	.65843	−22.73	1102	121.9	208.8	1.630	.7488	13.6
370	600.1	.68270	−21.12	1694	115.5	215.6	1.541	.7604	12.7
380	652.0	.72226	−18.15	2806	107.7	227.3	1.428	.7812	11.1
390	698.3	.75371	−15.41	3860	103.5	237.3	1.358	.7996	9.90
400	740.8	.77960	−12.83	4881	101.0	246.1	1.309	.8158	8.91

410	780.5	.80141	− 10.35	5883	99.5	254.0	1.274	.8303	8.07
420	818.2	.82007	− 7.97	6874	98.7	261.3	1.247	.8433	7.35
430	854.2	.83624	− 5.65	7860	98.4	268.0	1.225	.8550	6.74
440	888.9	.85039	− 3.38	8844	98.4	274.3	1.208	.8656	6.19
450	922.4	.86286	− 1.17	9829	98.7	280.2	1.193	.8753	5.72
475	1002	.88835	4.20	12310	100.0	293.6	1.166	.8957	4.74
500	1078	.90783	9.37	14832	101.8	305.7	1.147	.9120	4.00
525	1151	.92308	14.39	17405	104.0	316.7	1.132	.9253	3.41
550	1222	.93522	19.29	20036	106.4	327.0	1.121	.9361	2.94
575	1291	.94504	24.07	22727	108.9	336.6	1.112	.9451	2.55

Table 2—continued

P/MPa	4.0								
T_σ/K	357.049								
Liq	V_σ 120.9	Z_σ .16285	S_σ −47.66	H_σ −7592	C_P 285.7	w_σ 258.2	γ_σ 3.710	$(f/P)_\sigma$ .6862	μ_σ 3.78
Vap	355.1	.47846	−30.58	−1492	298.4	168.9	3.802	.6862	17.6
$\dfrac{T}{K}$	$\dfrac{V}{\text{cm}^3\,\text{mol}^{-1}}$	Z	$\dfrac{S}{\text{J K}^{-1}\,\text{mol}^{-1}}$	$\dfrac{H}{\text{J mol}^{-1}}$	$\dfrac{C_P}{\text{J K}^{-1}\,\text{mol}^{-1}}$	$\dfrac{w}{\text{m sec}^{-1}}$	γ	(f/P)	$\dfrac{\mu}{\text{K MPa}^{-1}}$
90	54.81	.29297	−174.77	−34059				.6882E − 09	
95	55.22	.27963	−170.06	−33623				.3596E − 08	
100	55.64	.26767	−166.04	−33232				.1570E − 07	
105	56.06	.25688	−162.39	−32858				.5898E − 07	
110	56.50	.24711	−158.95	−32488	74.1	1970	1.582	.1946E − 06	− .633
115	56.94	.23821	−155.64	−32116	75.0	1927	1.580	.5742E − 06	− .623
120	57.39	.23007	−152.42	−31738	76.3	1883	1.572	.1537E − 05	− .612
125	57.84	.22260	−149.28	−31353	77.6	1841	1.563	.3775E − 05	− .600
130	58.29	.21572	−146.21	−30962	78.8	1801	1.555	.8596E − 05	− .589
135	58.75	.20937	−143.22	−30565	79.9	1763	1.548	.1830E − 04	− .579
140	59.22	.20349	−140.29	−30163	80.9	1727	1.544	.3691E − 04	− .570
145	59.69	.19804	−137.44	−29756	81.8	1693	1.541	.6977E − 04	− .562
150	60.17	.19297	−134.65	−29345	82.5	1660	1.539	.1264E − 03	− .554
155	60.65	.18825	−131.94	−28931	83.1	1629	1.539	.2193E − 03	− .548
160	61.14	.18384	−129.29	−28514	83.7	1598	1.540	.3659E − 03	− .541

165	61.64	.17973	−126.71	−28095	84.2	1567	1.541	.5893E−03	−.535
170	62.15	.17588	−124.18	−27672	84.7	1537	1.543	.9190E−03	−.528
175	62.67	.17228	−121.72	−27248	85.2	1507	1.546	.1392E−02	−.521
180	63.19	.16890	−119.32	−26821	85.6	1478	1.549	.2052E−02	−.514
185	63.73	.16574	−116.97	−26391	86.1	1448	1.552	.2954E−02	−.507
190	64.28	.16277	−114.66	−25960	86.6	1418	1.555	.4157E−02	−.499
195	64.84	.15998	−112.41	−25526	87.1	1388	1.559	.5730E−02	−.490
200	65.42	.15737	−110.20	−25089	87.6	1359	1.562	.7751E−02	−.481
205	66.01	.15491	−108.02	−24649	88.2	1329	1.565	.1030E−01	−.471
210	66.61	.15261	−105.89	−24207	88.8	1299	1.569	.1348E−01	−.460
215	67.24	.15045	−103.79	−23761	89.5	1268	1.572	.1736E−01	−.449
220	67.88	.14843	−101.73	−23312	90.1	1238	1.576	.2206E−01	−.436
225	68.54	.14654	−99.70	−22860	90.9	1207	1.579	.2767E−01	−.423
230	69.21	.14478	−97.69	−22403	91.6	1177	1.583	.3430E−01	−.409
235	69.92	.14314	−95.71	−21943	92.5	1146	1.586	.4203E−01	−.393
240	70.64	.14161	−93.76	−21479	93.3	1115	1.590	.5097E−01	−.377
245	71.40	.14020	−91.82	−21010	94.3	1084	1.594	.6121E−01	−.359
250	72.18	.13890	−89.91	−20536	95.2	1053	1.598	.7284E−01	−.339
255	72.99	.13771	−88.01	−20057	96.3	1022	1.602	.8593E−01	−.318
260	73.84	.13663	−86.13	−19573	97.4	990.5	1.607	.1006	−.296
265	74.73	.13566	−84.27	−19084	98.5	959.1	1.612	.1168	−.271
270	75.65	.13480	−82.41	−18588	99.8	927.6	1.618	.1346	−.244
275	76.63	.13405	−80.57	−18086	101.1	896.0	1.624	.1542	−.215
280	77.65	.13342	−78.74	−17577	102.5	864.2	1.631	.1754	−.183
285	78.73	.13291	−76.91	−17061	104.0	832.3	1.639	.1984	−.147

Table 2—continued

P/MPa									
				4.0					
T_σ/K				357.049					
	V_σ $\overline{\text{cm}^3\,\text{mol}^{-1}}$	Z_σ	S_σ $\overline{\text{J K}^{-1}\,\text{mol}^{-1}}$	H_σ $\overline{\text{J mol}^{-1}}$	C_P $\overline{\text{J K}^{-1}\,\text{mol}^{-1}}$	w_σ $\overline{\text{m sec}^{-1}}$	γ_σ	$(f/P)_\sigma$	μ_σ $\overline{\text{K MPa}^{-1}}$
Liq	120.9	.16285	−47.66	−7592	285.7	258.2	3.710	.6862	3.78
Vap	355.1	.47846	−30.58	−1492	298.4	168.9	3.802	.6862	17.6
$\dfrac{T}{\text{K}}$	$\dfrac{V}{\text{cm}^3\,\text{mol}^{-1}}$	Z	$\dfrac{S}{\text{J K}^{-1}\,\text{mol}^{-1}}$	$\dfrac{H}{\text{J mol}^{-1}}$	$\dfrac{C_P}{\text{J K}^{-1}\,\text{mol}^{-1}}$	$\dfrac{w}{\text{m sec}^{-1}}$	γ	(f/P)	$\dfrac{\mu}{\text{K MPa}^{-1}}$
290	79.88	.13252	−75.09	−16537	105.6	800.2	1.648	.2231	−.107
295	81.10	.13226	−73.27	−16004	107.4	767.9	1.659	.2495	−.063
300	82.41	.13215	−71.45	−15463	109.3	735.3	1.671	.2776	−.012
305	83.81	.13220	−69.62	−14911	111.4	702.4	1.686	.3073	.045
310	85.32	.13242	−67.79	−14348	113.8	669.1	1.703	.3386	.110
315	86.97	.13283	−65.95	−13772	116.5	635.2	1.724	.3714	.187
320	88.79	.13349	−64.09	−13182	119.6	600.6	1.749	.4055	.278
325	90.80	.13442	−62.21	−12575	123.3	565.1	1.782	.4410	.388
330	93.08	.13569	−60.29	−11948	127.7	528.3	1.824	.4776	.524
335	95.69	.13742	−58.33	−11296	133.5	489.8	1.881	.5152	.699
340	98.77	.13976	−56.30	−10610	141.2	448.9	1.962	.5536	.933
345	102.6	.14304	−54.16	−9878	152.7	404.4	2.090	.5925	1.27
350	107.6	.14792	−51.84	−9070	172.8	354.0	2.325	.6317	1.80
355	115.4	.15646	−49.10	−8104	223.6	291.7	2.941	.6706	2.87
360	392.2	.52408	−28.57	−773	207.8	179.1	2.676	.6963	16.3
365	436.3	.57510	−26.09	126	160.0	191.3	2.078	.7117	14.7
370	471.0	.61244	−24.07	869	139.7	200.8	1.820	.7256	13.5
380	527.5	.66778	−20.63	2159	120.9	215.8	1.573	.7502	11.6
390	574.7	.70894	−17.61	3319	112.0	227.7	1.449	.7715	10.2
400	616.6	.74162	−14.85	4412	107.1	237.9	1.373	.7902	9.15

410	655.0	.76853	−12.24	5468	104.2	246.9	1.321	.8069	8.25
420	690.7	.79122	−9.75	6501	102.5	255.0	1.283	.8218	7.49
430	724.6	.81066	−7.35	7520	101.5	262.4	1.254	.8352	6.84
440	756.8	.82752	−5.02	8533	101.0	269.2	1.232	.8473	6.27
450	787.8	.84228	−2.75	9542	100.9	275.6	1.213	.8582	5.78
475	861.2	.87221	2.71	12070	101.6	290.0	1.179	.8815	4.77
500	930.1	.89489	7.96	14627	103.1	302.8	1.156	.9001	4.01
525	995.8	.91254	13.03	17227	105.0	314.4	1.140	.9152	3.41
550	1059	.92655	17.97	19879	107.2	325.0	1.127	.9275	2.94
575	1121	.93783	22.78	22588	109.5	335.0	1.117	.9377	2.55

175

Table 2—continued

P/MPa									
T_σ/K				4.5 363.563					
Liq Vap	V_σ 143.6 257.4	Z_σ .21375 .38326	S_σ −42.96 −33.76	H_σ −5830 −2486	$C_{P\sigma}$ 1038.5 1082.0	w_σ 190.0 158.6	γ_σ 12.95 13.27	$(f/P)_\sigma$ .6666 .6666	μ_σ 7.40 15.8
$\dfrac{T}{K}$	$\dfrac{V}{\text{cm}^3\,\text{mol}^{-1}}$	Z	$\dfrac{S}{\text{J K}^{-1}\,\text{mol}^{-1}}$	$\dfrac{H}{\text{J mol}^{-1}}$	$\dfrac{C_P}{\text{J K}^{-1}\,\text{mol}^{-1}}$	$\dfrac{w}{\text{m sec}^{-1}}$	γ	(f/P)	$\dfrac{\mu}{\text{K MPa}^{-1}}$
90	54.80	.32952	−174.81	−34036				.6345E−09	
95	55.20	.31451	−170.10	−33600				.3310E−08	
100	55.62	.30105	−166.08	−33208				.1443E−07	
105	56.05	.28892	−162.43	−32834				.5414E−07	
110	56.48	.27792	−158.99	−32465	74.1	1972	1.581	.1784E−06	−.633
115	56.92	.26790	−155.68	−32092	75.0	1929	1.579	.5259E−06	−.624
120	57.37	.25875	−152.47	−31714	76.2	1886	1.572	.1406E−05	−.612
125	57.82	.25034	−149.33	−31330	77.6	1843	1.563	.3450E−05	−.600
130	58.27	.24260	−146.26	−30939	78.8	1803	1.554	.7850E−05	−.589
135	58.73	.23546	−143.26	−30542	79.9	1766	1.548	.1670E−04	−.579
140	59.20	.22885	−140.34	−30140	80.9	1730	1.543	.3347E−04	−.570
145	59.67	.22271	−137.48	−29733	81.7	1696	1.540	.6357E−04	−.562
150	60.14	.21700	−134.70	−29323	82.5	1663	1.539	.1151E−03	−.555
155	60.62	.21169	−131.99	−28909	83.1	1632	1.538	.1996E−03	−.548
160	61.12	.20673	−129.34	−28492	83.7	1601	1.539	.3328E−03	−.542

165	61.61	.20210	−126.76	−28072	84.2	1570	1.540	.5357E−03	−.535
170	62.12	.19777	−124.24	−27650	84.7	1540	1.543	.8350E−03	−.529
175	62.64	.19372	−121.78	−27226	85.1	1510	1.545	.1264E−02	−.522
180	63.16	.18992	−119.37	−26799	85.6	1481	1.548	.1863E−02	−.515
185	63.70	.18635	−117.02	−26370	86.1	1451	1.551	.2681E−02	−.507
190	64.24	.18301	−114.72	−25938	86.5	1422	1.554	.3771E−02	−.499
195	64.80	.17987	−112.46	−25504	87.0	1392	1.558	.5197E−02	−.491
200	65.38	.17692	−110.25	−25068	87.6	1362	1.561	.7027E−02	−.482
205	65.96	.17416	−108.08	−24628	88.1	1332	1.564	.9337E−02	−.472
210	66.57	.17156	−105.95	−24186	88.7	1302	1.567	.1221E−01	−.461
215	67.18	.16913	−103.86	−23741	89.4	1272	1.571	.1573E−01	−.450
220	67.82	.16685	−101.79	−23292	90.1	1242	1.574	.1998E−01	−.438
225	68.47	.16472	−99.76	−22840	90.8	1212	1.577	.2505E−01	−.425
230	69.15	.16272	−97.76	−22385	91.5	1181	1.581	.3104E−01	−.411
235	69.85	.16087	−95.78	−21925	92.3	1151	1.584	.3803E−01	−.396
240	70.57	.15914	−93.83	−21461	93.2	1120	1.588	.4612E−01	−.379
245	71.32	.15754	−91.90	−20993	94.1	1089	1.591	.5537E−01	−.362
250	72.09	.15607	−89.99	−20520	95.1	1058	1.595	.6588E−01	−.343
255	72.90	.15472	−88.09	−20042	96.1	1027	1.599	.7771E−01	−.322
260	73.73	.15349	−86.22	−19559	97.2	996.3	1.604	.9092E−01	−.300
265	74.61	.15238	−84.36	−19070	98.3	965.2	1.608	.1056	−.276
270	75.53	.15140	−82.51	−18576	99.5	933.9	1.614	.1217	−.250
275	76.49	.15054	−80.67	−18075	100.8	902.6	1.620	.1394	−.222
280	77.50	.14980	−78.84	−17567	102.2	871.2	1.626	.1586	−.190
285	78.56	.14919	−77.02	−17053	103.6	839.7	1.634	.1793	−.156

Table 2—*continued*

P/MPa					4.5				
T_σ/K					363.563				
	V_σ	Z_σ	S_σ	H_σ	C_P	w_σ	γ_σ	$(f/P)_\sigma$	μ_σ
Liq	143.6	.21375	−42.96	−5830	1038.5	190.0	12.95	.6666	7.40
Vap	257.4	.38326	−33.76	−2486	1082.0	158.6	13.27	.6666	15.8
$\frac{T}{K}$	$\frac{V}{cm^3\,mol^{-1}}$	Z	$\frac{S}{J\,K^{-1}\,mol^{-1}}$	$\frac{H}{J\,mol^{-1}}$	$\frac{C_P}{J\,K^{-1}\,mol^{-1}}$	$\frac{w}{m\,sec^{-1}}$	γ	(f/P)	$\frac{\mu}{K\,MPa^{-1}}$
290	79.69	.14872	−75.20	−16531	105.2	808.0	1.642	.2016	−.118
295	80.89	.14840	−73.39	−16001	106.9	776.2	1.652	.2255	−.075
300	82.16	.14823	−71.58	−15462	108.7	744.1	1.663	.2509	−.028
305	83.53	.14822	−69.77	−14913	110.8	711.8	1.676	.2777	.027
310	85.00	.14841	−67.95	−14354	113.0	679.1	1.692	.3060	.088
315	86.60	.14880	−66.12	−13782	115.5	646.0	1.710	.3356	.160
320	88.35	.14943	−64.28	−13198	118.4	612.3	1.733	.3665	.244
325	90.27	.15034	−62.42	−12598	121.7	577.9	1.761	.3986	.344
330	92.43	.15159	−60.53	−11980	125.7	542.5	1.797	.4318	.466
335	94.88	.15329	−58.60	−11339	130.6	505.8	1.844	.4659	.619
340	97.72	.15555	−56.62	−10671	137.0	467.3	1.908	.5007	.817
345	101.1	.15863	−54.56	−9965	145.8	426.2	2.003	.5361	1.09
350	105.4	.16297	−52.38	−9205	159.4	381.2	2.155	.5719	1.49
355	111.2	.16961	−49.96	−8353	184.7	329.9	2.451	.6076	2.14
360	121.1	.18208	−46.98	−7289	257.7	265.5	3.337	.6428	3.55
365	295.8	.43860	−31.41	−1632	403.6	166.7	5.012	.6724	15.7
370	355.7	.52026	−27.75	−285	205.3	182.9	2.593	.6898	14.3
380	424.9	.60522	−23.30	1380	143.4	203.2	1.828	.7189	12.1
390	475.7	.66014	−19.86	2705	124.3	217.8	1.583	.7434	10.6
400	518.4	.70139	−16.84	3898	115.2	229.7	1.458	.7648	9.37

410	556.3	.73434	− 14.06	5022	110.1	239.8	1.380	.7836	8.40
420	591.0	.76157	− 11.45	6106	107.0	248.8	1.327	.8004	7.60
430	623.4	.78461	− 8.96	7165	105.1	256.9	1.288	.8155	6.92
440	653.9	.80439	− 6.56	8210	103.9	264.3	1.259	.8291	6.33
450	683.1	.82159	− 4.23	9246	103.3	271.2	1.235	.8414	5.82
475	751.4	.85616	1.35	11825	103.3	286.6	1.194	.8675	4.79
500	814.9	.88212	6.67	14418	104.3	300.0	1.167	.8884	4.02
525	875.2	.90221	11.80	17046	106.0	312.1	1.147	.9053	3.41
550	933.0	.91808	16.78	19721	108.0	323.2	1.133	.9191	2.93
575	988.9	.93083	21.62	22448	110.2	333.5	1.122	.9305	2.54

179

Table 2—continued

P/MPa **5.0**

T_σ/K

Liq
Vap

$\dfrac{T}{K}$	$\dfrac{V}{cm^3\,mol^{-1}}$ (V_σ)	Z (Z_σ)	$\dfrac{S}{J\,K^{-1}\,mol^{-1}}$ (S_σ)	$\dfrac{H}{J\,mol^{-1}}$ (H_σ)	$\dfrac{C_P}{J\,K^{-1}\,mol^{-1}}$ (C_P)	$\dfrac{w}{m\,sec^{-1}}$ (w_σ)	γ (γ_σ)	(f/P) ($(f/P)_\sigma$)	$\dfrac{\mu}{K\,MPa^{-1}}$ (μ_σ)
90	54.78	.36606	−174.85	−34012				.5924E − 09	
95	55.19	.34937	−170.14	−33576				.3085E − 08	
100	55.61	.33442	−166.12	−33185				.1343E − 07	
105	56.04	.32094	−162.48	−32811				.5031E − 07	
110	56.47	.30872	−159.04	−32441	74.0	1975	1.581	.1656E − 06	− .633
115	56.91	.29759	−155.73	−32069	75.0	1932	1.579	.4876E − 06	− .624
120	57.35	.28741	−152.51	−31691	76.2	1888	1.571	.1302E − 05	− .612
125	57.80	.27807	−149.37	−31307	77.6	1846	1.562	.3193E − 05	− .600
130	58.25	.26947	−146.30	−30916	78.8	1806	1.554	.7258E − 05	− .589
135	58.71	.26153	−143.31	−30519	79.9	1768	1.547	.1543E − 04	− .579
140	59.17	.25418	−140.39	−30117	80.9	1733	1.543	.3090E − 04	− .570
145	59.64	.24736	−137.53	−29710	81.7	1699	1.540	.5865E − 04	− .562
150	60.12	.24102	−134.75	−29300	82.4	1666	1.538	.1061E − 03	− .555
155	60.60	.23511	−132.03	−28886	83.1	1634	1.538	.1839E − 03	− .548
160	61.09	.22960	−129.39	−28469	83.6	1604	1.538	.3065E − 03	− .542

165	61.59	.22446	−126.81	−28050	84.1	1573	1.540	.4931E − 03	−.536
170	62.09	.21964	−124.29	−27628	84.6	1543	1.542	.7682E − 03	−.529
175	62.60	.21513	−121.83	−27203	85.1	1514	1.544	.1162E − 02	−.522
180	63.13	.21091	−119.42	−26777	85.5	1484	1.547	.1713E − 02	−.515
185	63.66	.20694	−117.07	−26348	86.0	1455	1.550	.2463E − 02	−.508
190	64.21	.20322	−114.77	−25917	86.5	1425	1.553	.3463E − 02	−.500
195	64.76	.19973	−112.52	−25483	87.0	1396	1.556	.4771E − 02	−.492
200	65.33	.19645	−110.31	−25047	87.5	1366	1.560	.6450E − 02	−.483
205	65.92	.19337	−108.14	−24608	88.1	1336	1.563	.8568E − 02	−.473
210	66.52	.19048	−106.01	−24166	88.7	1306	1.566	.1120E − 01	−.463
215	67.13	.18777	−103.92	−23721	89.3	1276	1.569	.1442E − 01	−.452
220	67.76	.18523	−101.86	−23273	90.0	1246	1.572	.1832E − 01	−.440
225	68.42	.18286	−99.83	−22821	90.7	1216	1.576	.2296E − 01	−.427
230	69.09	.18064	−97.83	−22366	91.4	1186	1.579	.2845E − 01	−.413
235	69.78	.17856	−95.85	−21907	92.2	1155	1.582	.3485E − 01	−.398
240	70.49	.17664	−93.90	−21443	93.1	1125	1.585	.4225E − 01	−.382
245	71.23	.17485	−91.97	−20976	94.0	1094	1.589	.5071E − 01	−.365
250	72.00	.17320	−90.07	−20503	94.9	1064	1.593	.6033E − 01	−.346
255	72.80	.17169	−88.18	−20026	95.9	1033	1.596	.7115E − 01	−.326
260	73.63	.17031	−86.30	−19544	97.0	1002	1.601	.8324E − 01	−.305
265	74.50	.16906	−84.45	−19056	98.1	971.2	1.605	.9664E − 01	−.281
270	75.40	.16794	−82.60	−18563	99.3	940.2	1.610	.1114	−.256
275	76.35	.16696	−80.77	−18063	100.5	909.2	1.615	.1275	−.228
280	77.35	.16612	−78.94	−17557	101.9	878.1	1.622	.1451	−.198
285	78.39	.16542	−77.13	−17045	103.3	846.9	1.628	.1641	−.165

Table 2—*continued*

182

P/MPa									5.0

T_σ/K — Liq / Vap

$\dfrac{T}{\mathrm{K}}$	$\dfrac{V}{\mathrm{cm^3\,mol^{-1}}}$ (V_σ)	Z (Z_σ)	$\dfrac{S}{\mathrm{J\,K^{-1}\,mol^{-1}}}$ (S_σ)	$\dfrac{H}{\mathrm{J\,mol^{-1}}}$ (H_σ)	$\dfrac{C_P}{\mathrm{J\,K^{-1}\,mol^{-1}}}$	$\dfrac{w}{\mathrm{m\,sec^{-1}}}$ (w_σ)	γ (γ_σ)	(f/P) ($(f/P)_\sigma$)	$\dfrac{\mu}{\mathrm{K\,MPa^{-1}}}$ (μ_σ)
290	79.50	.16486	−75.32	−16524	104.8	815.7	1.636	.1845	−.128
295	80.68	.16446	−73.51	−15996	106.4	784.2	1.645	.2063	−.087
300	81.92	.16422	−71.71	−15460	108.2	752.7	1.655	.2295	−.042
305	83.26	.16416	−69.91	−14914	110.1	720.9	1.667	.2541	.010
310	84.69	.16430	−68.10	−14358	112.3	688.8	1.681	.2800	.068
315	86.24	.16465	−66.28	−13791	114.6	656.4	1.698	.3071	.135
320	87.93	.16524	−64.46	−13211	117.3	623.6	1.718	.3354	.213
325	89.78	.16613	−62.62	−12617	120.3	590.2	1.743	.3648	.304
330	91.83	.16735	−60.75	−12007	123.9	556.0	1.773	.3952	.415
335	94.14	.16899	−58.86	−11377	128.2	520.7	1.813	.4264	.550
340	96.78	.17118	−56.92	−10723	133.6	484.2	1.865	.4585	.721
345	99.87	.17409	−54.92	−10038	140.7	445.7	1.938	.4911	.947
350	103.6	.17805	−52.83	−9311	150.8	404.6	2.046	.5240	1.26
355	108.4	.18370	−50.59	−8521	166.7	359.5	2.226	.5571	1.73
360	115.3	.19256	−48.07	−7620	197.7	308.0	2.588	.5900	2.52
365	127.6	.21027	−44.84	−6449	294.3	244.1	3.752	.6220	4.27
370	203.1	.33007	−35.90	−3157	1085.2	165.4	13.14	.6507	12.0
380	334.4	.52924	−26.46	370	190.7	189.7	2.374	.6870	12.5
390	393.0	.60607	−22.24	1992	143.4	207.6	1.795	.7153	10.8
400	438.0	.65855	−18.86	3328	126.1	221.4	1.575	.7394	9.55

410	476.4	.69873	−15.86	4541	117.3	232.8	1.456	.7606	8.53
420	510.6	.73115	−13.10	5687	112.3	242.8	1.381	.7793	7.69
430	542.1	.75814	−10.50	6793	109.2	251.6	1.328	.7961	6.98
440	571.5	.78108	−8.01	7874	107.2	259.7	1.290	.8112	6.38
450	599.3	.80087	−5.62	8940	106.1	267.0	1.261	.8249	5.85
475	663.7	.84024	.08	11575	105.1	283.3	1.210	.8538	4.80
500	723.0	.86956	5.48	14208	105.7	297.5	1.178	.8769	4.02
525	778.8	.89211	10.67	16865	107.0	310.1	1.155	.8955	3.41
550	832.1	.90986	15.69	19562	108.8	321.6	1.139	.9108	2.92
575	883.6	.92407	20.57	22308	110.9	332.2	1.126	.9234	2.53

183

Table 2—continued

	P/MPa		5.5							
	T_σ/K									
	Liq									
	Vap									
	V_σ	Z_σ	S_σ	H_σ	C_P	w_σ	γ_σ	$(f/P)_\sigma$	μ_σ	
$\dfrac{T}{K}$	$\dfrac{V}{\text{cm}^3\,\text{mol}^{-1}}$	Z	$\dfrac{S}{\text{J K}^{-1}\text{mol}^{-1}}$	$\dfrac{H}{\text{J mol}^{-1}}$	$\dfrac{C_P}{\text{J K}^{-1}\text{mol}^{-1}}$	$\dfrac{w}{\text{m sec}^{-1}}$	γ	(f/P)	$\dfrac{\mu}{\text{K MPa}^{-1}}$	
90	54.77	.40258	−174.89	−33988				.5586E − 09		
95	55.18	.38422	−170.18	−33552				.2904E − 08		
100	55.60	.36777	−166.16	−33161				.1263E − 07		
105	56.02	.35294	−162.52	−32787				.4723E − 07		
110	56.45	.33949	−159.08	−32418	74.0	1978	1.581	.1553E − 06	− .633	
115	56.89	.32725	−155.77	−32046	75.0	1935	1.579	.4566E − 06	− .624	
120	57.33	.31606	−152.55	−31668	76.2	1891	1.571	.1218E − 05	− .612	
125	57.78	.30578	−149.42	−31283	77.5	1849	1.562	.2984E − 05	− .600	
130	58.23	.29632	−146.35	−30893	78.8	1809	1.554	.6778E − 05	− .589	
135	58.69	.28759	−143.36	−30496	79.9	1771	1.547	.1440E − 04	− .579	
140	59.15	.27950	−140.43	−30094	80.9	1735	1.542	.2881E − 04	− .570	
145	59.62	.27199	−137.58	−29687	81.7	1701	1.539	.5465E − 04	− .563	
150	60.09	.26502	−134.80	−29277	82.4	1669	1.538	.9882E − 04	− .555	
155	60.57	.25852	−132.08	−28863	83.1	1637	1.537	.1712E − 03	− .549	
160	61.06	.25246	−129.44	−28446	83.6	1606	1.538	.2851E − 03	− .542	

165	61.56	.24679	−126.86	−28027	84.1	1576	1.539	.4584E−03	−.536
170	62.06	.24149	−124.34	−27605	84.6	1546	1.541	.7139E−03	−.530
175	62.57	.23653	−121.88	−27181	85.1	1517	1.544	.1080E−02	−.523
180	63.09	.23187	−119.48	−26755	85.5	1487	1.546	.1590E−02	−.516
185	63.63	.22751	−117.13	−26326	86.0	1458	1.549	.2286E−02	−.509
190	64.17	.22341	−114.83	−25895	86.4	1429	1.552	.3213E−02	−.501
195	64.72	.21957	−112.58	−25461	86.9	1399	1.555	.4425E−02	−.493
200	65.29	.21596	−110.37	−25025	87.5	1370	1.559	.5980E−02	−.484
205	65.87	.21256	−108.20	−24587	88.0	1340	1.562	.7941E−02	−.474
210	66.47	.20938	−106.07	−24145	88.6	1310	1.565	.1038E−01	−.464
215	67.08	.20639	−103.98	−23701	89.2	1280	1.568	.1336E−01	−.453
220	67.71	.20359	−101.92	−23253	89.9	1250	1.571	.1696E−01	−.441
225	68.36	.20097	−99.90	−22802	90.6	1220	1.574	.2126E−01	−.429
230	69.02	.19852	−97.90	−22347	91.3	1190	1.577	.2633E−01	−.415
235	69.71	.19623	−95.92	−21888	92.1	1160	1.580	.3225E−01	−.400
240	70.42	.19410	−93.97	−21425	93.0	1130	1.583	.3909E−01	−.385
245	71.16	.19212	−92.05	−20958	93.8	1099	1.587	.4692E−01	−.368
250	71.92	.19029	−90.14	−20487	94.8	1069	1.590	.5580E−01	−.350
255	72.71	.18861	−88.26	−20011	95.8	1038	1.594	.6580E−01	−.330
260	73.53	.18708	−86.39	−19529	96.8	1008	1.597	.7697E−01	−.309
265	74.39	.18569	−84.53	−19042	97.9	977.1	1.602	.8935E−01	−.286
270	75.28	.18444	−82.69	−18550	99.1	946.4	1.606	.1030	−.262
275	76.22	.18334	−80.86	−18052	100.3	915.7	1.611	.1179	−.235
280	77.20	.18238	−79.05	−17547	101.6	884.9	1.617	.1341	−.205
285	78.23	.18158	−77.24	−17036	103.0	854.1	1.623	.1517	−.173

Table 2—continued

P/MPa									5.5									
T_σ/K																		
Liq Vap																		
$\frac{T}{K}$	$\frac{V}{cm^3\,mol^{-1}}$	V_σ	Z	Z_σ	$\frac{S}{J\,K^{-1}\,mol^{-1}}$	S_σ	$\frac{H}{J\,mol^{-1}}$	H_σ	$\frac{C_P}{J\,K^{-1}\,mol^{-1}}$	C_P	$\frac{w}{m\,sec^{-1}}$	w_σ	γ	γ_σ	(f/P)	$(f/P)_\sigma$	$\frac{\mu}{K\,MPa^{-1}}$	μ_σ
290	79.32		.18093		−75.43		−16517		104.4		823.1		1.631		.1705		−.138	
295	80.47		.18045		−73.63		−15991		106.0		792.1		1.639		.1907		−.099	
300	81.69		.18014		−71.84		−15457		107.7		761.0		1.648		.2121		−.055	
305	83.00		.18002		−70.04		−14914		109.6		729.7		1.659		.2348		−.006	
310	84.40		.18010		−68.25		−14361		111.6		698.3		1.672		.2587		.049	
315	85.90		.18040		−66.44		−13798		113.8		666.5		1.687		.2838		.111	
320	87.53		.18095		−64.63		−13223		116.3		634.4		1.705		.3100		.184	
325	89.32		.18179		−62.81		−12634		119.1		601.9		1.726		.3372		.268	
330	91.28		.18297		−60.96		−12031		122.3		568.7		1.753		.3653		.369	
335	93.47		.18456		−59.10		−11410		126.2		534.8		1.786		.3943		.490	
340	95.94		.18666		−57.19		−10768		130.8		499.9		1.829		.4239		.640	
345	98.79		.18942		−55.24		−10100		136.8		463.5		1.887		.4542		.832	
350	102.2		.19309		−53.22		−9397		144.6		425.3		1.969		.4849		1.09	
355	106.3		.19809		−51.10		−8648		156.0		384.4		2.091		.5158		1.44	
360	111.7		.20531		−48.80		−7826		174.3		339.8		2.298		.5466		1.98	
365	119.6		.21683		−46.18		−6877		210.3		289.3		2.718		.5771		2.91	
370	134.4		.24025		−42.76		−5620		313.8		230.5		3.944		.6064		4.87	
380	246.3		.42874		−30.79		−1131		322.4		178.4		3.912		.6539		11.7	
390	321.3		.54499		−24.90		1133		175.4		197.8		2.156		.6870		10.9	
400	370.5		.61274		−20.96		2689		141.3		213.4		1.740		.7143		9.63	

410	410.1	.66167	−17.67	4020	126.6	226.2	1.553	.7378	8.60
420	444.4	.70001	−14.73	5242	118.7	237.1	1.446	.7585	7.74
430	475.4	.73135	−12.00	6403	113.9	246.6	1.375	.7770	7.01
440	504.0	.75767	−9.41	7526	110.9	255.3	1.326	.7936	6.40
450	530.7	.78017	−6.94	8625	109.1	263.1	1.289	.8086	5.86
475	592.1	.82453	−1.11	11321	107.0	280.4	1.227	.8403	4.80
500	648.0	.85725	4.37	13994	107.1	295.1	1.189	.8655	4.01
525	700.2	.88227	9.62	16682	108.1	308.2	1.164	.8860	3.40
550	749.9	.90189	14.68	19403	109.7	320.1	1.145	.9027	2.91
575	797.6	.91756	19.59	22167	111.5	331.0	1.131	.9165	2.52

Table 2—*continued*

P/MPa					6.0				
T_σ/K									
Liq Vap									
	V_σ	Z_σ	S_σ	H_σ	C_P	w_σ	γ_σ	$(f/P)_\sigma$	μ_σ
$\dfrac{T}{K}$	$\dfrac{V}{cm^3\,mol^{-1}}$	Z	$\dfrac{S}{J\,K^{-1}\,mol^{-1}}$	$\dfrac{H}{J\,mol^{-1}}$	$\dfrac{C_P}{J\,K^{-1}\,mol^{-1}}$	$\dfrac{w}{m\,sec^{-1}}$	γ	(f/P)	$\dfrac{\mu}{K\,MPa^{-1}}$
90	54.76	.43908	−174.93	−33964				.5311E−09	
95	55.17	.41906	−170.22	−33529				.2757E−08	
100	55.58	.40111	−166.21	−33137				.1197E−07	
105	56.01	.38493	−162.56	−32764				.4470E−07	
110	56.44	.37026	−159.12	−32394	74.0	1981	1.580	.1468E−06	−.633
115	56.87	.35690	−155.81	−32022	74.9	1937	1.578	.4312E−06	−.624
120	57.32	.34468	−152.60	−31644	76.2	1894	1.570	.1149E−05	−.612
125	57.76	.33347	−149.46	−31260	77.5	1851	1.561	.2813E−05	−.600
130	58.21	.32315	−146.40	−30869	78.8	1811	1.553	.6383E−05	−.589
135	58.67	.31362	−143.40	−30473	79.9	1774	1.547	.1355E−04	−.579
140	59.13	.30480	−140.48	−30071	80.9	1738	1.542	.2709E−04	−.571
145	59.60	.29661	−137.63	−29664	81.7	1704	1.539	.5135E−04	−.563
150	60.07	.28900	−134.84	−29254	82.4	1672	1.537	.9280E−04	−.556
155	60.55	.28191	−132.13	−28840	83.0	1640	1.537	.1606E−03	−.549
160	61.04	.27529	−129.49	−28424	83.6	1609	1.537	.2674E−03	−.543

165	61.53	.26910	−126.91	−28004	84.1	1579	1.539	.4297E−03	−.536
170	62.03	.26332	−124.39	−27583	84.6	1549	1.540	.6689E−03	−.530
175	62.54	.25790	−121.93	−27159	85.0	1520	1.543	.1011E−02	−.524
180	63.06	.25282	−119.53	−26733	85.5	1491	1.546	.1489E−02	−.517
185	63.59	.24806	−117.18	−26304	85.9	1461	1.548	.2139E−02	−.510
190	64.13	.24358	−114.88	−25873	86.4	1432	1.551	.3006E−02	−.502
195	64.68	.23938	−112.63	−25440	86.9	1403	1.554	.4138E−02	−.494
200	65.25	.23544	−110.43	−25004	87.4	1373	1.557	.5590E−02	−.485
205	65.83	.23173	−108.26	−24566	88.0	1344	1.561	.7421E−02	−.475
210	66.42	.22825	−106.13	−24125	88.5	1314	1.564	.9695E−02	−.465
215	67.03	.22499	−104.04	−23680	89.2	1284	1.567	.1248E−01	−.454
220	67.66	.22192	−101.99	−23233	89.8	1255	1.569	.1584E−01	−.443
225	68.30	.21905	−99.96	−22782	90.5	1225	1.572	.1985E−01	−.430
230	68.96	.21637	−97.96	−22328	91.2	1195	1.575	.2458E−01	−.417
235	69.64	.21386	−95.99	−21870	92.0	1165	1.578	.3010E−01	−.403
240	70.35	.21153	−94.05	−21407	92.8	1134	1.581	.3647E−01	−.387
245	71.08	.20936	−92.12	−20941	93.7	1104	1.584	.4376E−01	−.371
250	71.83	.20735	−90.22	−20470	94.6	1074	1.588	.5204E−01	−.353
255	72.62	.20550	−88.34	−19995	95.6	1044	1.591	.6136E−01	−.334
260	73.43	.20381	−86.47	−19514	96.6	1013	1.595	.7176E−01	−.313
265	74.28	.20227	−84.62	−19028	97.7	982.9	1.598	.8330E−01	−.291
270	75.16	.20089	−82.78	−18537	98.8	952.5	1.603	.9600E−01	−.267
275	76.08	.19966	−80.96	−18040	100.0	922.0	1.608	.1099	−.241
280	77.05	.19859	−79.15	−17537	101.3	891.6	1.613	.1250	−.212
285	78.07	.19768	−77.34	−17027	102.6	861.0	1.619	.1413	−.181

Table 2—continued

P/MPa										
					6.0					

T_σ/K	V_σ	Z_σ	S_σ	H_σ	C_P	w_σ	γ_σ	$(f/P)_\sigma$	μ_σ
$\dfrac{T}{K}$ Liq Vap	$\dfrac{V}{cm^3\,mol^{-1}}$	Z	$\dfrac{S}{J\,K^{-1}\,mol^{-1}}$	$\dfrac{H}{J\,mol^{-1}}$	$\dfrac{C_P}{J\,K^{-1}\,mol^{-1}}$	$\dfrac{w}{m\,sec^{-1}}$	γ	(f/P)	$\dfrac{\mu}{K\,MPa^{-1}}$
290	79.14	.19694	−75.54	−16510	104.1	830.5	1.625	.1589	−.147
295	80.27	.19637	−73.75	−15986	105.6	799.9	1.633	.1777	−.110
300	81.47	.19598	−71.96	−15454	107.2	769.2	1.641	.1976	−.068
305	82.75	.19579	−70.18	−14913	109.0	738.4	1.651	.2188	−.022
310	84.11	.19580	−68.39	−14363	110.9	707.4	1.663	.2411	.031
315	85.58	.19605	−66.60	−13804	113.0	676.3	1.676	.2644	.090
320	87.16	.19655	−64.80	−13233	115.4	644.9	1.692	.2888	.157
325	88.88	.19734	−62.99	−12649	118.0	613.2	1.711	.3142	.236
330	90.76	.19847	−61.17	−12052	121.0	581.0	1.735	.3405	.328
335	92.84	.20000	−59.32	−11439	124.4	548.2	1.763	.3675	.437
340	95.17	.20200	−57.45	−10807	128.5	514.6	1.799	.3952	.571
345	97.83	.20462	−55.54	−10152	133.5	480.0	1.846	.4236	.737
350	100.9	.20805	−53.57	−9469	140.0	444.0	1.910	.4523	.950
355	104.6	.21259	−51.53	−8749	148.6	406.2	2.000	.4813	1.24
360	109.2	.21881	−49.37	−7977	161.2	366.0	2.136	.5104	1.64
365	115.2	.22782	−47.02	−7125	181.5	322.4	2.363	.5393	2.25
370	124.2	.24227	−44.32	−6133	219.9	274.8	2.806	.5676	3.28
380	173.9	.33023	−36.14	−3059	390.6	190.5	4.729	.6193	8.03
390	257.4	.47625	−28.00	68	228.9	190.9	2.767	.6584	10.3
400	312.5	.56385	−23.20	1962	162.8	206.4	1.978	.6892	9.52

410	354.1	.62326	− 19.53	3451	138.4	220.1	1.680	.7152	8.57
420	388.9	.66826	− 16.35	4769	126.3	231.8	1.525	.7380	7.73
430	419.7	.70434	− 13.47	5994	119.4	242.1	1.431	.7582	7.01
440	447.7	.73425	− 10.77	7165	115.1	251.2	1.366	.7763	6.39
450	473.7	.75960	− 8.22	8301	112.3	259.5	1.320	.7925	5.86
475	532.5	.80907	− 2.25	11061	109.1	277.6	1.246	.8270	4.79
500	585.6	.84523	3.32	13778	108.5	293.0	1.202	.8544	3.99
525	634.9	.87273	8.63	16498	109.2	306.6	1.173	.8766	3.38
550	681.5	.89420	13.74	19243	110.5	318.8	1.152	.8947	2.89
575	726.1	.91131	18.69	22027	112.2	330.0	1.137	.9097	2.50

Table 2—continued

P/MPa									6.5
T_σ/K									
Liq Vap	V_σ	Z_σ	S_σ	H_σ	C_P	w_σ	γ_σ	$(f/P)_\sigma$	μ_σ
$\dfrac{T}{K}$	$\dfrac{V}{cm^3\,mol^{-1}}$	Z	$\dfrac{S}{J\,K^{-1}\,mol^{-1}}$	$\dfrac{H}{J\,mol^{-1}}$	$\dfrac{C_P}{J\,K^{-1}\,mol^{-1}}$	$\dfrac{w}{m\,sec^{-1}}$	γ	(f/P)	$\dfrac{\mu}{K\,MPa^{-1}}$
90	54.75	.47557	−174.97	−33941				.5085E − 09	
95	55.15	.45388	−170.26	−33505				.2635E − 08	
100	55.57	.43443	−166.25	−33114				.1142E − 07	
105	55.99	.41690	−162.60	−32740				.4261E − 07	
110	56.42	.40100	−159.17	−32371	74.0	1983	1.580	.1398E − 06	−.634
115	56.86	.38653	−155.86	−31999	74.9	1940	1.578	.4101E − 06	−.624
120	57.30	.37330	−152.64	−31621	76.2	1896	1.570	.1092E − 05	−.612
125	57.74	.36115	−149.51	−31237	77.5	1854	1.561	.2670E − 05	−.601
130	58.19	.34997	−146.44	−30846	78.8	1814	1.553	.6053E − 05	−.589
135	58.65	.33964	−143.45	−30449	79.9	1776	1.546	.1284E − 04	−.580
140	59.11	.33008	−140.52	−30048	80.8	1741	1.541	.2565E − 04	−.571
145	59.58	.32121	−137.67	−29641	81.7	1707	1.538	.4859E − 04	−.563
150	60.05	.31296	−134.89	−29231	82.4	1674	1.537	.8775E − 04	−.556
155	60.52	.30527	−132.18	−28817	83.0	1643	1.536	.1518E − 03	−.549
160	61.01	.29810	−129.54	−28401	83.6	1612	1.537	.2526E − 03	−.543

165	61.50	.29140	−126.96	−27982	84.1	1582	1.538	.4057E−03	−.537
170	62.00	.28513	−124.44	−27560	84.5	1552	1.540	.6312E−03	−.531
175	62.51	.27925	−121.98	−27137	85.0	1523	1.542	.9538E−03	−.524
180	63.03	.27375	−119.58	−26710	85.4	1494	1.545	.1404E−02	−.517
185	63.56	.26858	−117.23	−26282	85.9	1464	1.548	.2016E−02	−.510
190	64.10	.26373	−114.94	−25852	86.4	1435	1.550	.2832E−02	−.503
195	64.65	.25917	−112.69	−25419	86.8	1406	1.553	.3897E−02	−.495
200	65.21	.25489	−110.48	−24983	87.4	1377	1.556	.5262E−02	−.486
205	65.78	.25087	−108.32	−24545	87.9	1347	1.559	.6984E−02	−.477
210	66.37	.24710	−106.19	−24104	88.5	1318	1.562	.9121E−02	−.467
215	66.98	.24355	−104.11	−23660	89.1	1288	1.565	.1174E−01	−.456
220	67.60	.24022	−102.05	−23213	89.7	1259	1.568	.1489E−01	−.444
225	68.24	.23711	−100.03	−22763	90.4	1229	1.571	.1866E−01	−.432
230	68.90	.23419	−98.03	−22309	91.2	1199	1.574	.2310E−01	−.419
235	69.58	.23146	−96.06	−21851	91.9	1169	1.576	.2828E−01	−.405
240	70.28	.22892	−94.12	−21389	92.7	1139	1.579	.3426E−01	−.390
245	71.00	.22655	−92.20	−20924	93.6	1109	1.582	.4111E−01	−.373
250	71.75	.22437	−90.30	−20453	94.5	1079	1.585	.4888E−01	−.356
255	72.52	.22235	−88.42	−19979	95.4	1049	1.588	.5762E−01	−.337
260	73.33	.22049	−86.55	−19499	96.4	1019	1.592	.6738E−01	−.317
265	74.17	.21881	−84.71	−19014	97.5	988.7	1.595	.7819E−01	−.295
270	75.04	.21729	−82.87	−18524	98.6	958.5	1.599	.9011E−01	−.272
275	75.96	.21593	−81.05	−18028	99.8	928.3	1.604	.1031	−.247
280	76.91	.21474	−79.24	−17526	101.0	898.1	1.609	.1173	−.219
285	77.91	.21373	−77.44	−17017	102.3	867.9	1.614	.1326	−.189

Table 2—continued

P/MPa					6.5				
T_σ/K									
Liq Vap									
T/K	V_σ	Z_σ	S_σ	H_σ	C_P	w_σ	γ_σ	$(f/P)_\sigma$	μ_σ
	$\dfrac{V}{\text{cm}^3\,\text{mol}^{-1}}$	Z	$\dfrac{S}{\text{J K}^{-1}\,\text{mol}^{-1}}$	$\dfrac{H}{\text{J mol}^{-1}}$	$\dfrac{C_P}{\text{J K}^{-1}\,\text{mol}^{-1}}$	$\dfrac{w}{\text{m sec}^{-1}}$	γ	(f/P)	$\dfrac{\mu}{\text{K MPa}^{-1}}$
290	78.97	.21288	−75.65	−16502	103.7	837.7	1.620	.1491	−.156
295	80.08	.21222	−73.87	−15980	105.2	807.5	1.627	.1667	−.120
300	81.26	.21175	−72.09	−15450	106.8	777.2	1.635	.1854	−.080
305	82.51	.21148	−70.31	−14912	108.5	746.8	1.644	.2053	−.036
310	83.84	.21143	−68.53	−14365	110.3	716.4	1.654	.2262	.014
315	85.26	.21161	−66.75	−13808	112.3	685.8	1.667	.2481	.070
320	86.80	.21206	−64.96	−13241	114.5	655.0	1.681	.2710	.133
325	88.46	.21279	−63.17	−12662	117.0	624.0	1.698	.2948	.206
330	90.27	.21385	−61.36	−12071	119.7	592.7	1.718	.3195	.291
335	92.26	.21531	−59.54	−11465	122.8	560.8	1.743	.3449	.390
340	94.47	.21722	−57.69	−10842	126.5	528.5	1.774	.3710	.510
345	96.96	.21971	−55.81	−10198	130.9	495.3	1.813	.3977	.656
350	99.80	.22293	−53.89	−9531	136.3	461.2	1.863	.4248	.839
355	103.1	.22710	−51.91	−8833	143.2	425.7	1.932	.4522	1.07
360	107.1	.23264	−49.85	−8095	152.6	388.7	2.029	.4797	1.39
365	112.2	.24021	−47.66	−7301	166.1	349.6	2.175	.5072	1.84
370	118.9	.25117	−45.27	−6422	187.3	308.1	2.411	.5344	2.51
380	145.1	.29845	−39.21	−4148	280.2	225.0	3.453	.5861	5.21
390	204.5	.40998	−31.42	−1150	277.0	193.6	3.321	.6297	8.57
400	262.6	.51323	−25.62	1141	190.7	202.2	2.290	.6642	9.04

410	306.3	.58396	−21.43	2833	153.1	215.3	1.839	.6929	8.39
420	341.8	.63620	−17.98	4266	135.4	227.3	1.621	.7177	7.64
430	372.5	.67731	−14.92	5566	125.6	238.1	1.494	.7397	6.96
440	400.2	.71098	−12.11	6791	119.7	247.7	1.412	.7592	6.35
450	425.5	.73926	−9.46	7968	115.9	256.4	1.355	.7768	5.82
475	482.4	.79393	−3.34	10798	111.3	275.2	1.266	.8139	4.76
500	533.1	.83354	2.33	13561	110.0	291.2	1.215	.8435	3.97
525	579.9	.86350	7.70	16313	110.3	305.2	1.182	.8674	3.36
550	623.9	.88682	12.86	19083	111.4	317.7	1.159	.8869	2.87
575	665.9	.90534	17.84	21887	112.9	329.2	1.142	.9030	2.48

Table 2—continued

P/MPa									7.0
T_σ/K									
Liq Vap									
$\frac{T}{K}$	V_σ $\frac{V}{\text{cm}^3\text{mol}^{-1}}$	Z_σ Z	S_σ $\frac{S}{\text{J K}^{-1}\text{mol}^{-1}}$	H_σ $\frac{H}{\text{J mol}^{-1}}$	C_P $\frac{C_P}{\text{J K}^{-1}\text{mol}^{-1}}$	w_σ $\frac{w}{\text{m sec}^{-1}}$	γ_σ γ	$(f/P)_\sigma$ (f/P)	μ_σ $\frac{\mu}{\text{K MPa}^{-1}}$
90	54.74	.51205	−175.01	−33917				.4898E−09	
95	55.14	.48868	−170.30	−33481				.2533E−08	
100	55.56	.46774	−166.29	−33090				.1097E−07	
105	55.98	.44885	−162.65	−32717				.4086E−07	
110	56.41	.43173	−159.21	−32348	74.0	1986	1.579	.1338E−06	−.634
115	56.84	.41615	−155.90	−31976	74.9	1943	1.577	.3923E−06	−.624
120	57.28	.40189	−152.69	−31598	76.2	1899	1.570	.1043E−05	−.613
125	57.73	.38881	−149.55	−31214	77.5	1857	1.561	.2549E−05	−.601
130	58.18	.37677	−146.49	−30823	78.8	1817	1.552	.5774E−05	−.590
135	58.63	.36564	−143.49	−30426	79.9	1779	1.546	.1224E−04	−.580
140	59.09	.35534	−140.57	−30024	80.8	1743	1.541	.2443E−04	−.571
145	59.55	.34579	−137.72	−29618	81.7	1710	1.538	.4625E−04	−.563
150	60.02	.33690	−134.94	−29208	82.4	1677	1.536	.8346E−04	−.556
155	60.50	.32862	−132.23	−28795	83.0	1646	1.536	.1443E−03	−.550
160	60.98	.32089	−129.58	−28378	83.5	1615	1.536	.2400E−03	−.543

165	61.47	.31367	−127.01	−27959	84.0	1585	1.537	.3852E−03	−.537
170	61.97	.30692	−124.49	−27538	84.5	1555	1.539	.5991E−03	−.531
175	62.48	.30059	−122.03	−27114	85.0	1526	1.541	.9049E−03	−.525
180	63.00	.29465	−119.63	−26688	85.4	1497	1.544	.1331E−02	−.518
185	63.52	.28908	−117.29	−26260	85.9	1468	1.547	.1911E−02	−.511
190	64.06	.28385	−114.99	−25830	86.3	1439	1.549	.2683E−02	−.503
195	64.61	.27894	−112.74	−25397	86.8	1409	1.552	.3691E−02	−.495
200	65.17	.27433	−110.54	−24962	87.3	1380	1.555	.4983E−02	−.487
205	65.74	.26999	−108.38	−24524	87.8	1351	1.558	.6611E−02	−.478
210	66.33	.26592	−106.25	−24083	88.4	1322	1.561	.8632E−02	−.468
215	66.93	.26209	−104.17	−23640	89.0	1292	1.564	.1110E−01	−.457
220	67.55	.25850	−102.11	−23193	89.7	1263	1.566	.1409E−01	−.446
225	68.18	.25513	−100.09	−22743	90.3	1233	1.569	.1765E−01	−.434
230	68.84	.25198	−98.10	−22290	91.1	1203	1.572	.2184E−01	−.421
235	69.51	.24903	−96.13	−21832	91.8	1174	1.574	.2673E−01	−.407
240	70.20	.24628	−94.19	−21371	92.6	1144	1.577	.3238E−01	−.392
245	70.92	.24372	−92.27	−20906	93.5	1114	1.580	.3884E−01	−.376
250	71.67	.24135	−90.37	−20437	94.4	1084	1.583	.4617E−01	−.359
255	72.43	.23915	−88.50	−19962	95.3	1054	1.586	.5442E−01	−.341
260	73.23	.23714	−86.64	−19484	96.3	1024	1.589	.6363E−01	−.321
265	74.06	.23530	−84.79	−19000	97.3	994.3	1.592	.7384E−01	−.300
270	74.93	.23364	−82.96	−18510	98.4	964.4	1.596	.8508E−01	−.277
275	75.83	.23215	−81.15	−18015	99.6	934.5	1.600	.9737E−01	−.252
280	76.77	.23084	−79.34	−17514	100.8	904.6	1.605	.1107	−.226
285	77.76	.22971	−77.55	−17007	102.1	874.7	1.610	.1252	−.197

197

Table 2—continued

P/MPa							7.0											
T_σ/K																	
Liq Vap	V_σ	$\dfrac{V}{\text{cm}^3\,\text{mol}^{-1}}$	Z_σ	Z	S_σ	$\dfrac{S}{\text{J K}^{-1}\text{mol}^{-1}}$	H_σ	$\dfrac{H}{\text{J mol}^{-1}}$	C_P	$\dfrac{C_P}{\text{J K}^{-1}\text{mol}^{-1}}$	w_σ	$\dfrac{w}{\text{m sec}^{-1}}$	γ_σ	γ	$(f/P)_\sigma$	(f/P)	μ_σ	$\dfrac{\mu}{\text{K MPa}^{-1}}$
$\dfrac{T}{\text{K}}$																	
290		78.80		.22877		−75.76		−16494		103.4		844.8		1.615		.1407		−.165
295		79.89		.22801		−73.98		−15973		104.8		814.9		1.622		.1573		−.130
300		81.05		.22745		−72.21		−15445		106.4		785.0		1.629		.1750		−.092
305		82.27		.22710		−70.43		−14909		108.0		755.1		1.637		.1937		−.049
310		83.57		.22698		−68.66		−14365		109.8		725.1		1.647		.2135		−.002
315		84.96		.22709		−66.89		−13811		111.7		695.0		1.657		.2342		.051
320		86.46		.22746		−65.12		−13248		113.8		664.8		1.670		.2558		.110
325		88.06		.22813		−63.34		−12674		116.0		634.5		1.685		.2783		.178
330		89.81		.22913		−61.55		−12087		118.6		603.9		1.703		.3016		.257
335		91.72		.23051		−59.74		−11487		121.5		573.0		1.725		.3256		.348
340		93.82		.23233		−57.92		−10872		124.8		541.6		1.751		.3503		.456
345		96.17		.23469		−56.07		−10239		128.6		509.7		1.784		.3755		.586
350		98.82		.23771		−54.19		−9584		133.3		477.1		1.826		.4012		.745
355		101.9		.24159		−52.26		−8904		139.0		443.6		1.880		.4272		.945
360		105.4		.24661		−50.26		−8192		146.4		408.9		1.953		.4534		1.20
365		109.8		.25323		−48.18		−7437		156.2		373.0		2.055		.4797		1.55
370		115.3		.26226		−45.97		−6623		170.2		335.6		2.204		.5058		2.03
380		133.0		.29468		−40.83		−4696		221.4		259.5		2.765		.5562		3.75
390		170.4		.36791		−34.31		−2185		266.1		209.6		3.206		.6017		6.49
400		221.3		.46585		−28.11		265		216.2		203.4		2.577		.6395		8.04

410	265.5	.54511	−23.40	2171	169.7	212.8	2.021	.6709	7.99
420	301.5	.60437	−19.63	3735	145.7	224.1	1.731	.6978	7.45
430	332.3	.65057	−16.37	5121	132.6	234.9	1.567	.7215	6.84
440	359.6	.68804	−13.41	6405	124.8	244.7	1.463	.7425	6.27
450	384.5	.71930	−10.67	7626	119.8	253.7	1.393	.7614	5.76
475	439.6	.77918	−4.38	10531	113.6	273.2	1.287	.8012	4.72
500	488.3	.82222	1.38	13342	111.6	289.6	1.228	.8328	3.94
525	532.9	.85461	6.82	16128	111.5	304.0	1.191	.8584	3.33
550	574.7	.87974	12.02	18923	112.3	316.9	1.166	.8793	2.85
575	614.4	.89966	17.04	21747	113.7	328.6	1.147	.8965	2.46

Table 2—continued

P/MPa 7.5

$\dfrac{T}{K}$ Liq Vap	V_σ $\dfrac{V}{\text{cm}^3\,\text{mol}^{-1}}$	Z_σ Z	S_σ $\dfrac{S}{\text{J K}^{-1}\,\text{mol}^{-1}}$	H_σ $\dfrac{H}{\text{J mol}^{-1}}$	C_P $\dfrac{C_P}{\text{J K}^{-1}\,\text{mol}^{-1}}$	w_σ $\dfrac{w}{\text{m sec}^{-1}}$	γ_σ γ	$(f/P)_{b\sigma}$ (f/P)	μ_σ $\dfrac{\mu}{\text{K MPa}^{-1}}$
90	54.73	.54850	−175.05	−33893				.4742E−09	
95	55.13	.52347	−170.34	−33458				.2449E−08	
100	55.54	.50102	−166.33	−33067				.1058E−07	
105	55.96	.48079	−162.69	−32693				.3937E−07	
110	56.39	.46245	−159.25	−32324	74.0	1989	1.579	.1288E−06	−.634
115	56.83	.44575	−155.95	−31952	74.9	1945	1.577	.3772E−06	−.625
120	57.27	.43047	−152.73	−31574	76.2	1902	1.569	.1002E−05	−.613
125	57.71	.41645	−149.60	−31190	77.5	1859	1.560	.2446E−05	−.601
130	58.16	.40355	−146.53	−30800	78.7	1819	1.552	.5536E−05	−.590
135	58.61	.39163	−143.54	−30403	79.8	1782	1.545	.1172E−04	−.580
140	59.07	.38059	−140.62	−30001	80.8	1746	1.540	.2339E−04	−.571
145	59.53	.37035	−137.77	−29595	81.6	1712	1.537	.4424E−04	−.564
150	60.00	.36082	−134.99	−29185	82.4	1680	1.536	.7980E−04	−.557
155	60.48	.35195	−132.28	−28772	83.0	1649	1.535	.1379E−03	−.550
160	60.96	.34367	−129.63	−28356	83.5	1618	1.535	.2291E−03	−.544

165	61.45	.33593	−127.05	−27937	84.0	1588	1.537	.3677E−03	−.538
170	61.94	.32869	−124.54	−27516	84.5	1558	1.538	.5715E−03	−.532
175	62.45	.32190	−122.08	−27092	84.9	1529	1.541	.8629E−03	−.525
180	62.96	.31554	−119.69	−26666	85.4	1500	1.543	.1269E−02	−.519
185	63.49	.30956	−117.34	−26238	85.8	1471	1.546	.1821E−02	−.512
190	64.02	.30396	−115.05	−25808	86.3	1442	1.549	.2556E−02	−.504
195	64.57	.29869	−112.80	−25376	86.8	1413	1.551	.3514E−02	−.496
200	65.13	.29374	−110.60	−24941	87.3	1384	1.554	.4743E−02	−.488
205	65.70	.28908	−108.44	−24503	87.8	1355	1.557	.6291E−02	−.479
210	66.28	.28471	−106.31	−24063	88.4	1325	1.560	.8211E−02	−.469
215	66.88	.28060	−104.23	−23619	89.0	1296	1.562	.1056E−01	−.459
220	67.49	.27675	−102.18	−23173	89.6	1267	1.565	.1339E−01	−.447
225	68.13	.27313	−100.15	−22723	90.3	1237	1.568	.1677E−01	−.436
230	68.78	.26974	−98.16	−22270	91.0	1208	1.570	.2075E−01	−.423
235	69.44	.26657	−96.20	−21814	91.7	1178	1.573	.2540E−01	−.409
240	70.13	.26361	−94.26	−21353	92.5	1148	1.575	.3076E−01	−.394
245	70.85	.26085	−92.34	−20888	93.3	1119	1.578	.3689E−01	−.379
250	71.58	.25829	−90.45	−20420	94.2	1089	1.580	.4385E−01	−.362
255	72.35	.25592	−88.57	−19946	95.2	1059	1.583	.5167E−01	−.344
260	73.14	.25375	−86.72	−19468	96.1	1030	1.586	.6040E−01	−.325
265	73.96	.25176	−84.88	−18985	97.1	999.9	1.589	.7009E−01	−.304
270	74.81	.24995	−83.05	−18496	98.2	970.3	1.593	.8074E−01	−.282
275	75.71	.24833	−81.24	−18003	99.3	940.6	1.597	.9240E−01	−.258
280	76.64	.24690	−79.44	−17503	100.5	910.9	1.601	.1051	−.232
285	77.61	.24565	−77.65	−16997	101.8	881.3	1.605	.1188	−.204

Table 2—continued

P/MPa			7.5							
T_σ/K										
Liq Vap	V_σ	Z_σ	S_σ	H_σ		w_σ	γ_σ	$(f/P)_\sigma$	μ_σ	
$\dfrac{T}{K}$	$\dfrac{V}{cm^3\,mol^{-1}}$	Z	$\dfrac{S}{J\,K^{-1}\,mol^{-1}}$	$\dfrac{H}{J\,mol^{-1}}$	$\dfrac{C_P}{J\,K^{-1}\,mol^{-1}}$	$\dfrac{w}{m\,sec^{-1}}$	γ	(f/P)	$\dfrac{\mu}{K\,MPa^{-1}}$	
290	78.63	.24459	−75.87	−16485	103.1	851.8	1.610	.1335	−.173	
295	79.71	.24374	−74.09	−15966	104.5	822.2	1.616	.1493	−.139	
300	80.84	.24309	−72.32	−15440	106.0	792.7	1.623	.1660	−.103	
305	82.04	.24265	−70.56	−14906	107.5	763.1	1.630	.1838	−.062	
310	83.32	.24244	−68.80	−14364	109.2	733.6	1.639	.2025	−.017	
315	84.68	.24248	−67.03	−13814	111.1	704.0	1.649	.2221	.033	
320	86.13	.24279	−65.27	−13254	113.0	674.4	1.660	.2426	.089	
325	87.69	.24338	−63.50	−12683	115.2	644.6	1.674	.2640	.153	
330	89.37	.24431	−61.72	−12101	117.6	614.7	1.690	.2861	.226	
335	91.21	.24560	−59.94	−11507	120.2	584.6	1.709	.3089	.310	
340	93.22	.24732	−58.13	−10899	123.2	554.2	1.731	.3324	.408	
345	95.45	.24956	−56.31	−10274	126.7	523.3	1.759	.3564	.525	
350	97.93	.25241	−54.46	−9631	130.7	492.0	1.794	.3808	.666	
355	100.8	.25602	−52.57	−8965	135.6	460.1	1.838	.4057	.838	
360	104.0	.26063	−50.64	−8273	141.6	427.3	1.895	.4307	1.06	
365	107.9	.26656	−48.63	−7546	149.3	393.7	1.971	.4558	1.33	
370	112.5	.27434	−46.53	−6776	159.4	359.3	2.074	.4808	1.71	
380	126.2	.29950	−41.89	−5035	192.0	289.4	2.418	.5299	2.90	
390	151.3	.35003	−36.33	−2893	233.5	232.9	2.842	.5756	4.91	
400	190.5	.42954	−30.41	−553	224.1	212.0	2.669	.6156	6.71	

202

410	231.5	.50937	− 25.36	1490	184.8	213.9	2.187	.6494	7.31
420	267.1	.57377	− 21.28	3184	156.6	222.8	1.847	.6783	7.11
430	297.7	.62459	− 17.80	4661	140.1	232.9	1.645	.7037	6.64
440	324.7	.66573	− 14.70	6009	130.2	242.6	1.519	.7261	6.14
450	349.1	.69988	− 11.85	7278	124.0	251.7	1.434	.7463	5.67
475	402.8	.76491	− 5.39	10262	116.1	271.5	1.309	.7887	4.67
500	449.7	.81132	.47	13122	113.2	288.4	1.243	.8223	3.90
525	492.4	.84610	5.98	15942	112.6	303.1	1.201	.8495	3.30
550	532.3	.87300	11.23	18763	113.2	316.2	1.173	.8718	2.82
575	570.0	.89428	16.28	21607	114.4	328.1	1.153	.8902	2.44

204

Table 2—continued

P/MPa									8.0	
T_σ/K										
Liq Vap										
$\frac{T}{K}$	$\frac{V}{cm^3\,mol^{-1}}$	Z	$\frac{S}{J\,K^{-1}\,mol^{-1}}$	$\frac{H}{J\,mol^{-1}}$	$\frac{C_P}{J\,K^{-1}\,mol^{-1}}$	$\frac{w}{m\,sec^{-1}}$	γ	(f/P)	$\frac{\mu}{K\,MPa^{-1}}$
90	54.71	.58495	−175.09	−33869				.4610E−09	
95	55.12	.55824	−170.39	−33434				.2377E−08	
100	55.53	.53430	−166.37	−33043				.1026E−07	
105	55.95	.51271	−162.73	−32670				.3812E−07	
110	56.38	.49315	−159.30	−32301	74.0	1991	1.578	.1246E−06	−.634
115	56.81	.47533	−155.99	−31929	74.9	1948	1.576	.3643E−06	−.625
120	57.25	.45903	−152.78	−31551	76.2	1904	1.569	.9669E−06	−.613
125	57.69	.44408	−149.64	−31167	77.5	1862	1.560	.2357E−05	−.601
130	58.14	.43031	−146.58	−30776	78.7	1822	1.551	.5332E−05	−.590
135	58.59	.41759	−143.58	−30380	79.8	1784	1.545	.1128E−04	−.580
140	59.05	.40582	−140.66	−29978	80.8	1749	1.540	.2249E−04	−.572
145	59.51	.39489	−137.81	−29572	81.6	1715	1.537	.4251E−04	−.564
150	59.98	.38473	−135.03	−29162	82.3	1683	1.535	.7663E−04	−.557
155	60.45	.37526	−132.32	−28749	83.0	1651	1.534	.1323E−03	−.550
160	60.93	.36642	−129.68	−28333	83.5	1621	1.535	.2198E−03	−.544

165	61.42	.35817	−127.10	−27914	84.0	1591	1.536	.3525E−03	−.538
170	61.91	.35044	−124.59	−27493	84.5	1561	1.538	.5477E−03	−.532
175	62.42	.34319	−122.14	−27070	84.9	1532	1.540	.8265E−03	−.526
180	62.93	.33640	−119.74	−26644	85.3	1503	1.542	.1215E−02	−.519
185	63.45	.33003	−117.39	−26216	85.8	1474	1.545	.1742E−02	−.512
190	63.99	.32404	−115.10	−25786	86.2	1445	1.548	.2445E−02	−.505
195	64.53	.31841	−112.85	−25354	86.7	1416	1.550	.3361E−02	−.497
200	65.09	.31312	−110.65	−24919	87.2	1387	1.553	.4534E−02	−.489
205	65.65	.30815	−108.49	−24482	87.7	1358	1.556	.6012E−02	−.480
210	66.23	.30348	−106.37	−24042	88.3	1329	1.558	.7845E−02	−.470
215	66.83	.29909	−104.29	−23599	88.9	1300	1.561	.1009E−01	−.460
220	67.44	.29497	−102.24	−23153	89.5	1271	1.564	.1279E−01	−.449
225	68.07	.29110	−100.22	−22704	90.2	1241	1.566	.1601E−01	−.437
230	68.72	.28747	−98.23	−22251	90.9	1212	1.568	.1981E−01	−.425
235	69.38	.28407	−96.27	−21795	91.6	1183	1.571	.2424E−01	−.411
240	70.07	.28090	−94.33	−21335	92.4	1153	1.573	.2935E−01	−.397
245	70.77	.27795	−92.42	−20871	93.2	1124	1.576	.3519E−01	−.381
250	71.50	.27520	−90.52	−20402	94.1	1094	1.578	.4182E−01	−.365
255	72.26	.27266	−88.65	−19930	95.0	1064	1.581	.4927E−01	−.347
260	73.04	.27031	−86.80	−19452	96.0	1035	1.583	.5759E−01	−.329
265	73.86	.26817	−84.96	−18970	97.0	1005	1.586	.6682E−01	−.308
270	74.70	.26622	−83.14	−18483	98.0	976.0	1.590	.7697E−01	−.287
275	75.58	.26446	−81.33	−17990	99.1	946.6	1.593	.8807E−01	−.263
280	76.50	.26290	−79.53	−17491	100.3	917.2	1.597	.1001	−.238
285	77.47	.26153	−77.75	−16987	101.5	887.9	1.601	.1132	−.211

Table 2—continued

P/MPa									8.0									
T_σ/K																	
Liq Vap	V_σ		Z_σ		S_σ		H_σ		C_P		w_σ		γ_σ		$(f/P)_\sigma$		μ_σ	
$\dfrac{T}{K}$	$\dfrac{V}{\text{cm}^3\,\text{mol}^{-1}}$		Z		$\dfrac{S}{\text{J K}^{-1}\,\text{mol}^{-1}}$		$\dfrac{H}{\text{J mol}^{-1}}$		$\dfrac{C_P}{\text{J K}^{-1}\,\text{mol}^{-1}}$		$\dfrac{w}{\text{m sec}^{-1}}$		γ		(f/P)		$\dfrac{\mu}{\text{K MPa}^{-1}}$	
290	78.47		.26037		−75.97		−16476		102.8		858.6		1.606		.1272		−.181	
295	79.53		.25941		−74.20		−15959		104.2		829.4		1.611		.1422		−.149	
300	80.65		.25866		−72.44		−15434		105.6		800.2		1.617		.1582		−.113	
305	81.82		.25813		−70.68		−14903		107.1		771.0		1.624		.1751		−.074	
310	83.07		.25784		−68.93		−14363		108.7		741.9		1.632		.1929		−.031	
315	84.40		.25780		−67.17		−13815		110.5		712.8		1.641		.2116		.016	
320	85.81		.25802		−65.42		−13258		112.4		683.6		1.651		.2312		.070	
325	87.33		.25854		−63.66		−12691		114.4		654.4		1.663		.2515		.130	
330	88.96		.25939		−61.90		−12114		116.6		625.2		1.677		.2726		.198	
335	90.73		.26059		−60.13		−11525		119.1		595.7		1.694		.2944		.275	
340	92.66		.26222		−58.34		−10922		121.9		566.1		1.714		.3168		.366	
345	94.78		.26433		−56.54		−10305		125.0		536.3		1.738		.3397		.471	
350	97.13		.26701		−54.72		−9672		128.6		506.1		1.767		.3631		.596	
355	99.76		.27040		−52.86		−9018		132.8		475.4		1.803		.3868		.748	
360	102.8		.27466		−50.97		−8342		137.9		444.3		1.849		.4108		.933	
365	106.2		.28005		−49.03		−7638		144.1		412.6		1.907		.4349		1.17	
370	110.3		.28695		−47.02		−6898		151.9		380.3		1.984		.4590		1.46	
380	121.6		.30780		−42.69		−5274		174.8		315.3		2.216		.5066		2.36	
390	140.1		.34561		−37.74		−3369		206.3		258.1		2.533		.5519		3.82	
400	169.3		.40736		−32.31		−1224		216.0		226.5		2.584		.5928		5.46	

410	204.6	.48005	− 27.22	837	193.2	219.5	2.281	.6285	6.43
420	238.3	.54588	− 22.90	2630	166.3	224.1	1.952	.6593	6.61
430	268.2	.60006	− 19.22	4193	147.7	232.4	1.725	.6863	6.35
440	294.7	.64439	− 15.97	5606	135.9	241.5	1.576	.7102	5.96
450	318.6	.68123	− 13.01	6924	128.3	250.4	1.477	.7315	5.54
475	370.8	.75119	− 6.37	9990	118.6	270.3	1.333	.7765	4.59
500	416.2	.80087	− .40	12901	114.9	287.5	1.257	.8121	3.85
525	457.2	.83798	5.17	15757	113.8	302.4	1.211	.8409	3.26
550	495.4	.86661	10.47	18604	114.1	315.8	1.180	.8645	2.79
575	531.4	.88921	15.56	21468	115.1	327.9	1.158	.8840	2.41

Table 2—continued

P/MPa					8.5				
T_σ/K									
Liq Vap									
T/K	V_σ / V (cm³ mol⁻¹)	Z_σ / Z	S_σ / S (J K⁻¹ mol⁻¹)	H_σ / H (J mol⁻¹)	C_P (J K⁻¹ mol⁻¹)	w_σ / w (m sec⁻¹)	γ_σ / γ	$(f/P)_\sigma$ / (f/P)	μ_σ / μ (K MPa⁻¹)
90	54.70	.62138	−175.13	−33845				.4501E−09	
95	55.10	.59299	−170.43	−33410				.2317E−08	
100	55.52	.56756	−166.42	−33020				.9984E−08	
105	55.94	.54462	−162.77	−32646				.3704E−07	
110	56.36	.52383	−159.34	−32277	74.0	1994	1.578	.1209E−06	−.634
115	56.79	.50490	−156.03	−31905	74.9	1951	1.576	.3532E−06	−.625
120	57.23	.48758	−152.82	−31528	76.2	1907	1.568	.9365E−06	−.613
125	57.67	.47169	−149.68	−31144	77.5	1864	1.559	.2281E−05	−.601
130	58.12	.45706	−146.62	−30753	78.7	1825	1.551	.5155E−05	−.590
135	58.57	.44354	−143.63	−30357	79.8	1787	1.544	.1090E−04	−.580
140	59.03	.43103	−140.71	−29955	80.8	1751	1.540	.2171E−04	−.572
145	59.49	.41942	−137.86	−29549	81.6	1718	1.536	.4101E−04	−.564
150	59.95	.40862	−135.08	−29139	82.3	1685	1.534	.7388E−04	−.557
155	60.43	.39855	−132.37	−28726	82.9	1654	1.534	.1275E−03	−.551
160	60.91	.38916	−129.73	−28310	83.5	1624	1.534	.2117E−03	−.545

165	61.39	.38038	− 127.15	− 27892	84.0	1594	1.535	.3393E − 03	− .539
170	61.89	.37217	− 124.64	− 27471	84.4	1564	1.537	.5269E − 03	− .532
175	62.39	.36447	− 122.19	− 27047	84.9	1535	1.539	.7947E − 03	− .526
180	62.90	.35724	− 119.79	− 26622	85.3	1506	1.542	.1168E − 02	− .520
185	63.42	.35046	− 117.45	− 26194	85.7	1477	1.544	.1674E − 02	− .513
190	63.95	.34410	− 115.15	− 25765	86.2	1449	1.547	.2348E − 02	− .506
195	64.49	.33811	− 112.91	− 25332	86.7	1420	1.549	.3227E − 02	− .498
200	65.04	.33249	− 110.71	− 24898	87.2	1391	1.552	.4352E − 02	− .490
205	65.61	.32720	− 108.55	− 24461	87.7	1362	1.555	.5769E − 02	− .481
210	66.19	.32222	− 106.43	− 24021	88.2	1333	1.557	.7525E − 02	− .471
215	66.78	.31755	− 104.35	− 23578	88.8	1304	1.560	.9672E − 02	− .461
220	67.39	.31316	− 102.30	− 23133	89.4	1275	1.562	.1226E − 01	− .450
225	68.01	.30904	− 100.28	− 22684	90.1	1245	1.564	.1535E − 01	− .439
230	68.66	.30517	− 98.29	− 22232	90.8	1216	1.567	.1898E − 01	− .426
235	69.32	.30155	− 96.33	− 21776	91.5	1187	1.569	.2322E − 01	− .413
240	70.00	.29817	− 94.40	− 21316	92.3	1158	1.571	.2811E − 01	− .399
245	70.70	.29501	− 92.49	− 20853	93.1	1128	1.574	.3370E − 01	− .384
250	71.42	.29207	− 90.60	− 20385	94.0	1099	1.576	.4004E − 01	− .368
255	72.17	.28935	− 88.73	− 19913	94.9	1070	1.578	.4717E − 01	− .351
260	72.95	.28684	− 86.88	− 19436	95.8	1040	1.581	.5513E − 01	− .332
265	73.76	.28454	− 85.04	− 18955	96.8	1011	1.584	.6395E − 01	− .312
270	74.59	.28244	− 83.22	− 18468	97.8	981.7	1.587	.7365E − 01	− .291
275	75.46	.28054	− 81.42	− 17977	98.9	952.5	1.590	.8427E − 01	− .268
280	76.37	.27885	− 79.63	− 17479	100.1	923.4	1.593	.9581E − 01	− .244
285	77.32	.27736	− 77.84	− 16976	101.2	894.3	1.597	.1083	− .217

Table 2—*continued*

P/MPa																			
T_σ/K																			**8.5**
Liq Vap	V_σ		Z_σ		S_σ		H_σ		C_P		w_σ		γ_σ		$(f/P)_\sigma$		μ_σ		
$\dfrac{T}{K}$	$\dfrac{V}{\text{cm}^3\,\text{mol}^{-1}}$		Z		$\dfrac{S}{\text{J K}^{-1}\text{mol}^{-1}}$		$\dfrac{H}{\text{J mol}^{-1}}$		$\dfrac{C_P}{\text{J K}^{-1}\text{mol}^{-1}}$		$\dfrac{w}{\text{m sec}^{-1}}$		γ		(f/P)		$\dfrac{\mu}{\text{K MPa}^{-1}}$		
290	78.32		.27608		−76.07		−16467		102.5		865.3		1.602		.1217		−.188		
295	79.36		.27502		−74.31		−15951		103.8		836.4		1.606		.1360		−.157		
300	80.45		.27417		−72.55		−15428		105.2		807.5		1.612		.1513		−.123		
305	81.61		.27355		−70.80		−14898		106.7		778.7		1.618		.1675		−.086		
310	82.83		.27317		−69.05		−14361		108.3		750.0		1.625		.1845		−.045		
315	84.13		.27304		−67.31		−13816		109.9		721.3		1.633		.2024		.001		
320	85.51		.27318		−65.56		−13261		111.7		692.6		1.643		.2211		.051		
325	86.98		.27362		−63.82		−12698		113.7		664.0		1.653		.2406		.108		
330	88.57		.27438		−62.06		−12124		115.8		635.3		1.666		.2608		.171		
335	90.27		.27549		−60.31		−11540		118.1		606.5		1.681		.2816		.244		
340	92.13		.27702		−58.54		−10943		120.6		577.6		1.698		.3031		.327		
345	94.15		.27900		−56.76		−10333		123.5		548.6		1.719		.3250		.423		
350	96.38		.28153		−54.96		−9708		126.7		519.4		1.743		.3474		.536		
355	98.86		.28471		−53.13		−9065		130.5		489.9		1.774		.3702		.670		
360	101.7		.28868		−51.28		−8402		134.8		460.0		1.811		.3933		.831		
365	104.8		.29363		−49.39		−7715		140.0		429.9		1.858		.4165		1.03		
370	108.5		.29984		−47.44		−7000		146.3		399.4		1.916		.4398		1.27		
380	118.1		.31781		−43.32		−5457		163.6		338.4		2.084		.4859		1.98		
390	132.8		.34800		−38.78		−3706		187.1		282.5		2.314		.5304		3.08		
400	155.1		.39652		−33.80		−1741		202.5		244.4		2.437		.5717		4.45		

410	184.1	.45908	− 28.87	258	193.7	229.2	2.290	.6086	5.52
420	214.6	.52230	− 24.44	2094	173.0	228.4	2.025	.6409	5.99
430	243.0	.57784	− 20.60	3727	154.4	233.9	1.796	.6694	5.97
440	268.8	.62449	− 17.21	5201	141.4	241.7	1.633	.6946	5.71
450	292.1	.66362	− 14.14	6568	132.6	250.0	1.520	.7172	5.37
475	342.9	.73812	− 7.32	9718	121.1	269.7	1.357	.7645	4.50
500	386.8	.79093	− 1.24	12680	116.6	286.9	1.272	.8021	3.79
525	426.4	.83028	4.40	15571	115.1	302.0	1.221	.8325	3.21
550	463.0	.86058	9.75	18445	115.0	315.6	1.188	.8574	2.75
575	497.5	.88446	14.88	21330	115.8	327.8	1.164	.8779	2.38

Table 2—continued

P/MPa					9.0				
T_σ/K									
Liq Vap	V_σ	Z_σ	S_σ	H_σ	C_P	w_σ	γ_σ	$(f/P)_\sigma$	μ_σ
$\dfrac{T}{K}$	$\dfrac{V}{cm^3\,mol^{-1}}$	Z	$\dfrac{S}{J\,K^{-1}\,mol^{-1}}$	$\dfrac{H}{J\,mol^{-1}}$	$\dfrac{C_P}{J\,K^{-1}\,mol^{-1}}$	$\dfrac{w}{m\,sec^{-1}}$	γ	(f/P)	$\dfrac{\mu}{K\,MPa^{-1}}$
90	54.69	.65779	− 175.17	− 33822				.4409E − 09	
95	55.09	.62774	− 170.47	− 33387				.2266E − 08	
100	55.50	.60080	− 166.46	− 32996				.9750E − 08	
105	55.92	.57651	− 162.82	− 32623				.3612E − 07	
110	56.35	.55450	− 159.38	− 32254	73.9	1997	1.577	.1178E − 06	− .634
115	56.78	.53445	− 156.08	− 31882	74.9	1953	1.576	.3436E − 06	− .625
120	57.22	.51611	− 152.86	− 31504	76.1	1909	1.568	.9102E − 06	− .613
125	57.66	.49928	− 149.73	− 31120	77.5	1867	1.559	.2215E − 05	− .601
130	58.10	.48379	− 146.67	− 30730	78.7	1827	1.551	.5001E − 05	− .590
135	58.55	.46948	− 143.67	− 30334	79.8	1790	1.544	.1056E − 04	− .581
140	59.01	.45623	− 140.75	− 29932	80.8	1754	1.539	.2103E − 04	− .572
145	59.47	.44393	− 137.91	− 29526	81.6	1720	1.536	.3970E − 04	− .564
150	59.93	.43249	− 135.13	− 29116	82.3	1688	1.534	.7147E − 04	− .557
155	60.40	.42183	− 132.42	− 28703	82.9	1657	1.533	.1233E − 03	− .551
160	60.88	.41188	− 129.78	− 28287	83.5	1626	1.534	.2045E − 03	− .545

165	61.37	.40258	−127.20	−27869	83.9	1597	1.535	.3277E−03	−.539
170	61.86	.39388	−124.69	−27448	84.4	1567	1.536	.5086E−03	−.533
175	62.36	.38572	−122.24	−27025	84.8	1538	1.538	.7669E−03	−.527
180	62.87	.37807	−119.84	−26600	85.3	1509	1.541	.1126E−02	−.520
185	63.39	.37088	−117.50	−26172	85.7	1481	1.543	.1614E−02	−.514
190	63.91	.36413	−115.21	−25743	86.1	1452	1.546	.2263E−02	−.506
195	64.45	.35779	−112.96	−25311	86.6	1423	1.548	.3109E−02	−.499
200	65.00	.35183	−110.76	−24877	87.1	1394	1.551	.4191E−02	−.491
205	65.57	.34622	−108.61	−24440	87.6	1365	1.554	.5554E−02	−.482
210	66.14	.34094	−106.49	−24000	88.2	1337	1.556	.7243E−02	−.473
215	66.73	.33599	−104.41	−23558	88.8	1308	1.558	.9307E−02	−.463
220	67.34	.33133	−102.36	−23113	89.4	1279	1.561	.1180E−01	−.452
225	67.96	.32695	−100.34	−22664	90.0	1250	1.563	.1476E−01	−.440
230	68.60	.32284	−98.36	−22212	90.7	1220	1.565	.1825E−01	−.428
235	69.25	.31900	−96.40	−21757	91.4	1191	1.567	.2232E−01	−.415
240	69.93	.31540	−94.47	−21298	92.2	1162	1.569	.2702E−01	−.401
245	70.63	.31204	−92.56	−20835	93.0	1133	1.572	.3239E−01	−.386
250	71.34	.30891	−90.67	−20368	93.8	1104	1.574	.3847E−01	−.371
255	72.09	.30601	−88.80	−19896	94.7	1075	1.576	.4531E−01	−.354
260	72.86	.30333	−86.95	−19420	95.7	1045	1.578	.5295E−01	−.336
265	73.66	.30087	−85.12	−18940	96.6	1016	1.581	.6141E−01	−.316
270	74.48	.29862	−83.31	−18454	97.7	987.3	1.584	.7073E−01	−.296
275	75.35	.29658	−81.51	−17963	98.7	958.4	1.586	.8091E−01	−.273
280	76.24	.29476	−79.72	−17467	99.8	929.5	1.590	.9198E−01	−.249
285	77.18	.29315	−77.94	−16965	101.0	900.7	1.593	.1039	−.224

Table 2—continued

P/MPa									9.0

Tσ/K

Liq Vap	V_σ	Z_σ	S_σ	H_σ	C_P	w_σ	γ_σ	$(f/P)_\sigma$	μ_σ
$\frac{T}{K}$	$\frac{V}{cm^3\,mol^{-1}}$	Z	$\frac{S}{J\,K^{-1}\,mol^{-1}}$	$\frac{H}{J\,mol^{-1}}$	$\frac{C_P}{J\,K^{-1}\,mol^{-1}}$	$\frac{w}{m\,sec^{-1}}$	γ	(f/P)	$\frac{\mu}{K\,MPa^{-1}}$
290	78.16	.29175	−76.17	−16457	102.2	871.9	1.597	.1168	−.196
295	79.19	.29057	−74.41	−15942	103.5	843.3	1.602	.1306	−.166
300	80.27	.28962	−72.66	−15421	104.9	814.8	1.607	.1452	−.133
305	81.40	.28890	−70.92	−14894	106.3	786.3	1.612	.1607	−.097
310	82.60	.28843	−69.18	−14358	107.8	757.9	1.619	.1771	−.057
315	83.87	.28821	−67.44	−13815	109.4	729.7	1.626	.1942	−.014
320	85.22	.28826	−65.70	−13264	111.1	701.4	1.635	.2122	.034
325	86.65	.28861	−63.97	−12703	113.0	673.2	1.644	.2309	.087
330	88.19	.28928	−62.23	−12134	115.0	645.1	1.655	.2503	.147
335	89.84	.29031	−60.48	−11553	117.1	616.9	1.668	.2703	.215
340	91.63	.29172	−58.73	−10962	119.5	588.7	1.684	.2909	.291
345	93.57	.29359	−56.96	−10358	122.2	560.4	1.702	.3120	.380
350	95.70	.29597	−55.19	−9740	125.1	532.0	1.723	.3336	.482
355	98.04	.29895	−53.39	−9106	128.4	503.5	1.749	.3555	.602
360	100.7	.30265	−51.57	−8455	132.3	474.8	1.779	.3778	.744
365	103.6	.30723	−49.71	−7783	136.7	445.9	1.817	.4002	.915
370	107.0	.31290	−47.82	−7086	141.9	416.9	1.864	.4227	1.12
380	115.4	.32878	−43.86	−5603	155.6	359.1	1.990	.4675	1.69
390	127.5	.35401	−39.59	−3959	173.7	305.1	2.161	.5111	2.56
400	145.3	.39325	−34.98	−2137	189.1	263.8	2.291	.5523	3.67

410	169.0	.44609	−30.27	−231	189.1	242.0	2.242	.5898	4.69
420	195.6	.50409	−25.86	1600	175.7	235.6	2.057	.6233	5.32
430	222.0	.55876	−21.92	3275	159.5	237.6	1.851	.6530	5.51
440	246.5	.60653	−18.41	4801	146.3	243.4	1.685	.6795	5.40
450	269.1	.64735	−15.24	6214	136.8	250.7	1.563	.7032	5.15
475	318.5	.72579	−8.24	9445	123.7	269.5	1.381	.7529	4.40
500	361.0	.78154	−2.06	12460	118.3	286.7	1.288	.7924	3.72
525	399.2	.82302	3.66	15387	116.3	301.9	1.232	.8243	3.16
550	434.4	.85492	9.05	18288	116.0	315.6	1.195	.8504	2.71
575	467.5	.88003	14.22	21193	116.6	328.0	1.170	.8721	2.35

Table 2—*continued*

P/MPa										9.5								
T_σ/K																		
Liq Vap	V_σ	Z_σ	S_σ	H_σ	C_P	w_σ	γ_σ	$(f/P)_\sigma$	μ_σ								
$\dfrac{T}{K}$	$\dfrac{V}{cm^3\,mol^{-1}}$	Z	$\dfrac{S}{J\,K^{-1}\,mol^{-1}}$	$\dfrac{H}{J\,mol^{-1}}$	$\dfrac{C_P}{J\,K^{-1}\,mol^{-1}}$	$\dfrac{w}{m\,sec^{-1}}$	γ	(f/P)	$\dfrac{\mu}{K\,MPa^{-1}}$
90	54.68	.69419	−175.21	−33798				.4333E−09	
95	55.08	.66246	−170.51	−33363				.2223E−08	
100	55.49	.63403	−166.50	−32972				.9550E−08	
105	55.91	.60839	−162.86	−32599				.3534E−07	
110	56.33	.58515	−159.43	−32230	73.9	1999	1.577	.1150E−06	−.635
115	56.76	.56398	−156.12	−31859	74.9	1956	1.575	.3353E−06	−.625
120	57.20	.54463	−152.91	−31481	76.1	1912	1.567	.8874E−06	−.613
125	57.64	.52686	−149.77	−31097	77.5	1870	1.558	.2158E−05	−.602
130	58.08	.51050	−146.71	−30707	78.7	1830	1.550	.4867E−05	−.591
135	58.53	.49539	−143.72	−30310	79.8	1792	1.544	.1027E−04	−.581
140	58.98	.48140	−140.80	−29909	80.8	1757	1.539	.2043E−04	−.572
145	59.44	.46842	−137.95	−29503	81.6	1723	1.535	.3855E−04	−.565
150	59.91	.45634	−135.17	−29093	82.3	1691	1.533	.6935E−04	−.558
155	60.38	.44509	−132.47	−28681	82.9	1660	1.533	.1195E−03	−.551
160	60.85	.43458	−129.83	−28265	83.4	1629	1.533	.1983E−03	−.545

165	61.34	.42476	− 127.25	− 27846	83.9	1599	1.534	.3175E − 03	−.539
170	61.83	.41557	− 124.74	− 27426	84.4	1570	1.536	.4925E − 03	−.533
175	62.33	.40695	− 122.29	− 27003	84.8	1541	1.538	.7422E − 03	−.527
180	62.84	.39887	− 119.89	− 26578	85.2	1512	1.540	.1089E − 02	−.521
185	63.35	.39128	− 117.55	− 26150	85.7	1484	1.542	.1561E − 02	−.514
190	63.88	.38415	− 115.26	− 25721	86.1	1455	1.545	.2188E − 02	−.507
195	64.42	.37745	− 113.02	− 25289	86.6	1426	1.547	.3004E − 02	−.500
200	64.96	.37114	− 110.82	− 24855	87.1	1398	1.550	.4049E − 02	−.492
205	65.53	.36522	− 108.66	− 24419	87.6	1369	1.552	.5364E − 02	−.483
210	66.10	.35964	− 106.55	− 23979	88.1	1340	1.555	.6993E − 02	−.474
215	66.69	.35440	− 104.47	− 23537	88.7	1311	1.557	.8983E − 02	−.464
220	67.29	.34947	− 102.42	− 23092	89.3	1283	1.559	.1138E − 01	−.453
225	67.90	.34484	− 100.41	− 22644	89.9	1254	1.561	.1424E − 01	−.442
230	68.54	.34049	− 98.42	− 22193	90.6	1225	1.564	.1760E − 01	−.430
235	69.19	.33641	− 96.47	− 21738	91.3	1196	1.566	.2152E − 01	−.417
240	69.86	.33260	− 94.54	− 21279	92.1	1167	1.568	.2605E − 01	−.404
245	70.55	.32904	− 92.63	− 20817	92.9	1138	1.570	.3122E − 01	−.389
250	71.27	.32572	− 90.74	− 20350	93.7	1109	1.572	.3708E − 01	−.373
255	72.00	.32264	− 88.88	− 19880	94.6	1080	1.574	.4367E − 01	−.357
260	72.77	.31979	− 87.03	− 19404	95.5	1051	1.576	.5102E − 01	−.339
265	73.56	.31716	− 85.20	− 18924	96.5	1022	1.578	.5916E − 01	−.320
270	74.38	.31476	− 83.39	− 18440	97.5	992.9	1.581	.6812E − 01	−.300
275	75.23	.31258	− 81.59	− 17950	98.5	964.1	1.583	.7792E − 01	−.278
280	76.12	.31062	− 79.81	− 17454	99.6	935.5	1.586	.8858E − 01	−.255
285	77.04	.30888	− 78.04	− 16953	100.8	906.9	1.590	.1001	−.230

Table 2—continued

P/MPa						9.5			
T_σ/K									
	V_σ	Z_σ	S_σ	H_σ	C_P	w_σ	γ_σ	$(f/P)_\sigma$	μ_σ
Liq Vap $\dfrac{T}{K}$	$\dfrac{V}{cm^3\,mol^{-1}}$	Z	$\dfrac{S}{J\,K^{-1}\,mol^{-1}}$	$\dfrac{H}{J\,mol^{-1}}$	$\dfrac{C_P}{J\,K^{-1}\,mol^{-1}}$	$\dfrac{w}{m\,sec^{-1}}$	γ	(f/P)	$\dfrac{\mu}{K\,MPa^{-1}}$
290	78.01	.30736	−76.27	−16447	102.0	878.5	1.593	.1125	−.203
295	79.02	.30607	−74.52	−15934	103.2	850.1	1.597	.1257	−.174
300	80.08	.30501	−72.77	−15414	104.5	821.9	1.602	.1398	−.142
305	81.20	.30419	−71.03	−14888	105.9	793.7	1.607	.1547	−.107
310	82.37	.30362	−69.30	−14355	107.4	765.7	1.613	.1705	−.070
315	83.62	.30331	−67.57	−13814	108.9	737.8	1.619	.1870	−.028
320	84.93	.30327	−65.84	−13265	110.6	710.0	1.627	.2043	.018
325	86.34	.30353	−64.11	−12708	112.4	682.2	1.636	.2223	.068
330	87.83	.30411	−62.38	−12141	114.3	654.6	1.646	.2409	.124
335	89.43	.30503	−60.65	−11565	116.3	626.9	1.657	.2602	.188
340	91.16	.30635	−58.91	−10978	118.5	599.3	1.671	.2801	.259
345	93.02	.30809	−57.16	−10380	121.0	571.8	1.686	.3005	.341
350	95.06	.31032	−55.40	−9768	123.6	544.1	1.705	.3213	.434
355	97.28	.31312	−53.63	−9143	126.7	516.5	1.727	.3425	.542
360	99.75	.31659	−51.83	−8501	130.1	488.7	1.753	.3639	.669
365	102.5	.32084	−50.01	−7841	133.9	460.9	1.784	.3856	.819
370	105.6	.32606	−48.16	−7161	138.4	433.1	1.821	.4074	.998
380	113.2	.34033	−44.33	−5723	149.6	378.0	1.920	.4510	1.47
390	123.6	.36208	−40.26	−4157	164.0	326.0	2.050	.4937	2.17
400	138.2	.39480	−35.93	−2445	177.8	283.4	2.166	.5345	3.07

410	157.7	.43949	−31.45	−633	182.3	256.8	2.171	.5723	3.99
420	180.6	.49137	−27.13	1162	175.1	245.2	2.052	.6066	4.67
430	204.5	.54340	−23.15	2852	162.5	243.4	1.884	.6374	5.01
440	227.6	.59100	−19.56	4414	150.2	246.7	1.726	.6649	5.04
450	249.2	.63273	−16.30	5865	140.5	252.6	1.601	.6896	4.90
475	297.0	.71433	−9.14	9174	126.2	270.0	1.406	.7416	4.27
500	338.1	.77274	−2.85	12240	120.0	286.9	1.303	.7829	3.64
525	375.0	.81622	2.94	15204	117.5	302.1	1.243	.8163	3.11
550	409.0	.84965	8.38	18131	116.9	315.8	1.203	.8437	2.67
575	440.8	.87594	13.59	21057	117.3	328.3	1.175	.8663	2.31

Table 2—continued

P/MPa									
T_σ/K									
					10				
Liq Vap									
V_σ	Z_σ	S_σ	H_σ	C_P	w_σ	γ_σ	$(f/P)_\sigma$	μ_σ	
$\dfrac{T}{\text{K}}$	$\dfrac{V}{\text{cm}^3\,\text{mol}^{-1}}$	$\dfrac{Z}{}$	$\dfrac{S}{\text{J K}^{-1}\text{mol}^{-1}}$	$\dfrac{H}{\text{J mol}^{-1}}$	$\dfrac{C_P}{\text{J K}^{-1}\text{mol}^{-1}}$	$\dfrac{w}{\text{m sec}^{-1}}$	γ	(f/P)	$\dfrac{\mu}{\text{K MPa}^{-1}}$
90	54.67	.73057	−175.25	−33774				4269E − 09	
95	55.07	.69717	−170.55	−33339				2186E − 08	
100	55.48	.66724	−166.54	−32949				9380E − 08	
105	55.89	.64025	−162.90	−32576				3466E − 07	
110	56.32	.61578	−159.47	−32207	73.9	2002	1.577	1127E − 06	− .635
115	56.75	.59350	−156.16	−31835	74.9	1958	1.575	3282E − 06	− .625
120	57.18	.57312	−152.95	−31458	76.1	1914	1.567	8676E − 06	− .614
125	57.62	.55442	−149.82	−31074	77.4	1872	1.558	2107E − 05	− .602
130	58.06	.53720	−146.76	−30683	78.7	1832	1.550	4749E − 05	− .591
135	58.51	.52129	−143.76	−30287	79.8	1795	1.543	1002E − 04	− .581
140	58.96	.50656	−140.85	−29886	80.7	1759	1.538	1991E − 04	− .572
145	59.42	.49289	−138.00	−29480	81.6	1726	1.535	3754E − 04	− .565
150	59.89	.48018	−135.22	−29071	82.3	1693	1.533	6749E − 04	− .558
155	60.35	.46832	−132.51	−28658	82.9	1662	1.532	1163E − 03	− .552
160	60.83	.45726	−129.87	−28242	83.4	1632	1.532	1927E − 03	− .546

165	61.31	.44692	−127.30	−27824	83.9	1602	1.533	.3084E−03	−.540
170	61.80	.43724	−124.79	−27403	84.3	1573	1.535	.4782E−03	−.534
175	62.30	.42817	−122.34	−26980	84.8	1544	1.537	.7204E−03	−.528
180	62.80	.41965	−119.94	−26555	85.2	1515	1.539	.1057E−02	−.522
185	63.32	.41166	−117.60	−26128	85.6	1487	1.542	.1514E−02	−.515
190	63.84	.40415	−115.31	−25699	86.1	1458	1.544	.2121E−02	−.508
195	64.38	.39708	−113.07	−25268	86.5	1430	1.546	.2911E−02	−.500
200	64.93	.39044	−110.88	−24834	87.0	1401	1.549	.3922E−02	−.492
205	65.48	.38419	−108.72	−24397	87.5	1373	1.551	.5194E−02	−.484
210	66.05	.37831	−106.61	−23959	88.1	1344	1.554	.6770E−02	−.475
215	66.64	.37278	−104.53	−23517	88.6	1315	1.556	.8694E−02	−.465
220	67.24	.36758	−102.48	−23072	89.2	1286	1.558	.1101E−01	−.455
225	67.85	.36269	−100.47	−22624	89.9	1258	1.560	.1378E−01	−.444
230	68.48	.35811	−98.49	−22173	90.5	1229	1.562	.1703E−01	−.432
235	69.13	.35380	−96.53	−21719	91.3	1200	1.564	.2081E−01	−.419
240	69.80	.34977	−94.60	−21261	92.0	1171	1.566	.2518E−01	−.406
245	70.48	.34601	−92.70	−20799	92.8	1142	1.568	.3017E−01	−.391
250	71.19	.34250	−90.82	−20333	93.6	1113	1.569	.3583E−01	−.376
255	71.92	.33923	−88.95	−19863	94.5	1084	1.571	.4219E−01	−.360
260	72.68	.33621	−87.11	−19388	95.4	1056	1.573	.4929E−01	−.342
265	73.46	.33342	−85.28	−18909	96.3	1027	1.575	.5715E−01	−.324
270	74.27	.33086	−83.48	−18425	97.3	998.4	1.578	.6580E−01	−.304
275	75.12	.32853	−81.68	−17936	98.3	969.8	1.580	.7526E−01	−.283
280	75.99	.32644	−79.90	−17441	99.4	941.4	1.583	.8553E−01	−.260
285	76.91	.32457	−78.13	−16942	100.5	913.1	1.586	.9664E−01	−.236

Table 2—continued

P/MPa						10			
T_σ/K									
Liq Vap									
T/K	V_σ / V (cm³ mol⁻¹)	Z_σ / Z	S_σ / S (J K⁻¹ mol⁻¹)	H_σ / H (J mol⁻¹)	C_P / C_P (J K⁻¹ mol⁻¹)	w_σ / w (m sec⁻¹)	γ_σ / γ	$(f/P)_\sigma$ / (f/P)	μ_σ / μ (K MPa⁻¹)
290	77.86	.32293	−76.37	−16436	101.7	884.9	1.589	.1086	−.210
295	78.86	.32152	−74.62	−15924	102.9	856.8	1.593	.1214	−.181
300	79.90	.32035	−72.88	−15407	104.2	828.8	1.597	.1350	−.151
305	81.00	.31942	−71.15	−14882	105.6	801.0	1.602	.1494	−.117
310	82.16	.31875	−69.42	−14351	107.0	773.3	1.607	.1645	−.081
315	83.37	.31834	−67.70	−13812	108.5	745.8	1.613	.1805	−.041
320	84.66	.31821	−65.97	−13266	110.1	718.3	1.620	.1972	.002
325	86.03	.31838	−64.26	−12711	111.8	691.0	1.628	.2145	.050
330	87.49	.31886	−62.54	−12148	113.6	663.8	1.637	.2326	.103
335	89.04	.31968	−60.81	−11575	115.5	636.7	1.647	.2512	.163
340	90.71	.32089	−59.09	−10993	117.6	609.6	1.659	.2704	.230
345	92.51	.32251	−57.35	−10399	119.9	582.7	1.672	.2901	.305
350	94.46	.32460	−55.61	−9794	122.4	555.7	1.688	.3102	.391
355	96.58	.32722	−53.86	−9175	125.1	528.8	1.707	.3307	.489
360	98.92	.33047	−52.09	−8542	128.2	501.9	1.729	.3515	.603
365	101.5	.33444	−50.30	−7893	131.6	475.1	1.755	.3725	.736
370	104.4	.33927	−48.48	−7225	135.5	448.3	1.786	.3937	.893
380	111.3	.35225	−44.75	−5825	144.9	395.4	1.865	.4361	1.30
390	120.4	.37142	−40.83	−4319	156.7	345.4	1.966	.4779	1.87
400	132.8	.39941	−36.71	−2690	168.7	302.6	2.064	.5182	2.61

410	149.2	.43762	− 32.45	− 964	175.1	272.6	2.095	.5561	3.42
420	168.9	.48355	− 28.24	781	172.4	256.6	2.025	.5909	4.09
430	190.2	.53194	− 24.28	2464	163.5	251.0	1.895	.6224	4.50
440	211.5	.57820	− 20.65	4046	153.0	251.7	1.755	.6509	4.66
450	232.0	.62003	− 17.32	5527	143.6	255.7	1.633	.6765	4.61
475	278.0	.70384	− 10.00	8907	128.6	271.2	1.429	.7306	4.13
500	317.8	.76457	− 3.61	12023	121.7	287.5	1.319	.7737	3.55
525	353.5	.80991	2.24	15022	118.7	302.6	1.253	.8085	3.05
550	386.3	.84477	7.74	17975	117.8	316.3	1.211	.8371	2.62
575	417.0	.87218	12.98	20922	118.0	328.9	1.181	.8607	2.27

Table 2—continued

P/MPa												
T_σ/K							**11**					
Liq Vap												
T/K	V_σ / $\dfrac{V}{\text{cm}^3\,\text{mol}^{-1}}$	Z_σ / Z	S_σ / $\dfrac{S}{\text{J K}^{-1}\,\text{mol}^{-1}}$	H_σ / $\dfrac{H}{\text{J mol}^{-1}}$	C_P / $\dfrac{C_P}{\text{J K}^{-1}\,\text{mol}^{-1}}$	w_σ / $\dfrac{w}{\text{m sec}^{-1}}$	γ_σ / γ	$(f/P)_\sigma$ / (f/P)	μ_σ / $\dfrac{\mu}{\text{K MPa}^{-1}}$
90	54.65	.80330	-175.33	-33726				.4175E-09	
95	55.04	.76655	-170.63	-33292				.2131E-08	
100	55.45	.73362	-166.62	-32902				.9116E-08	
105	55.87	.70392	-162.98	-32529				.3359E-07	
110	56.29	.67700	-159.55	-32160	73.9	2007	1.576	.1090E-06	-.635
115	56.72	.65249	-156.25	-31788	74.8	1963	1.574	.3166E-06	-.626
120	57.15	.63007	-153.04	-31411	76.1	1920	1.566	.8352E-06	-.614
125	57.59	.60949	-149.91	-31027	77.4	1877	1.557	.2025E-05	-.602
130	58.03	.59054	-146.84	-30637	78.7	1837	1.549	.4556E-05	-.591
135	58.47	.57304	-143.85	-30241	79.8	1800	1.542	.9593E-05	-.581
140	58.92	.55683	-140.94	-29840	80.7	1764	1.537	.1904E-04	-.573
145	59.38	.54179	-138.09	-29434	81.5	1731	1.534	.3585E-04	-.565
150	59.84	.52779	-135.31	-29025	82.2	1699	1.532	.6437E-04	-.559
155	60.31	.51475	-132.61	-28612	82.8	1668	1.531	.1108E-03	-.552
160	60.78	.50257	-129.97	-28196	83.4	1637	1.531	.1834E-03	-.546

165	61.26	.49119	−127.40	−27778	83.8	1608	1.532	.2932E−03	−.541
170	61.75	.48053	−124.89	−27358	84.3	1579	1.534	.4542E−03	−.535
175	62.24	.47053	−122.44	−26936	84.7	1550	1.536	.6835E−03	−.529
180	62.74	.46116	−120.04	−26511	85.1	1521	1.538	.1002E−02	−.523
185	63.25	.45235	−117.71	−26084	85.6	1493	1.540	.1434E−02	−.516
190	63.77	.44407	−115.42	−25655	86.0	1465	1.542	.2007E−02	−.509
195	64.30	.43629	−113.18	−25224	86.4	1436	1.545	.2754E−02	−.502
200	64.85	.42897	−110.99	−24791	86.9	1408	1.547	.3708E−02	−.494
205	65.40	.42207	−108.83	−24355	87.4	1380	1.549	.4907E−02	−.486
210	65.97	.41559	−106.72	−23917	87.9	1351	1.551	.6392E−02	−.477
215	66.54	.40948	−104.64	−23476	88.5	1323	1.553	.8204E−02	−.467
220	67.14	.40374	−102.60	−23032	89.1	1294	1.555	.1039E−01	−.457
225	67.74	.39834	−100.59	−22585	89.7	1265	1.557	.1299E−01	−.447
230	68.37	.39326	−98.61	−22134	90.4	1237	1.559	.1604E−01	−.435
235	69.01	.38850	−96.66	−21681	91.1	1208	1.561	.1960E−01	−.423
240	69.66	.38403	−94.74	−21223	91.8	1180	1.562	.2371E−01	−.410
245	70.34	.37985	−92.84	−20762	92.6	1151	1.564	.2840E−01	−.396
250	71.04	.37595	−90.96	−20297	93.4	1123	1.565	.3371E−01	−.381
255	71.76	.37231	−89.10	−19828	94.2	1094	1.567	.3968E−01	−.366
260	72.50	.36894	−87.26	−19355	95.1	1066	1.569	.4634E−01	−.349
265	73.27	.36582	−85.44	−18877	96.0	1037	1.570	.5371E−01	−.331
270	74.07	.36295	−83.64	−18395	97.0	1009	1.572	.6183E−01	−.312
275	74.90	.36032	−81.85	−17908	98.0	981.0	1.574	.7070E−01	−.292
280	75.75	.35794	−80.08	−17415	99.0	953.0	1.576	.8033E−01	−.270
285	76.65	.35580	−78.31	−16917	100.1	925.2	1.579	.9075E−01	−.247

Table 2—continued

P/MPa

T_σ/K

11

$\dfrac{T}{K}$	$\dfrac{V_\sigma}{cm^3\,mol^{-1}}$ $\dfrac{V}{}$	Z_σ Z	$\dfrac{S_\sigma}{J\,K^{-1}\,mol^{-1}}$ $\dfrac{S}{}$	$\dfrac{H_\sigma}{J\,mol^{-1}}$ $\dfrac{H}{}$	$\dfrac{C_P}{J\,K^{-1}\,mol^{-1}}$ $\dfrac{C_P}{}$	$\dfrac{w_\sigma}{m\,sec^{-1}}$ $\dfrac{w}{}$	γ_σ γ	$(f/P)_\sigma$ (f/P)	μ_σ $\dfrac{\mu}{K\,MPa^{-1}}$
290	77.58	.35391	−76.56	−16414	101.2	897.4	1.582	.1019	−.222
295	78.55	.35226	−74.82	−15905	102.4	869.9	1.585	.1139	−.196
300	79.56	.35086	−73.09	−15390	103.6	842.4	1.588	.1267	−.167
305	80.62	.34972	−71.37	−14869	104.9	815.2	1.592	.1402	−.136
310	81.74	.34884	−69.65	−14341	106.2	788.1	1.596	.1544	−.103
315	82.91	.34822	−67.94	−13806	107.6	761.2	1.601	.1694	−.066
320	84.15	.34789	−66.23	−13264	109.1	734.5	1.607	.1850	−.026
325	85.45	.34786	−64.53	−12715	110.7	707.9	1.613	.2013	.017
330	86.84	.34814	−62.83	−12157	112.3	681.6	1.620	.2183	.065
335	88.31	.34876	−61.13	−11591	114.1	655.3	1.628	.2358	.118
340	89.88	.34974	−59.42	−11016	116.0	629.3	1.637	.2538	.177
345	91.56	.35111	−57.71	−10431	118.0	603.4	1.648	.2723	.242
350	93.37	.35293	−56.00	−9836	120.1	577.6	1.660	.2913	.316
355	95.32	.35523	−54.28	−9230	122.5	552.0	1.674	.3106	.399
360	97.43	.35808	−52.55	−8611	125.0	526.5	1.691	.3302	.493
365	99.75	.36155	−50.81	−7979	127.8	501.2	1.709	.3501	.600
370	102.3	.36573	−49.05	−7332	130.9	476.1	1.731	.3701	.724
380	108.2	.37673	−45.46	−5989	138.0	426.9	1.784	.4105	1.03
390	115.6	.39230	−41.77	−4567	146.5	380.1	1.850	.4505	1.44
400	125.2	.41401	−37.95	−3057	155.5	338.2	1.917	.4897	1.97

410	137.3	.44309	− 34.02	− 1464	162.4	304.9	1.958	.5271	2.57
420	152.1	.47914	− 30.07	175	164.4	282.7	1.944	.5623	3.15
430	168.9	.51961	− 26.23	1807	161.2	270.5	1.874	.5949	3.60
440	186.6	.56114	− 22.59	3388	154.8	265.7	1.777	.6247	3.88
450	204.4	.60103	− 19.19	4900	147.5	265.7	1.674	.6518	3.99
475	246.3	.68615	− 11.65	8388	132.8	275.7	1.471	.7097	3.80

Table 2—continued

○ **12**

P/MPa									
T_σ/K									
Liq Vap									
$\dfrac{T}{K}$	$\dfrac{V}{cm^3\,mol^{-1}}$ (V_σ)	Z (Z_σ)	$\dfrac{S}{J\,K^{-1}\,mol^{-1}}$ (S_σ)	$\dfrac{H}{J\,mol^{-1}}$ (H_σ)	$\dfrac{C_P}{J\,K^{-1}\,mol^{-1}}$ (C_P)	$\dfrac{w}{m\,sec^{-1}}$ (w_σ)	γ (γ_σ)	(f/P) ($(f/P)_\sigma$)	$\dfrac{\mu}{K\,MPa^{-1}}$ (μ_σ)
90	54.62	.87596	−175.41	−33679				.4117E−09	
95	55.02	.83587	−170.71	−33245				.2095E−08	
100	55.42	.79994	−166.70	−32854				.8932E−08	
105	55.84	.76753	−163.07	−32482				.3283E−07	
110	56.26	.73816	−159.64	−32113	73.9	2012	1.575	.1062E−06	−.635
115	56.68	.71142	−156.34	−31741	74.8	1968	1.573	.3079E−06	−.626
120	57.12	.68695	−153.13	−31364	76.1	1925	1.565	.8107E−06	−.614
125	57.55	.66450	−149.99	−30981	77.4	1882	1.556	.1962E−05	−.602
130	57.99	.64382	−146.93	−30591	78.6	1842	1.548	.4406E−05	−.591
135	58.43	.62472	−143.94	−30195	79.7	1805	1.541	.9263E−05	−.582
140	58.88	.60704	−141.03	−29793	80.7	1770	1.536	.1836E−04	−.573
145	59.34	.59062	−138.18	−29388	81.5	1736	1.533	.3452E−04	−.566
150	59.79	.57534	−135.41	−28979	82.2	1704	1.531	.6190E−04	−.559
155	60.26	.56110	−132.70	−28566	82.8	1673	1.530	.1064E−03	−.553
160	60.73	.54781	−130.06	−28151	83.3	1643	1.530	.1759E−03	−.547

165	61.21	.53538	− 127.49	− 27733	83.8	1613	1.531	.2810E − 03	− .541
170	61.69	.52374	− 124.98	− 27313	84.2	1585	1.532	.4349E − 03	− .536
175	62.18	.51283	− 122.54	− 26891	84.7	1556	1.534	.6539E − 03	− .530
180	62.68	.50259	− 120.15	− 26466	85.1	1527	1.536	.9578E − 03	− .524
185	63.19	.49296	− 117.81	− 26040	85.5	1499	1.538	.1370E − 02	− .517
190	63.70	.48392	− 115.52	− 25612	85.9	1471	1.540	.1916E − 02	− .511
195	64.23	.47541	− 113.29	− 25181	86.4	1443	1.543	.2626E − 02	− .504
200	64.77	.46740	− 111.09	− 24748	86.8	1415	1.545	.3534E − 02	− .496
205	65.32	.45986	− 108.94	− 24313	87.3	1386	1.547	.4674E − 02	− .488
210	65.88	.45277	− 106.83	− 23875	87.8	1358	1.549	.6085E − 02	− .479
215	66.45	.44608	− 104.76	− 23434	88.4	1330	1.551	.7805E − 02	− .470
220	67.04	.43979	− 102.72	− 22991	89.0	1302	1.553	.9877E − 02	− .460
225	67.64	.43387	− 100.72	− 22544	89.6	1273	1.554	.1234E − 01	− .449
230	68.25	.42831	− 98.74	− 22095	90.2	1245	1.556	.1524E − 01	− .438
235	68.89	.42308	− 96.79	− 21642	90.9	1217	1.557	.1861E − 01	− .426
240	69.54	.41817	− 94.87	− 21186	91.6	1188	1.559	.2250E − 01	− .414
245	70.20	.41357	− 92.97	− 20726	92.4	1160	1.560	.2694E − 01	− .400
250	70.89	.40927	− 91.10	− 20262	93.2	1132	1.561	.3197E − 01	− .386
255	71.60	.40526	− 89.25	− 19794	94.0	1104	1.563	.3762E − 01	− .371
260	72.33	.40153	− 87.41	− 19322	94.8	1076	1.564	.4392E − 01	− .355
265	73.09	.39807	− 85.60	− 18845	95.7	1048	1.566	.5090E − 01	− .338
270	73.87	.39488	− 83.80	− 18364	96.7	1020	1.567	.5857E − 01	− .320
275	74.68	.39195	− 82.02	− 17879	97.6	992.0	1.569	.6696E − 01	− .300
280	75.52	.38928	− 80.25	− 17388	98.6	964.4	1.570	.7607E − 01	− .280
285	76.39	.38687	− 78.49	− 16892	99.7	936.9	1.572	.8592E − 01	− .258

Table 2—continued

12

P/MPa									
T_σ/K									
	V_σ	Z_σ	S_σ	H_σ	C_P	w_σ	γ_σ	$(f/P)_b$	μ_σ
Liq Vap T/K	$\dfrac{V}{\text{cm}^3\,\text{mol}^{-1}}$	Z	$\dfrac{S}{\text{J K}^{-1}\,\text{mol}^{-1}}$	$\dfrac{H}{\text{J mol}^{-1}}$	$\dfrac{C_P}{\text{J K}^{-1}\,\text{mol}^{-1}}$	$\dfrac{w}{\text{m sec}^{-1}}$	γ	(f/P)	$\dfrac{\mu}{\text{K MPa}^{-1}}$
290	77.30	.38471	−76.75	−16391	100.8	909.6	1.575	.9650E−01	−.234
295	78.25	.38282	−75.02	−15884	101.9	882.5	1.577	.1078	−.209
300	79.23	.38118	−73.30	−15372	103.1	855.6	1.580	.1199	−.182
305	80.26	.37980	−71.58	−14854	104.3	828.9	1.583	.1326	−.153
310	81.34	.37870	−69.88	−14329	105.6	802.4	1.586	.1461	−.122
315	82.47	.37787	−68.18	−13798	106.9	776.1	1.590	.1602	−.088
320	83.66	.37733	−66.48	−13260	108.3	750.0	1.595	.1750	−.052
325	84.91	.37708	−64.79	−12715	109.7	724.1	1.600	.1905	−.012
330	86.23	.37715	−63.11	−12163	111.3	698.4	1.605	.2065	.031
335	87.63	.37755	−61.42	−11602	112.9	673.0	1.612	.2230	.079
340	89.12	.37830	−59.74	−11034	114.6	647.8	1.619	.2401	.131
345	90.70	.37944	−58.05	−10457	116.4	622.8	1.627	.2577	.189
350	92.39	.38098	−56.36	−9870	118.3	598.0	1.637	.2756	.253
355	94.20	.38297	−54.67	−9274	120.3	573.4	1.647	.2940	.324
360	96.15	.38546	−52.97	−8666	122.5	549.1	1.660	.3126	.404
365	98.25	.38850	−51.27	−8048	124.9	525.1	1.674	.3316	.493
370	100.5	.39216	−49.55	−7417	127.5	501.3	1.689	.3507	.594
380	105.8	.40166	−46.08	−6115	133.2	454.8	1.727	.3892	.838
390	112.1	.41477	−42.54	−4751	139.7	410.7	1.771	.4277	1.15
400	119.9	.43250	−38.91	−3319	146.6	370.3	1.818	.4657	1.54

12 MPa

410	129.5	.45575	−35.21	−1821	152.8	336.1	1.854	.5024	1.99
420	141.0	.48470	−31.48	−272	156.4	310.3	1.859	.5375	2.46
430	154.4	.51822	−27.79	1294	156.3	293.2	1.826	.5704	2.88
440	169.0	.55425	−24.23	2844	153.3	283.7	1.763	.6010	3.20
450	184.1	.59057	−20.84	4353	148.4	279.6	1.686	.6293	3.39
475	221.6	.67330	−13.16	7902	135.9	283.1	1.502	.6901	3.42

Table 2—continued

		P/MPa				13				
T_σ/K										
Liq Vap		V_σ	Z_σ	S_σ	H_σ	C_P	w_σ	γ_σ	$(f/P)_\sigma$	μ_σ
$\dfrac{T}{K}$		$\dfrac{V}{cm^3\,mol^{-1}}$	Z	$\dfrac{S}{J\,K^{-1}\,mol^{-1}}$	$\dfrac{H}{J\,mol^{-1}}$	$\dfrac{C_P}{J\,K^{-1}\,mol^{-1}}$	$\dfrac{w}{m\,sec^{-1}}$	γ	(f/P)	$\dfrac{\mu}{K\,MPa^{-1}}$
90		54.60	.94857	−175.49	−33631				.4088E−09	
95		54.99	.90513	−170.79	−33197				.2073E−08	
100		55.40	.86619	−166.79	−32807				.8813E−08	
105		55.81	.83109	−163.15	−32435				.3230E−07	
110		56.23	.79926	−159.72	−32066	73.9	2017	1.574	.1043E−06	−.636
115		56.65	.77028	−156.42	−31695	74.8	1973	1.572	.3016E−06	−.626
120		57.08	.74377	−153.21	−31318	76.1	1929	1.564	.7924E−06	−.614
125		57.52	.71944	−150.08	−30934	77.4	1887	1.555	.1914E−05	−.603
130		57.95	.69704	−147.02	−30544	78.6	1847	1.547	.4292E−05	−.592
135		58.40	.67634	−144.03	−30148	79.7	1810	1.540	.9008E−05	−.582
140		58.84	.65717	−141.12	−29747	80.7	1775	1.535	.1783E−04	−.574
145		59.29	.63938	−138.27	−29342	81.5	1741	1.532	.3347E−04	−.566
150		59.75	.62282	−135.50	−28933	82.2	1709	1.530	.5995E−04	−.560
155		60.21	.60739	−132.79	−28520	82.8	1678	1.529	.1029E−03	−.554
160		60.68	.59298	−130.16	−28105	83.3	1648	1.529	.1700E−03	−.548

165	61.15	.57950	−127.59	−27688	83.8	1619	1.530	.2712E−03	−.542
170	61.63	.56688	−125.08	−27268	84.2	1590	1.531	.4193E−03	−.536
175	62.12	.55505	−122.64	−26846	84.6	1562	1.533	.6300E−03	−.531
180	62.62	.54394	−120.25	−26422	85.0	1533	1.535	.9219E−03	−.525
185	63.12	.53350	−117.91	−25996	85.4	1505	1.537	.1317E−02	−.519
190	63.64	.52368	−115.63	−25568	85.8	1477	1.539	.1841E−02	−.512
195	64.16	.51445	−113.39	−25137	86.3	1449	1.541	.2522E−02	−.505
200	64.69	.50576	−111.20	−24705	86.7	1421	1.543	.3392E−02	−.498
205	65.24	.49757	−109.05	−24270	87.2	1393	1.545	.4483E−02	−.490
210	65.79	.48986	−106.95	−23832	87.7	1365	1.547	.5832E−02	−.481
215	66.36	.48259	−104.88	−23392	88.3	1337	1.548	.7478E−02	−.472
220	66.94	.47575	−102.84	−22950	88.8	1309	1.550	.9457E−02	−.463
225	67.53	.46931	−100.84	−22504	89.4	1281	1.552	.1181E−01	−.452
230	68.14	.46325	−98.86	−22055	90.1	1253	1.553	.1458E−01	−.441
235	68.77	.45755	−96.92	−21603	90.8	1225	1.554	.1780E−01	−.430
240	69.41	.45220	−95.00	−21148	91.5	1197	1.555	.2151E−01	−.418
245	70.07	.44718	−93.11	−20688	92.2	1169	1.557	.2574E−01	−.405
250	70.75	.44248	−91.24	−20226	93.0	1141	1.558	.3054E−01	−.391
255	71.45	.43809	−89.39	−19759	93.8	1113	1.559	.3592E−01	−.376
260	72.17	.43399	−87.56	−19288	94.6	1085	1.560	.4192E−01	−.361
265	72.91	.43019	−85.75	−18813	95.5	1058	1.561	.4857E−01	−.344
270	73.68	.42667	−83.96	−18333	96.4	1030	1.562	.5587E−01	−.327
275	74.47	.42343	−82.18	−17849	97.3	1003	1.563	.6386E−01	−.308
280	75.30	.42047	−80.42	−17360	98.3	975.4	1.565	.7253E−01	−.289
285	76.15	.41777	−78.67	−16866	99.3	948.3	1.566	.8190E−01	−.268

Table 2—continued

			13					
P/MPa								
T_σ/K								
V_σ	Z_σ	S_σ	H_σ	C_P	w_σ	γ_σ	$(f/P)_\sigma$	μ_σ
Liq Vap								
$\dfrac{T}{K}$ / $\dfrac{V}{\text{cm}^3\,\text{mol}^{-1}}$	Z	$\dfrac{S}{\text{J K}^{-1}\,\text{mol}^{-1}}$	$\dfrac{H}{\text{J mol}^{-1}}$	$\dfrac{C_P}{\text{J K}^{-1}\,\text{mol}^{-1}}$	$\dfrac{w}{\text{m sec}^{-1}}$	γ	(f/P)	$\dfrac{\mu}{\text{K MPa}^{-1}}$
290 77.04	.41534	−76.93	−16367	100.3	921.5	1.568	.9198E−01	−.246
295 77.96	.41319	−75.21	−15862	101.4	894.8	1.570	.1027	−.222
300 78.92	.41130	−73.50	−15353	102.6	868.4	1.572	.1142	−.197
305 79.92	.40969	−71.79	−14837	103.7	842.1	1.574	.1264	−.169
310 80.96	.40835	−70.09	−14315	104.9	816.1	1.577	.1392	−.140
315 82.06	.40730	−68.41	−13787	106.2	790.4	1.580	.1526	−.109
320 83.20	.40653	−66.72	−13253	107.5	764.8	1.584	.1667	−.075
325 84.40	.40607	−65.05	−12712	108.9	739.6	1.588	.1814	−.038
330 85.67	.40592	−63.37	−12164	110.3	714.6	1.592	.1967	.001
335 87.01	.40609	−61.70	−11609	111.8	689.8	1.597	.2124	.044
340 88.42	.40661	−60.04	−11047	113.3	665.3	1.603	.2287	.091
345 89.91	.40750	−58.37	−10476	115.0	641.1	1.609	.2455	.143
350 91.50	.40877	−56.70	−9896	116.7	617.1	1.617	.2626	.199
355 93.20	.41047	−55.03	−9308	118.5	593.5	1.625	.2801	.261
360 95.00	.41263	−53.36	−8711	120.5	570.1	1.634	.2980	.330
365 96.94	.41528	−51.69	−8103	122.5	547.1	1.645	.3160	.406
370 99.03	.41848	−50.00	−7485	124.7	524.3	1.656	.3344	.491
380 103.7	.42677	−46.62	−6215	129.5	480.1	1.683	.3714	.690
390 109.3	.43805	−43.18	−4894	134.8	438.1	1.715	.4085	.938
400 115.9	.45301	−39.70	−3518	140.4	399.3	1.748	.4454	1.24

410	123.8	.47228	− 36.17	− 2087	145.6	365.2	1.776	.4813	1.59
420	133.3	.49616	− 32.61	− 610	149.5	337.6	1.787	.5160	1.97
430	144.2	.52420	− 29.07	895	151.0	317.3	1.773	.5490	2.33
440	156.2	.55517	− 25.60	2403	150.2	304.0	1.733	.5800	2.63
450	169.1	.58750	− 22.26	3892	147.4	296.3	1.678	.6089	2.85
475	202.2	.66544	− 14.54	7460	137.8	293.1	1.522	.6721	3.04

Table 2—continued

P/MPa						14			
T_σ/K									
Liq Vap	V_σ	Z_σ	S_σ	H_σ	C_P	w_σ	γ_σ	$(f/P)_\sigma$	μ_σ
$\dfrac{T}{K}$	$\dfrac{V}{cm^3\,mol^{-1}}$	Z	$\dfrac{S}{J\,K^{-1}\,mol^{-1}}$	$\dfrac{H}{J\,mol^{-1}}$	$\dfrac{C_P}{J\,K^{-1}\,mol^{-1}}$	$\dfrac{w}{m\,sec^{-1}}$	γ	(f/P)	$\dfrac{\mu}{K\,MPa^{-1}}$
90	54.58	1.0211	-175.56	-33584				.4084E-09	
95	54.97	.97433	-170.87	-33150				.2064E-08	
100	55.37	.93239	-166.87	-32760				.8748E-08	
105	55.78	.89458	-163.23	-32388				.3198E-07	
110	56.20	.86030	-159.81	-32019	73.8	2022	1.573	.1030E-06	-.636
115	56.62	.82908	-156.51	-31648	74.8	1978	1.571	.2971E-06	-.627
120	57.05	.80053	-153.30	-31271	76.0	1934	1.563	.7792E-06	-.615
125	57.48	.77433	-150.17	-30887	77.4	1892	1.554	.1878E-05	-.603
130	57.92	.75019	-147.11	-30497	78.6	1852	1.546	.4204E-05	-.592
135	58.36	.72789	-144.12	-30102	79.7	1815	1.539	.8811E-05	-.583
140	58.80	.70724	-141.21	-29701	80.6	1780	1.534	.1741E-04	-.574
145	59.25	.68807	-138.36	-29296	81.4	1746	1.531	.3265E-04	-.567
150	59.71	.67023	-135.59	-28887	82.1	1714	1.529	.5840E-04	-.560
155	60.17	.65360	-132.89	-28475	82.7	1683	1.528	.1001E-03	-.554
160	60.63	.63808	-130.25	-28060	83.2	1654	1.528	.1652E-03	-.548

236

165	61.10	.62355	−127.68	−27642	83.7	1624	1.528	.2633E−03	−.543
170	61.58	.60995	−125.18	−27223	84.1	1596	1.530	.4067E−03	−.537
175	62.07	.59719	−122.73	−26801	84.5	1567	1.531	.6105E−03	−.532
180	62.56	.58521	−120.35	−26377	85.0	1539	1.533	.8927E−03	−.526
185	63.06	.57396	−118.01	−25951	85.4	1511	1.535	.1274E−02	−.520
190	63.57	.56337	−115.73	−25524	85.8	1483	1.537	.1780E−02	−.513
195	64.09	.55341	−113.50	−25094	86.2	1456	1.539	.2437E−02	−.506
200	64.62	.54402	−111.31	−24662	86.7	1428	1.541	.3274E−02	−.499
205	65.16	.53519	−109.16	−24227	87.1	1400	1.543	.4325E−02	−.491
210	65.71	.52686	−107.06	−23790	87.6	1372	1.545	.5624E−02	−.483
215	66.27	.51901	−104.99	−23351	88.2	1344	1.546	.7206E−02	−.474
220	66.84	.51162	−102.96	−22909	88.7	1316	1.548	.9109E−02	−.465
225	67.43	.50465	−100.96	−22463	89.3	1289	1.549	.1137E−01	−.455
230	68.04	.49809	−98.99	−22015	89.9	1261	1.550	.1403E−01	−.444
235	68.65	.49192	−97.05	−21564	90.6	1233	1.551	.1712E−01	−.433
240	69.29	.48612	−95.13	−21109	91.3	1205	1.552	.2068E−01	−.421
245	69.94	.48067	−93.24	−20651	92.0	1177	1.553	.2474E−01	−.409
250	70.61	.47557	−91.37	−20189	92.8	1150	1.554	.2934E−01	−.395
255	71.30	.47079	−89.53	−19723	93.5	1122	1.555	.3450E−01	−.381
260	72.01	.46633	−87.71	−19253	94.4	1095	1.556	.4025E−01	−.366
265	72.74	.46218	−85.90	−18780	95.2	1067	1.556	.4661E−01	−.351
270	73.49	.45833	−84.11	−18301	96.1	1040	1.557	.5361E−01	−.334
275	74.27	.45477	−82.34	−17819	97.0	1013	1.558	.6126E−01	−.316
280	75.08	.45150	−80.58	−17331	98.0	986.2	1.559	.6956E−01	−.297
285	75.91	.44851	−78.84	−16839	98.9	959.5	1.560	.7853E−01	−.277

237

Table 2—continued

P/MPa									
T_σ/K					**14**				
Liq Vap	V_σ	Z_σ	S_σ	H_σ	C_P	w_σ	γ_σ	$(f/P)_\sigma$	μ_σ
$\dfrac{T}{\text{K}}$	$\dfrac{V}{\text{cm}^3\,\text{mol}^{-1}}$	Z	$\dfrac{S}{\text{J K}^{-1}\,\text{mol}^{-1}}$	$\dfrac{H}{\text{J mol}^{-1}}$	$\dfrac{C_P}{\text{J K}^{-1}\,\text{mol}^{-1}}$	$\dfrac{w}{\text{m sec}^{-1}}$	γ	(f/P)	$\dfrac{\mu}{\text{K MPa}^{-1}}$
290	76.78	.44581	−77.11	−16342	99.9	933.0	1.562	.8817E−01	−.256
295	77.68	.44339	−75.39	−15839	101.0	906.8	1.563	.9848E−01	−.234
300	78.61	.44124	−73.69	−15332	102.1	880.7	1.565	.1095	−.210
305	79.59	.43938	−71.99	−14819	103.2	855.0	1.566	.1211	−.184
310	80.60	.43781	−70.30	−14300	104.4	829.4	1.569	.1333	−.157
315	81.66	.43652	−68.63	−13775	105.6	804.2	1.571	.1462	−.127
320	82.77	.43552	−66.95	−13244	106.8	779.2	1.574	.1597	−.096
325	83.93	.43483	−65.29	−12707	108.1	754.4	1.577	.1738	−.062
330	85.14	.43445	−63.63	−12163	109.4	730.0	1.580	.1884	−.026
335	86.42	.43440	−61.97	−11612	110.8	705.9	1.584	.2035	.013
340	87.77	.43468	−60.32	−11055	112.3	682.0	1.589	.2191	.056
345	89.19	.43532	−58.67	−10490	113.8	658.5	1.594	.2352	.102
350	90.69	.43633	−57.02	−9917	115.4	635.2	1.599	.2516	.152
355	92.29	.43774	−55.37	−9336	117.0	612.3	1.606	.2684	.207
360	93.98	.43958	−53.72	−8746	118.8	589.8	1.613	.2855	.267
365	95.78	.44187	−52.07	−8148	120.6	567.6	1.620	.3029	.333
370	97.71	.44467	−50.42	−7540	122.5	545.7	1.629	.3206	.406
380	102.0	.45192	−47.10	−6295	126.6	503.3	1.649	.3562	.573
390	106.9	.46175	−43.75	−5007	131.1	463.0	1.672	.3922	.776
400	112.7	.47462	−40.38	−3673	135.8	425.6	1.696	.4280	1.02

410	119.6	.49101	−36.97	−2293	140.3	392.1	1.717	.4632	1.30
420	127.5	.51117	−33.54	−871	144.0	363.8	1.728	.4974	1.60
430	136.6	.53496	−30.12	582	146.2	341.5	1.723	.5302	1.91
440	146.8	.56168	−26.76	2047	146.6	325.4	1.698	.5613	2.18
450	157.7	.59022	−23.47	3508	145.4	314.7	1.659	.5906	2.40
475	186.8	.66225	−15.78	7064	138.6	305.2	1.531	.6555	2.68

Table 2—continued

		P/MPa	15							
	T_σ/K									
Liq Vap	V_σ	Z_σ	S_σ	H_σ	C_P	w_σ	γ_σ	$(f/P)_\sigma$	μ_σ	
$\dfrac{T}{K}$	$\dfrac{V}{cm^3\,mol^{-1}}$	Z	$\dfrac{S}{J\,K^{-1}\,mol^{-1}}$	$\dfrac{H}{J\,mol^{-1}}$	$\dfrac{C_P}{J\,K^{-1}\,mol^{-1}}$	$\dfrac{w}{m\,sec^{-1}}$	γ	(f/P)	$\dfrac{\mu}{K\,MPa^{-1}}$	
90	54.56	1.0936	−175.64	−33536				.4100E−09		
95	54.95	1.0435	−170.95	−33102				.2065E−08		
100	55.35	.99853	−166.95	−32713				.8727E−08		
105	55.76	.95801	−163.32	−32340				.3181E−07		
110	56.17	.92128	−159.89	−31972	73.8	2027	1.571	.1022E−06	−.636	
115	56.59	.88783	−156.59	−31601	74.7	1983	1.570	.2942E−06	−.627	
120	57.02	.85723	−153.38	−31224	76.0	1939	1.562	.7700E−06	−.615	
125	57.45	.82914	−150.25	−30841	77.3	1897	1.553	.1853E−05	−.603	
130	57.88	.80328	−147.20	−30451	78.6	1857	1.545	.4140E−05	−.592	
135	58.32	.77938	−144.21	−30055	79.7	1820	1.538	.8662E−05	−.583	
140	58.76	.75725	−141.30	−29655	80.6	1784	1.533	.1709E−04	−.575	
145	59.21	.73670	−138.45	−29249	81.4	1751	1.530	.3200E−04	−.567	
150	59.66	.71758	−135.68	−28841	82.1	1719	1.528	.5718E−04	−.561	
155	60.12	.69975	−132.98	−28429	82.7	1689	1.527	.9791E−04	−.555	
160	60.58	.68311	−130.35	−28014	83.2	1659	1.527	.1614E−03	−.549	

165	61.05	.66753	-127.78	-27597	83.7	1630	1.527	.2570E-03	-.544
170	61.53	.65294	-125.27	-27177	84.1	1601	1.528	.3965E-03	-.538
175	62.01	.63926	-122.83	-26756	84.5	1573	1.530	.5946E-03	-.533
180	62.50	.62641	-120.44	-26332	84.9	1545	1.532	.8687E-03	-.527
185	63.00	.61434	-118.11	-25907	85.3	1517	1.533	.1239E-02	-.521
190	63.50	.60298	-115.83	-25480	85.7	1489	1.535	.1729E-02	-.515
195	64.02	.59228	-113.60	-25050	86.1	1462	1.537	.2366E-02	-.508
200	64.54	.58221	-111.42	-24618	86.6	1434	1.539	.3177E-02	-.501
205	65.08	.57272	-109.27	-24184	87.0	1407	1.541	.4194E-02	-.493
210	65.62	.56377	-107.17	-23748	87.5	1379	1.542	.5450E-02	-.485
215	66.18	.55534	-105.10	-23309	88.1	1351	1.544	.6979E-02	-.476
220	66.75	.54739	-103.07	-22867	88.6	1324	1.545	.8818E-02	-.467
225	67.33	.53990	-101.08	-22423	89.2	1296	1.546	.1100E-01	-.458
230	67.93	.53284	-99.11	-21975	89.8	1268	1.547	.1357E-01	-.447
235	68.54	.52619	-97.17	-21525	90.5	1241	1.548	.1655E-01	-.436
240	69.17	.51994	-95.26	-21071	91.1	1213	1.549	.1998E-01	-.425
245	69.81	.51406	-93.37	-20613	91.8	1186	1.550	.2390E-01	-.413
250	70.47	.50854	-91.51	-20152	92.6	1158	1.550	.2833E-01	-.400
255	71.15	.50337	-89.67	-19687	93.3	1131	1.551	.3330E-01	-.386
260	71.85	.49854	-87.85	-19219	94.1	1104	1.552	.3884E-01	-.372
265	72.57	.49403	-86.05	-18746	95.0	1077	1.552	.4496E-01	-.356
270	73.31	.48984	-84.26	-18269	95.8	1050	1.553	.5170E-01	-.340
275	74.08	.48596	-82.50	-17788	96.7	1023	1.553	.5906E-01	-.323
280	74.87	.48238	-80.75	-17302	97.6	996.7	1.554	.6705E-01	-.305
285	75.68	.47910	-79.01	-16811	98.6	970.4	1.555	.7568E-01	-.286

Table 2—continued

P/MPa 15

Tσ/K

Liq Vap $\frac{T}{K}$	V_σ / V $\frac{V}{cm^3\,mol^{-1}}$	Z_σ / Z	S_σ / S $\frac{S}{J\,K^{-1}\,mol^{-1}}$	H_σ / H $\frac{H}{J\,mol^{-1}}$	C_P $\frac{C_P}{J\,K^{-1}\,mol^{-1}}$	w_σ / w $\frac{w}{m\,sec^{-1}}$	γ_σ / γ	$(f/P)_\sigma$ / (f/P)	μ_σ / μ $\frac{\mu}{K\,MPa^{-1}}$
290	76.53	.47611	−77.29	−16316	99.6	944.3	1.556	.8495E−01	−.266
295	77.41	.47342	−75.58	−15815	100.6	918.4	1.557	.9487E−01	−.245
300	78.32	.47101	−73.88	−15310	101.6	892.8	1.558	.1054	−.222
305	79.27	.46890	−72.19	−14799	102.7	867.4	1.559	.1166	−.198
310	80.26	.46707	−70.51	−14283	103.8	842.3	1.561	.1284	−.172
315	81.28	.46554	−68.84	−13761	105.0	817.5	1.562	.1408	−.145
320	82.36	.46431	−67.18	−13233	106.2	793.0	1.564	.1538	−.115
325	83.48	.46339	−65.52	−12699	107.4	768.7	1.567	.1673	−.084
330	84.65	.46277	−63.87	−12159	108.6	744.8	1.569	.1814	−.051
335	85.88	.46248	−62.23	−11612	110.0	721.2	1.572	.1959	−.014
340	87.17	.46252	−60.59	−11059	111.3	698.0	1.576	.2109	.024
345	88.52	.46291	−58.95	−10499	112.7	675.0	1.580	.2264	.066
350	89.95	.46366	−57.32	−9932	114.2	652.4	1.584	.2422	.111
355	91.46	.46479	−55.69	−9357	115.7	630.2	1.589	.2584	.160
360	93.05	.46632	−54.06	−8775	117.3	608.3	1.594	.2750	.214
365	94.74	.46829	−52.43	−8184	118.9	586.8	1.600	.2918	.271
370	96.54	.47070	−50.80	−7585	120.6	565.7	1.607	.3088	.334
380	100.5	.47705	−47.54	−6361	124.3	524.8	1.621	.3433	.478
390	105.0	.48566	−44.26	−5099	128.1	486.0	1.638	.3782	.648
400	110.2	.49688	−40.97	−3798	132.2	449.8	1.656	.4131	.849

410	116.1	.51105	− 37.65	− 2456	136.1	417.0	1.671	.4475	1.08
420	123.0	.52839	− 34.33	− 1077	139.5	388.5	1.681	.4811	1.33
430	130.8	.54887	− 31.02	331	142.0	365.0	1.679	.5136	1.58
440	139.5	.57208	− 27.74	1758	143.1	347.0	1.664	.5448	1.83
450	149.0	.59728	− 24.52	3189	143.0	334.0	1.635	.5743	2.03
475	174.6	.66315	− 16.89	6717	138.6	318.8	1.532	.6404	2.35

Table 2—*continued*

P/MPa					20				
T_σ/K									
Liq Vap									
$\dfrac{T}{K}$	V_σ $\dfrac{V}{cm^3\,mol^{-1}}$	Z_σ Z	S_σ $\dfrac{S}{J\,K^{-1}\,mol^{-1}}$	H_σ $\dfrac{H}{J\,mol^{-1}}$	C_P $\dfrac{C_P}{J\,K^{-1}\,mol^{-1}}$	w_σ $\dfrac{w}{m\,sec^{-1}}$	γ_σ γ	$(f/P)_\sigma$ (f/P)	μ_σ $\dfrac{\mu}{K\,MPa^{-1}}$
90	54.45	1.4553	−176.02	−33298				.4426E−09	
95	54.83	1.3884	−171.34	−32865				.2192E−08	
100	55.22	1.3284	−167.35	−32476				.9126E−08	
105	55.62	1.2743	−163.73	−32105				.3282E−07	
110	56.03	1.2253	−160.30	−31737	73.7	2051	1.566	.1042E−06	−.638
115	56.44	1.1806	−157.01	−31367	74.6	2007	1.564	.2966E−06	−.628
120	56.86	1.1398	−153.81	−30990	75.9	1963	1.557	.7682E−06	−.616
125	57.28	1.1023	−150.68	−30607	77.2	1921	1.548	.1831E−05	−.605
130	57.71	1.0678	−147.63	−30218	78.5	1881	1.540	.4057E−05	−.594
135	58.14	1.0359	−144.65	−29823	79.5	1843	1.534	.8420E−05	−.585
140	58.57	1.0063	−141.74	−29423	80.5	1809	1.528	.1649E−04	−.576
145	59.01	.97888	−138.90	−29018	81.3	1775	1.525	.3067E−04	−.569
150	59.45	.95334	−136.13	−28610	81.9	1744	1.522	.5445E−04	−.563
155	59.89	.92951	−133.43	−28199	82.5	1714	1.521	.9268E−04	−.557
160	60.35	.90725	−130.81	−27785	83.0	1685	1.521	.1519E−03	−.552

165	60.80	.88641	−128.24	−27369	83.5	1656	1.521	.2407E−03	−.547
170	61.26	.86688	−125.75	−26951	83.9	1628	1.522	.3695E−03	−.542
175	61.73	.84856	−123.31	−26530	84.3	1600	1.523	.5516E−03	−.537
180	62.21	.83134	−120.93	−26108	84.6	1573	1.524	.8024E−03	−.532
185	62.69	.81514	−118.61	−25684	85.0	1546	1.526	.1140E−02	−.526
190	63.18	.79988	−116.34	−25258	85.4	1519	1.527	.1585E−02	−.520
195	63.68	.78551	−114.11	−24830	85.8	1492	1.529	.2160E−02	−.514
200	64.18	.77196	−111.94	−24400	86.2	1465	1.530	.2891E−02	−.508
205	64.70	.75917	−109.80	−23968	86.6	1439	1.531	.3805E−02	−.501
210	65.22	.74710	−107.71	−23534	87.1	1412	1.532	.4930E−02	−.494
215	65.76	.73570	−105.65	−23097	87.6	1385	1.533	.6295E−02	−.486
220	66.30	.72493	−103.64	−22658	88.1	1358	1.534	.7932E−02	−.478
225	66.86	.71476	−101.65	−22217	88.6	1332	1.534	.9872E−02	−.469
230	67.42	.70515	−99.70	−21772	89.2	1305	1.534	.1214E−01	−.460
235	68.00	.69608	−97.77	−21325	89.8	1279	1.534	.1478E−01	−.451
240	68.59	.68752	−95.87	−20874	90.4	1252	1.534	.1781E−01	−.441
245	69.20	.67944	−94.00	−20420	91.1	1226	1.534	.2126E−01	−.430
250	69.82	.67182	−92.16	−19964	91.7	1200	1.534	.2515E−01	−.419
255	70.46	.66464	−90.33	−19503	92.4	1174	1.534	.2951E−01	−.407
260	71.11	.65789	−88.53	−19039	93.2	1148	1.533	.3436E−01	−.395
265	71.78	.65155	−86.75	−18571	93.9	1122	1.533	.3972E−01	−.382
270	72.46	.64560	−84.99	−18100	94.7	1097	1.532	.4561E−01	−.368
275	73.17	.64003	−83.24	−17625	95.5	1071	1.532	.5203E−01	−.354
280	73.90	.63483	−81.52	−17145	96.3	1046	1.531	.5900E−01	−.340
285	74.64	.63000	−79.80	−16661	97.1	1021	1.530	.6651E−01	−.324

Table 2—continued

P/MPa					20				
T_σ/K									
Liq Vap $\frac{T}{K}$	V_σ $\frac{V}{cm^3\,mol^{-1}}$	Z_σ Z	S_σ $\frac{S}{J\,K^{-1}\,mol^{-1}}$	H_σ $\frac{H}{J\,mol^{-1}}$	C_P $\frac{C_P}{J\,K^{-1}\,mol^{-1}}$	w_σ $\frac{w}{m\,sec^{-1}}$	γ_σ γ	$(f/P)_\sigma$ (f/P)	μ_σ $\frac{\mu}{K\,MPa^{-1}}$
290	75.41	.62551	−78.11	−16174	98.0	996.8	1.530	.7459E−01	−.308
295	76.20	.62137	−76.42	−15681	98.9	972.5	1.529	.8321E−01	−.291
300	77.02	.61756	−74.75	−15185	99.8	948.6	1.528	.9239E−01	−.273
305	77.86	.61409	−73.10	−14683	100.7	924.9	1.528	.1021	−.254
310	78.73	.61094	−71.45	−14177	101.7	901.6	1.528	.1124	−.234
315	79.63	.60812	−69.82	−13667	102.6	878.6	1.527	.1231	−.214
320	80.57	.60562	−68.19	−13151	103.6	855.9	1.527	.1344	−.192
325	81.53	.60344	−66.58	−12630	104.6	833.6	1.527	.1462	−.169
330	82.53	.60158	−64.97	−12105	105.6	811.7	1.527	.1584	−.145
335	83.57	.60005	−63.38	−11574	106.7	790.2	1.527	.1711	−.120
340	84.64	.59883	−61.79	−11038	107.7	769.0	1.527	.1841	−.094
345	85.76	.59795	−60.21	−10496	108.8	748.3	1.527	.1976	−.066
350	86.92	.59739	−58.63	−9950	109.9	727.9	1.528	.2115	−.036
355	88.13	.59716	−57.07	−9397	111.1	708.0	1.528	.2256	−.005
360	89.39	.59728	−55.51	−8839	112.2	688.5	1.529	.2401	.028
365	90.70	.59774	−53.95	−8275	113.4	669.4	1.530	.2549	.062
370	92.07	.59856	−52.40	−7705	114.5	650.8	1.531	.2699	.099
380	94.99	.60129	−49.31	−6548	116.9	614.9	1.533	.3005	.179
390	98.18	.60555	−46.24	−5367	119.4	581.0	1.536	.3316	.267
400	101.7	.61142	−43.19	−4160	121.9	549.1	1.538	.3630	.366

410	105.5	.61897	−40.15	−2929	124.3	519.5	1.541	.3945	.473
420	109.7	.62827	−37.13	−1674	126.7	492.4	1.542	.4257	.589
430	114.3	.63932	−34.12	−397	128.8	468.0	1.540	.4565	.711
440	119.3	.65208	−31.14	901	130.7	446.6	1.536	.4866	.836
450	124.7	.66640	−28.19	2215	132.1	428.3	1.527	.5158	.958
475	139.7	.70738	−20.99	5543	133.7	395.8	1.488	.5841	1.22

Table 2—*continued*

	P/MPa						25			
	T_σ/K									
Liq Vap	V_σ	Z_σ	S_σ	H_σ	C_P	w_σ	γ_σ	$(f/P)_\sigma$	μ_σ	
$\dfrac{T}{\text{K}}$	$\dfrac{V}{\text{cm}^3\,\text{mol}^{-1}}$	Z	$\dfrac{S}{\text{J K}^{-1}\text{mol}^{-1}}$	$\dfrac{H}{\text{J mol}^{-1}}$	$\dfrac{C_P}{\text{J K}^{-1}\text{mol}^{-1}}$	$\dfrac{w}{\text{m sec}^{-1}}$	γ	(f/P)	$\dfrac{\mu}{\text{K MPa}^{-1}}$
90	54.34	1.8156	−176.39	−33059				.5093E − 09	
95	54.72	1.7319	−171.72	−32628				.2480E − 08	
100	55.10	1.6568	−167.74	−32240				.1017E − 07	
105	55.50	1.5892	−164.13	−31869				.3610E − 07	
110	55.89	1.5279	−160.71	−31502	73.6	2074	1.560	.1131E − 06	− .639
115	56.30	1.4720	−157.42	−31132	74.5	2030	1.559	.3186E − 06	− .630
120	56.71	1.4209	−154.22	−30756	75.8	1985	1.552	.8169E − 06	− .618
125	57.12	1.3740	−151.10	−30374	77.1	1943	1.543	.1929E − 05	− .606
130	57.54	1.3308	−148.05	−29985	78.4	1903	1.535	.4237E − 05	− .595
135	57.96	1.2909	−145.07	−29590	79.4	1866	1.528	.8724E − 05	− .586
140	58.38	1.2539	−142.17	−29191	80.4	1831	1.523	.1696E − 04	− .578
145	58.81	1.2195	−139.33	−28787	81.2	1799	1.520	.3133E − 04	− .571
150	59.24	1.1875	−136.57	−28379	81.8	1768	1.517	.5526E − 04	− .565
155	59.68	1.1577	−133.88	−27969	82.4	1738	1.516	.9350E − 04	− .560
160	60.12	1.1298	−131.25	−27556	82.9	1709	1.515	.1524E − 03	− .555

165	60.56	1.1037	−128.70	−27140	83.3	1681	1.515	.2402E−03	−.550
170	61.01	1.0792	−126.21	−26723	83.7	1653	1.516	.3670E−03	−.545
175	61.47	1.0562	−123.77	−26304	84.0	1626	1.517	.5453E−03	−.540
180	61.93	1.0345	−121.40	−25883	84.4	1600	1.518	.7899E−03	−.536
185	62.40	1.0142	−119.09	−25460	84.7	1573	1.519	.1118E−02	−.531
190	62.87	.99501	−116.82	−25035	85.1	1547	1.520	.1548E−02	−.525
195	63.36	.97692	−114.61	−24609	85.5	1521	1.521	.2102E−02	−.520
200	63.84	.95985	−112.44	−24181	85.9	1495	1.521	.2804E−02	−.514
205	64.34	.94372	−110.31	−23750	86.3	1469	1.522	.3678E−02	−.508
210	64.85	.92848	−108.23	−23318	86.7	1443	1.523	.4751E−02	−.501
215	65.36	.91406	−106.18	−22883	87.2	1417	1.523	.6050E−02	−.495
220	65.88	.90043	−104.17	−22446	87.6	1391	1.523	.7602E−02	−.487
225	66.41	.88752	−102.20	−22007	88.1	1365	1.523	.9437E−02	−.479
230	66.95	.87531	−100.26	−21565	88.7	1340	1.523	.1158E−01	−.471
235	67.51	.86375	−98.34	−21120	89.2	1314	1.522	.1406E−01	−.463
240	68.07	.85282	−96.46	−20673	89.8	1289	1.522	.1691E−01	−.454
245	68.64	.84246	−94.60	−20222	90.4	1263	1.521	.2014E−01	−.445
250	69.23	.83267	−92.77	−19768	91.0	1238	1.520	.2378E−01	−.435
255	69.83	.82342	−90.96	−19312	91.7	1213	1.519	.2786E−01	−.424
260	70.44	.81467	−89.17	−18851	92.4	1188	1.518	.3238E−01	−.414
265	71.07	.80641	−87.41	−18388	93.1	1164	1.516	.3737E−01	−.403
270	71.71	.79862	−85.66	−17921	93.8	1139	1.515	.4284E−01	−.391
275	72.37	.79128	−83.93	−17450	94.5	1115	1.513	.4880E−01	−.379
280	73.04	.78437	−82.22	−16976	95.3	1091	1.512	.5526E−01	−.367
285	73.73	.77788	−80.53	−16498	96.0	1067	1.510	.6223E−01	−.353

Table 2—continued

P/MPa   **25**

T_σ/K

Liq Vap $\dfrac{T}{\mathrm{K}}$	V_σ $\dfrac{V}{\mathrm{cm^3\,mol^{-1}}}$	Z_σ Z	S_σ $\dfrac{S}{\mathrm{J\,K^{-1}\,mol^{-1}}}$	H_σ $\dfrac{H}{\mathrm{J\,mol^{-1}}}$	C_P $\dfrac{C_P}{\mathrm{J\,K^{-1}\,mol^{-1}}}$	w_σ $\dfrac{w}{\mathrm{m\,sec^{-1}}}$	γ_σ γ	$(f/P)_\sigma$ (f/P)	μ_σ $\dfrac{\mu}{\mathrm{K\,MPa^{-1}}}$
290	74.44	.77180	−78.85	−16016	96.8	1044	1.509	.6970E−01	−.340
295	75.16	.76611	−77.19	−15530	97.6	1021	1.507	.7767E−01	−.326
300	75.91	.76081	−75.54	−15040	98.4	998.4	1.505	.8615E−01	−.311
305	76.67	.75588	−73.91	−14545	99.2	976.0	1.504	.9512E−01	−.296
310	77.46	.75131	−72.29	−14047	100.1	953.9	1.502	.1046	−.280
315	78.27	.74710	−70.68	−13545	100.9	932.3	1.501	.1145	−.264
320	79.10	.74324	−69.09	−13038	101.8	911.0	1.499	.1249	−.247
325	79.95	.73972	−67.50	−12526	102.7	890.0	1.497	.1358	−.229
330	80.83	.73654	−65.93	−12011	103.6	869.5	1.496	.1470	−.211
335	81.74	.73369	−64.36	−11491	104.5	849.3	1.494	.1587	−.192
340	82.68	.73117	−62.81	−10966	105.4	829.6	1.493	.1708	−.172
345	83.64	.72898	−61.26	−10437	106.3	810.2	1.492	.1832	−.151
350	84.64	.72711	−59.73	−9903	107.2	791.3	1.490	.1960	−.130
355	85.66	.72556	−58.20	−9364	108.2	772.8	1.489	.2091	−.107
360	86.72	.72434	−56.68	−8821	109.1	754.8	1.488	.2225	−.084
365	87.82	.72343	−55.17	−8273	110.1	737.2	1.486	.2362	−.060
370	88.95	.72284	−53.66	−7721	111.0	720.0	1.485	.2501	−.035
380	91.32	.72262	−50.68	−6601	112.9	686.9	1.483	.2785	.018
390	93.86	.72367	−47.72	−5462	114.9	655.7	1.481	.3076	.075
400	96.58	.72601	−44.79	−4303	116.8	626.3	1.479	.3370	.136

410	99.49	.72963	−41.88	−3126	118.7	598.9	1.476	.3667	.201
420	102.6	.73453	−38.99	−1929	120.6	573.3	1.474	.3963	.270
430	105.9	.74071	−36.14	−714	122.4	549.9	1.470	.4258	.341
440	109.5	.74811	−33.30	518	124.0	528.5	1.466	.4549	.415
450	113.2	.75668	−30.50	1766	125.6	509.2	1.460	.4834	.489
475	123.6	.78251	−23.63	4944	128.5	470.7	1.437	.5515	.666

Table 2—*continued*

P/MPa										30									
T_σ/K																			
Liq Vap	V_σ		Z_σ		S_σ		H_σ		C_P		w_σ		γ_σ		$(f/P)_\sigma$		μ_σ		
$\dfrac{T}{K}$	$\dfrac{V}{cm^3\,mol^{-1}}$		Z		$\dfrac{S}{J\,K^{-1}\,mol^{-1}}$		$\dfrac{H}{J\,mol^{-1}}$		$\dfrac{C_P}{J\,K^{-1}\,mol^{-1}}$		$\dfrac{w}{m\,sec^{-1}}$		γ		(f/P)		$\dfrac{\mu}{K\,MPa^{-1}}$	
90	54.24		2.1746		−176.76		−32820								.6099E−09			
95	54.61		2.0741		−172.10		−32390								.2922E−08			
100	54.98		1.9840		−168.13		−32003								.1180E−07			
105	55.37		1.9027		−164.52		−31633								.4132E−07			
110	55.76		1.8291		−161.11		−31267		73.5		2095		1.553		.1279E−06		−.641	
115	56.16		1.7620		−157.82		−30897		74.4		2051		1.553		.3562E−06		−.631	
120	56.56		1.7007		−154.63		−30522		75.7		2007		1.546		.9041E−06		−.619	
125	56.96		1.6443		−151.51		−30140		77.0		1964		1.538		.2115E−05		−.607	
130	57.37		1.5924		−148.46		−29752		78.3		1925		1.530		.4606E−05		−.597	
135	57.79		1.5445		−145.49		−29357		79.3		1888		1.523		.9408E−05		−.587	
140	58.20		1.5000		−142.59		−28958		80.3		1853		1.518		.1816E−04		−.580	
145	58.62		1.4587		−139.76		−28555		81.0		1821		1.514		.3330E−04		−.573	
150	59.04		1.4203		−137.00		−28148		81.7		1790		1.512		.5837E−04		−.567	
155	59.47		1.3844		−134.31		−27738		82.2		1761		1.510		.9818E−04		−.562	
160	59.90		1.3508		−131.69		−27326		82.7		1732		1.510		.1592E−03		−.557	

165	60.33	1.3194	−129.14	−26911		83.1	1705	1.509	.2495E−03	−.552
170	60.77	1.2899	−126.65	−26495		83.5	1678	1.510	.3793E−03	−.548
175	61.22	1.2622	−124.23	−26076		83.8	1651	1.510	.5611E−03	−.544
180	61.67	1.2362	−121.86	−25656		84.2	1625	1.511	.8092E−03	−.539
185	62.12	1.2116	−119.55	−25235		84.5	1599	1.512	.1140E−02	−.535
190	62.58	1.1885	−117.29	−24811		84.9	1574	1.512	.1573E−02	−.530
195	63.05	1.1667	−115.08	−24386		85.2	1548	1.513	.2129E−02	−.525
200	63.52	1.1460	−112.92	−23959		85.6	1523	1.513	.2830E−02	−.520
205	64.00	1.1265	−110.80	−23530		86.0	1497	1.514	.3700E−02	−.514
210	64.49	1.1081	−108.73	−23100		86.4	1472	1.514	.4764E−02	−.508
215	64.98	1.0906	−106.69	−22667		86.8	1447	1.514	.6049E−02	−.502
220	65.49	1.0741	−104.69	−22232		87.2	1422	1.513	.7581E−02	−.495
225	66.00	1.0584	−102.72	−21794		87.7	1397	1.513	.9386E−02	−.488
230	66.52	1.0435	−100.79	−21354		88.2	1372	1.512	.1149E−01	−.481
235	67.05	1.0294	−98.89	−20912		88.7	1347	1.511	.1392E−01	−.473
240	67.58	1.0161	−97.01	−20467		89.3	1323	1.510	.1670E−01	−.465
245	68.13	1.0034	−95.17	−20019		89.9	1298	1.509	.1985E−01	−.457
250	68.69	.99138	−93.35	−19568		90.5	1274	1.507	.2339E−01	−.448
255	69.26	.97999	−91.55	−19114		91.1	1250	1.506	.2735E−01	−.439
260	69.84	.96919	−89.77	−18657		91.7	1226	1.504	.3174E−01	−.430
265	70.43	.95896	−88.02	−18197		92.4	1202	1.502	.3657E−01	−.420
270	71.03	.94926	−86.29	−17734		93.0	1179	1.500	.4185E−01	−.410
275	71.65	.94009	−84.57	−17267		93.7	1155	1.498	.4760E−01	−.399
280	72.28	.93141	−82.88	−16797		94.4	1132	1.496	.5383E−01	−.388
285	72.92	.92321	−81.20	−16323		95.1	1110	1.494	.6053E−01	−.377

Table 2—continued

P/MPa							**30**			
T_σ/K										
Liq Vap	V_σ	Z_σ	S_σ	H_σ	C_P	w_σ	γ_σ	$(f/P)_\sigma$	μ_σ
$\dfrac{T}{K}$	$\dfrac{V}{cm^3\,mol^{-1}}$	Z	$\dfrac{S}{J\,K^{-1}\,mol^{-1}}$	$\dfrac{H}{J\,mol^{-1}}$	$\dfrac{C_P}{J\,K^{-1}\,mol^{-1}}$	$\dfrac{w}{m\,sec^{-1}}$	γ	(f/P)	$\dfrac{\mu}{K\,MPa^{-1}}$
290	73.58	.91547	−79.54	−15846	95.8	1087	1.491	.6771E−01	−.366
295	74.25	.90817	−77.90	−15364	96.6	1065	1.489	.7537E−01	−.354
300	74.94	.90131	−76.27	−14880	97.3	1044	1.487	.8351E−01	−.341
305	75.64	.89486	−74.65	−14391	98.1	1022	1.484	.9211E−01	−.328
310	76.36	.88880	−73.05	−13899	98.9	1001	1.482	.1012	−.315
315	77.10	.88315	−71.46	−13402	99.7	980.5	1.479	.1107	−.302
320	77.85	.87786	−69.89	−12902	100.4	960.2	1.477	.1206	−.288
325	78.63	.87295	−68.32	−12398	101.2	940.3	1.475	.1310	−.273
330	79.42	.86840	−66.77	−11890	102.1	920.8	1.472	.1418	−.259
335	80.24	.86420	−65.23	−11377	102.9	901.7	1.470	.1530	−.243
340	81.07	.86035	−63.70	−10861	103.7	882.9	1.467	.1645	−.228
345	81.93	.85683	−62.18	−10340	104.5	864.6	1.465	.1764	−.212
350	82.80	.85365	−60.67	−9816	105.3	846.7	1.463	.1886	−.195
355	83.71	.85078	−59.17	−9287	106.2	829.2	1.461	.2011	−.178
360	84.63	.84824	−57.68	−8754	107.0	812.1	1.458	.2140	−.160
365	85.58	.84601	−56.20	−8217	107.8	795.5	1.456	.2270	−.142
370	86.56	.84409	−54.73	−7676	108.7	779.3	1.454	.2403	−.123
380	88.59	.84116	−51.81	−6580	110.4	748.1	1.449	.2676	−.085
390	90.73	.83942	−48.92	−5468	112.1	718.6	1.445	.2955	−.044
400	92.99	.83883	−46.06	−4339	113.7	690.8	1.441	.3238	−.001

410	95.38	.83937	−43.23	−3194	115.4	664.8	1.436	.3525	.044
420	97.89	.84102	−40.43	−2032	117.0	640.4	1.432	.3812	.091
430	100.5	.84372	−37.66	−854	118.6	617.8	1.427	.4098	.140
440	103.3	.84746	−34.92	340	120.1	596.9	1.422	.4383	.189
450	106.3	.85217	−32.20	1548	121.6	577.7	1.416	.4663	.239
475	114.2	.86776	−25.54	4628	124.7	537.3	1.399	.5340	.361

Table 2—continued

P/MPa = 35

Liq Vap T/K	$\dfrac{V}{\text{cm}^3\,\text{mol}^{-1}}$	Z	$\dfrac{S}{\text{J K}^{-1}\text{mol}^{-1}}$	$\dfrac{H}{\text{J mol}^{-1}}$	$\dfrac{C_P}{\text{J K}^{-1}\text{mol}^{-1}}$	$\dfrac{w}{\text{m sec}^{-1}}$	γ	(f/P)	$\dfrac{\mu}{\text{K MPa}^{-1}}$
90	54.14	2.5324	−177.11	−32582				.7509E−09	
95	54.50	2.4150	−172.47	−32152				.3537E−08	
100	54.87	2.3098	−168.51	−31766				.1408E−07	
105	55.25	2.2150	−164.90	−31397				.4862E−07	
110	55.63	2.1290	−161.50	−31031	73.3	2116	1.547	.1487E−06	−.642
115	56.02	2.0507	−158.22	−30662	74.3	2071	1.546	.4094E−06	−.632
120	56.42	1.9791	−155.03	−30287	75.6	2027	1.540	.1029E−05	−.620
125	56.81	1.9133	−151.91	−29906	77.0	1985	1.532	.2384E−05	−.609
130	57.21	1.8527	−148.87	−29518	78.2	1945	1.524	.5146E−05	−.598
135	57.62	1.7967	−145.90	−29124	79.3	1909	1.518	.1043E−04	−.589
140	58.03	1.7448	−143.00	−28726	80.2	1874	1.513	.1998E−04	−.581
145	58.44	1.6965	−140.17	−28323	80.9	1842	1.509	.3639E−04	−.574
150	58.85	1.6516	−137.42	−27916	81.6	1812	1.507	.6337E−04	−.569
155	59.27	1.6096	−134.73	−27507	82.1	1783	1.505	.1060E−03	−.563
160	59.69	1.5704	−132.12	−27095	82.6	1755	1.504	.1708E−03	−.559

165	60.11	1.5337	−129.57	−26681	83.0	1727	1.504	.2663E − 03	−.555
170	60.54	1.4992	−127.09	−26266	83.3	1701	1.504	.4030E − 03	−.551
175	60.98	1.4668	−124.67	−25848	83.7	1675	1.504	.5933E − 03	−.547
180	61.41	1.4363	−122.31	−25429	84.0	1649	1.505	.8519E − 03	−.542
185	61.86	1.4075	−120.00	−25008	84.3	1624	1.505	.1196E − 02	−.538
190	62.31	1.3804	−117.75	−24586	84.6	1599	1.506	.1643E − 02	−.534
195	62.76	1.3548	−115.55	−24162	85.0	1574	1.506	.2216E − 02	−.529
200	63.22	1.3306	−113.39	−23736	85.3	1549	1.506	.2935E − 02	−.524
205	63.68	1.3077	−111.28	−23309	85.7	1525	1.506	.3825E − 02	−.519
210	64.15	1.2860	−109.21	−22879	86.1	1500	1.506	.4910E − 02	−.514
215	64.63	1.2655	−107.18	−22448	86.5	1475	1.505	.6216E − 02	−.508
220	65.12	1.2460	−105.19	−22014	86.9	1451	1.505	.7768E − 02	−.502
225	65.61	1.2275	−103.23	−21579	87.4	1427	1.504	.9592E − 02	−.496
230	66.11	1.2100	−101.30	−21141	87.8	1403	1.503	.1171E − 01	−.489
235	66.62	1.1933	−99.41	−20700	88.3	1379	1.501	.1416E − 01	−.482
240	67.13	1.1775	−97.54	−20257	88.9	1355	1.500	.1694E − 01	−.475
245	67.65	1.1624	−95.71	−19812	89.4	1331	1.498	.2010E − 01	−.467
250	68.19	1.1482	−93.90	−19363	90.0	1307	1.496	.2364E − 01	−.459
255	68.73	1.1346	−92.11	−18912	90.5	1284	1.494	.2758E − 01	−.451
260	69.28	1.1217	−90.34	−18458	91.1	1261	1.492	.3195E − 01	−.443
265	69.84	1.1095	−88.60	−18001	91.8	1238	1.490	.3675E − 01	−.434
270	70.41	1.0978	−86.88	−17540	92.4	1215	1.487	.4199E − 01	−.425
275	70.99	1.0868	−85.18	−17077	93.0	1193	1.485	.4769E − 01	−.416
280	71.59	1.0763	−83.50	−16610	93.7	1171	1.482	.5385E − 01	−.406
285	72.19	1.0663	−81.83	−16140	94.4	1149	1.479	.6047E − 01	−.396

Table 2—continued

P/MPa					35				
T_σ/K	V_σ	Z_σ	S_σ	H_σ	C_P	w_σ	γ_σ	$(f/P)_\sigma$	μ_σ
Liq Vap $\dfrac{T}{K}$	$\dfrac{V}{cm^3\,mol^{-1}}$	Z	$\dfrac{S}{J\,K^{-1}\,mol^{-1}}$	$\dfrac{H}{J\,mol^{-1}}$	$\dfrac{C_P}{J\,K^{-1}\,mol^{-1}}$	$\dfrac{w}{m\,sec^{-1}}$	γ	(f/P)	$\dfrac{\mu}{K\,MPa^{-1}}$
290	72.81	1.0569	−80.19	−15666	95.1	1127	1.476	.6755E−01	−.386
295	73.44	1.0479	−78.55	−15189	95.8	1106	1.474	.7510E−01	−.376
300	74.08	1.0394	−76.94	−14708	96.5	1085	1.471	.8311E−01	−.365
305	74.73	1.0314	−75.34	−14224	97.2	1065	1.468	.9156E−01	−.354
310	75.40	1.0239	−73.75	−13736	97.9	1044	1.465	.1005	−.343
315	76.08	1.0167	−72.18	−13245	98.6	1025	1.462	.1098	−.332
320	76.78	1.0100	−70.62	−12750	99.4	1005	1.459	.1196	−.320
325	77.49	1.0036	−69.07	−12251	100.1	986.0	1.456	.1297	−.308
330	78.21	.99769	−67.54	−11749	100.9	967.3	1.453	.1403	−.295
335	78.95	.99211	−66.02	−11242	101.6	948.9	1.450	.1512	−.283
340	79.71	.98690	−64.51	−10732	102.4	931.0	1.447	.1625	−.270
345	80.48	.98205	−63.01	−10218	103.2	913.4	1.444	.1742	−.257
350	81.28	.97754	−61.52	−9701	103.9	896.3	1.441	.1861	−.243
355	82.08	.97336	−60.04	−9179	104.7	879.5	1.438	.1984	−.229
360	82.91	.96952	−58.57	−8654	105.5	863.2	1.436	.2109	−.215
365	83.76	.96599	−57.11	−8125	106.2	847.2	1.433	.2237	−.201
370	84.62	.96277	−55.66	−7592	107.0	831.7	1.430	.2367	−.186
380	86.41	.95725	−52.78	−6514	108.5	801.8	1.424	.2634	−.156
390	88.28	.95290	−49.94	−5421	110.1	773.6	1.419	.2907	−.124
400	90.24	.94967	−47.14	−4312	111.6	746.9	1.413	.3185	−.092

410	92.28	.94751	− 44.36	− 3188	113.1	721.9	1.408	.3466	− .058
420	94.42	.94637	− 41.62	− 2050	114.6	698.4	1.403	.3749	− .023
430	96.65	.94623	− 38.90	− 896	116.1	676.4	1.397	.4032	.012
440	98.98	.94701	− 36.22	272	117.5	656.0	1.392	.4313	.049
450	101.4	.94869	− 33.56	1454	118.9	637.0	1.386	.4591	.085
475	107.9	.95638	− 27.05	4466	122.0	596.0	1.370	.5267	.175

Table 2—continued

P/MPa						40			
T_σ/K									
Liq									
Vap									
	V_σ	Z_σ	S_σ	H_σ	C_P	w_σ	γ_σ	$(f/P)_\sigma$	μ_σ
$\dfrac{T}{K}$	$\dfrac{V}{cm^3\,mol^{-1}}$	Z	$\dfrac{S}{J\,K^{-1}\,mol^{-1}}$	$\dfrac{H}{J\,mol^{-1}}$	$\dfrac{C_P}{J\,K^{-1}\,mol^{-1}}$	$\dfrac{w}{m\,sec^{-1}}$	γ	(f/P)	$\dfrac{\mu}{K\,MPa^{-1}}$
95	54.40	2.7548	−172.83	−31914				.4369E−08	
100	54.76	2.6345	−168.88	−31529				.1713E−07	
105	55.13	2.5260	−165.28	−31161				.5836E−07	
110	55.51	2.4277	−161.88	−30795	73.2	2135	1.540	.1763E−06	−.644
115	55.89	2.3381	−158.61	−30427	74.2	2091	1.540	.4800E−06	−.634
120	56.28	2.2562	−155.42	−30053	75.5	2047	1.534	.1194E−05	−.622
125	56.67	2.1810	−152.31	−29672	76.9	2005	1.526	.2741E−05	−.610
130	57.06	2.1116	−149.27	−29284	78.1	1965	1.519	.5865E−05	−.599
135	57.46	2.0476	−146.30	−28891	79.2	1929	1.513	.1179E−04	−.590
140	57.86	1.9882	−143.41	−28493	80.1	1895	1.508	.2242E−04	−.582
145	58.26	1.9330	−140.58	−28090	80.9	1863	1.504	.4056E−04	−.576
150	58.66	1.8816	−137.83	−27684	81.5	1832	1.501	.7017E−04	−.570
155	59.07	1.8335	−135.15	−27275	82.0	1804	1.500	.1166E−03	−.565
160	59.49	1.7886	−132.54	−26864	82.5	1776	1.499	.1870E−03	−.561
165	59.90	1.7466	−129.99	−26451	82.8	1749	1.498	.2900E−03	−.557

170	60.32	1.7071	−127.52	−26036	83.2	1723	1.498	.4366E−03	−.553
175	60.74	1.6699	−125.10	−25619	83.5	1698	1.498	.6399E−03	−.549
180	61.17	1.6350	−122.74	−25201	83.8	1673	1.499	.9148E−03	−.545
185	61.60	1.6020	−120.44	−24781	84.1	1648	1.499	.1279E−02	−.541
190	62.04	1.5709	−118.19	−24359	84.4	1623	1.499	.1751E−02	−.537
195	62.48	1.5415	−116.00	−23936	84.8	1599	1.499	.2352E−02	−.533
200	62.93	1.5137	−113.85	−23512	85.1	1575	1.499	.3104E−02	−.529
205	63.38	1.4874	−111.74	−23085	85.4	1550	1.499	.4032E−02	−.524
210	63.83	1.4624	−109.68	−22657	85.8	1526	1.498	.5160E−02	−.519
215	64.30	1.4388	−107.65	−22227	86.2	1503	1.497	.6513E−02	−.514
220	64.77	1.4163	−105.67	−21795	86.6	1479	1.496	.8117E−02	−.508
225	65.24	1.3950	−103.72	−21361	87.0	1455	1.495	.9997E−02	−.502
230	65.72	1.3747	−101.80	−20925	87.5	1432	1.494	.1218E−01	−.496
235	66.21	1.3555	−99.91	−20486	88.0	1408	1.492	.1468E−01	−.490
240	66.71	1.3372	−98.05	−20045	88.5	1385	1.490	.1753E−01	−.483
245	67.21	1.3198	−96.22	−19601	89.0	1362	1.488	.2075E−01	−.476
250	67.72	1.3032	−94.42	−19155	89.5	1339	1.486	.2436E−01	−.469
255	68.24	1.2874	−92.64	−18706	90.1	1316	1.484	.2837E−01	−.462
260	68.76	1.2724	−90.89	−18254	90.7	1294	1.481	.3279E−01	−.454
265	69.30	1.2581	−89.15	−17799	91.3	1272	1.478	.3765E−01	−.446
270	69.84	1.2445	−87.44	−17341	91.9	1250	1.475	.4295E−01	−.438
275	70.40	1.2315	−85.75	−16880	92.5	1228	1.473	.4870E−01	−.430
280	70.96	1.2192	−84.08	−16416	93.1	1207	1.470	.5491E−01	−.421
285	71.53	1.2074	−82.43	−15949	93.8	1185	1.466	.6157E−01	−.413
290	72.11	1.1963	−80.79	−15479	94.4	1165	1.463	.6869E−01	−.404

Table 2—continued

P/MPa									40									
T_σ/K																		
$\frac{T}{K}$ Liq Vap	V_σ $\frac{V}{cm^3\,mol^{-1}}$		Z_σ Z		S_σ $\frac{S}{J\,K^{-1}mol^{-1}}$		H_σ $\frac{H}{J\,mol^{-1}}$		C_P $\frac{C_P}{J\,K^{-1}mol^{-1}}$		w_σ $\frac{w}{m\,sec^{-1}}$		γ_σ γ		$(f/P)_\sigma$ (f/P)		μ_σ $\frac{\mu}{K\,MPa^{-1}}$	

$\frac{T}{K}$	V_σ	$\frac{V}{cm^3\,mol^{-1}}$	Z_σ	Z	S_σ	$\frac{S}{J\,K^{-1}mol^{-1}}$	H_σ	$\frac{H}{J\,mol^{-1}}$	C_P	$\frac{C_P}{J\,K^{-1}mol^{-1}}$	w_σ	$\frac{w}{m\,sec^{-1}}$	γ_σ	γ	$(f/P)_\sigma$	(f/P)	μ_σ	$\frac{\mu}{K\,MPa^{-1}}$
295		72.70		1.1856		−79.17		−15005		95.1		1144		1.460		.7627E−01		−.394
300		73.30		1.1755		−77.57		−14528		95.8		1124		1.457		.8429E−01		−.385
305		73.92		1.1659		−75.98		−14047		96.4		1104		1.454		.9276E−01		−.375
310		74.54		1.1568		−74.40		−13564		97.1		1084		1.450		.1017		−.366
315		75.17		1.1481		−72.84		−13076		97.8		1065		1.447		.1110		−.356
320		75.82		1.1399		−71.30		−12585		98.5		1046		1.444		.1208		−.346
325		76.48		1.1321		−69.76		−12091		99.2		1028		1.440		.1309		−.335
330		77.15		1.1248		−68.24		−11593		99.9		1010		1.437		.1414		−.325
335		77.84		1.1178		−66.74		−11091		100.7		992.1		1.434		.1523		−.314
340		78.53		1.1112		−65.24		−10586		101.4		974.8		1.430		.1636		−.303
345		79.24		1.1050		−63.75		−10077		102.1		957.9		1.427		.1752		−.292
350		79.97		1.0992		−62.28		−9565		102.8		941.3		1.424		.1871		−.280
355		80.70		1.0937		−60.82		−9049		103.6		925.2		1.420		.1992		−.269
360		81.45		1.0885		−59.36		−8530		104.3		909.4		1.417		.2117		−.257
365		82.22		1.0837		−57.92		−8006		105.0		894.0		1.414		.2244		−.245
370		83.00		1.0792		−56.49		−7480		105.7		879.0		1.411		.2374		−.233
380		84.61		1.0712		−53.65		−6415		107.2		850.2		1.405		.2639		−.208
390		86.28		1.0643		−50.84		−5336		108.6		822.8		1.398		.2910		−.183
400		88.02		1.0586		−48.07		−4242		110.1		797.0		1.392		.3187		−.157
410		89.82		1.0540		−45.34		−3135		111.5		772.7		1.386		.3466		−.130

420	91.69	1.0503	−42.64	−2013	112.9	749.8	1.381	.3748	−.103
430	93.63	1.0476	−39.96	−877	114.3	728.4	1.375	.4030	−.075
440	95.65	1.0458	−37.32	273	115.6	708.3	1.369	.4310	−.047
450	97.73	1.0449	−34.71	1436	117.0	689.6	1.363	.4589	−.019
475	103.3	1.0459	−28.30	4400	120.1	648.6	1.348	.5267	.051

Table 2—continued

P/MPa					45				
T_σ/K									
Liq Vap	V_σ	Z_σ	S_σ	H_σ	C_P	w_σ	γ_σ	$(f/P)_b$	μ_σ
$\dfrac{T}{K}$	$\dfrac{V}{\text{cm}^3\,\text{mol}^{-1}}$	Z	$\dfrac{S}{\text{J K}^{-1}\,\text{mol}^{-1}}$	$\dfrac{H}{\text{J mol}^{-1}}$	$\dfrac{C_P}{\text{J K}^{-1}\,\text{mol}^{-1}}$	$\dfrac{w}{\text{m sec}^{-1}}$	γ	(f/P)	$\dfrac{\mu}{\text{K MPa}^{-1}}$
95	54.30	3.0934	−173.18	−31676				.5478E−08	
100	54.65	2.9580	−169.24	−31292				.2116E−07	
105	55.02	2.8358	−165.65	−30924				.7111E−07	
110	55.39	2.7251	−162.26	−30560	73.1	2154	1.532	.2122E−06	−.645
115	55.76	2.6243	−158.99	−30192	74.1	2110	1.533	.5713E−06	−.635
120	56.14	2.5321	−155.81	−29818	75.5	2066	1.527	.1406E−05	−.623
125	56.52	2.4474	−152.70	−29437	76.8	2023	1.520	.3198E−05	−.611
130	56.91	2.3694	−149.66	−29050	78.0	1984	1.513	.6785E−05	−.600
135	57.30	2.2972	−146.70	−28657	79.1	1948	1.507	.1353E−04	−.591
140	57.69	2.2303	−143.80	−28259	80.0	1914	1.502	.2554E−04	−.583
145	58.09	2.1682	−140.98	−27857	80.8	1882	1.499	.4589E−04	−.577
150	58.48	2.1103	−138.23	−27452	81.4	1852	1.496	.7889E−04	−.571
155	58.89	2.0562	−135.55	−27044	81.9	1824	1.494	.1303E−03	−.567
160	59.29	2.0056	−132.95	−26633	82.4	1797	1.493	.2077E−03	−.562
165	59.70	1.9582	−130.41	−26220	82.7	1770	1.493	.3206E−03	−.559

170	60.11	1.9136	− 127.93	− 25806	83.1	1745	1.493	.4803E − 03	− .555
175	60.52	1.8717	− 125.52	− 25389	83.4	1720	1.493	.7005E − 03	− .551
180	60.94	1.8323	− 123.17	− 24972	83.7	1695	1.493	.9971E − 03	− .548
185	61.36	1.7951	− 120.87	− 24553	84.0	1671	1.493	.1388E − 02	− .544
190	61.78	1.7600	− 118.63	− 24132	84.3	1646	1.493	.1893E − 02	− .540
195	62.21	1.7268	− 116.43	− 23710	84.6	1623	1.493	.2533E − 02	− .536
200	62.65	1.6954	− 114.29	− 23286	84.9	1599	1.492	.3333E − 02	− .532
205	63.09	1.6656	− 112.19	− 22861	85.2	1575	1.492	.4315E − 02	− .528
210	63.53	1.6374	− 110.13	− 22434	85.6	1552	1.491	.5504E − 02	− .523
215	63.98	1.6106	− 108.11	− 22005	86.0	1529	1.490	.6927E − 02	− .518
220	64.43	1.5852	− 106.13	− 21575	86.4	1505	1.489	.8609E − 02	− .513
225	64.89	1.5610	− 104.19	− 21142	86.8	1482	1.487	.1057E − 01	− .508
230	65.36	1.5380	− 102.27	− 20707	87.2	1459	1.486	.1285E − 01	− .502
235	65.83	1.5162	− 100.39	− 20270	87.7	1436	1.484	.1545E − 01	− .497
240	66.31	1.4953	− 98.54	− 19830	88.2	1414	1.482	.1841E − 01	− .490
245	66.79	1.4755	− 96.72	− 19388	88.7	1391	1.479	.2174E − 01	− .484
250	67.28	1.4566	− 94.92	− 18943	89.2	1369	1.477	.2547E − 01	− .478
255	67.78	1.4386	− 93.15	− 18496	89.7	1347	1.474	.2960E − 01	− .471
260	68.28	1.4215	− 91.41	− 18046	90.3	1325	1.471	.3416E − 01	− .464
265	68.80	1.4051	− 89.68	− 17594	90.8	1304	1.468	.3915E − 01	− .457
270	69.32	1.3895	− 87.98	− 17138	91.4	1282	1.465	.4458E − 01	− .449
275	69.84	1.3746	− 86.29	− 16679	92.0	1261	1.462	.5047E − 01	− .442
280	70.38	1.3604	− 84.63	− 16218	92.6	1240	1.459	.5681E − 01	− .434
285	70.92	1.3468	− 82.99	− 15753	93.2	1220	1.455	.6361E − 01	− .426
290	71.47	1.3339	− 81.36	− 15285	93.9	1199	1.452	.7086E − 01	− .418

Table 2—continued

P/MPa						45			
T_σ/K									
Liq Vap	V_σ	Z_σ	S_σ	H_σ	C_P	w_σ	γ_σ	$(f/P)_\sigma$	μ_σ
$\dfrac{T}{K}$	$\dfrac{V}{\text{cm}^3\,\text{mol}^{-1}}$	Z	$\dfrac{S}{\text{J K}^{-1}\,\text{mol}^{-1}}$	$\dfrac{H}{\text{J mol}^{-1}}$	$\dfrac{C_P}{\text{J K}^{-1}\,\text{mol}^{-1}}$	$\dfrac{w}{\text{m sec}^{-1}}$	γ	(f/P)	$\dfrac{\mu}{\text{K MPa}^{-1}}$
295	72.03	1.3215	−79.75	−14814	94.5	1179	1.448	.7857E−01	−.410
300	72.60	1.3098	−78.16	−14340	95.2	1160	1.445	.8672E−01	−.402
305	73.18	1.2986	−76.58	−13863	95.8	1141	1.441	.9532E−01	−.393
310	73.76	1.2879	−75.01	−13382	96.5	1122	1.438	.1044	−.384
315	74.36	1.2777	−73.46	−12898	97.2	1103	1.434	.1138	−.376
320	74.97	1.2679	−71.93	−12410	97.8	1085	1.430	.1237	−.367
325	75.58	1.2587	−70.41	−11919	98.5	1067	1.427	.1339	−.357
330	76.21	1.2499	−68.90	−11425	99.2	1049	1.423	.1446	−.348
335	76.85	1.2415	−67.40	−10928	99.9	1032	1.420	.1556	−.339
340	77.49	1.2336	−65.92	−10426	100.6	1015	1.416	.1669	−.329
345	78.15	1.2260	−64.44	−9922	101.3	998.9	1.412	.1786	−.320
350	78.82	1.2189	−62.98	−9414	102.0	982.8	1.409	.1906	−.310
355	79.50	1.2121	−61.53	−8902	102.6	967.1	1.405	.2028	−.300
360	80.19	1.2056	−60.09	−8387	103.3	951.8	1.402	.2154	−.290
365	80.90	1.1995	−58.66	−7869	104.0	936.9	1.399	.2282	−.280
370	81.61	1.1938	−57.24	−7347	104.7	922.3	1.395	.2412	−.270
380	83.08	1.1833	−54.43	−6293	106.1	894.3	1.388	.2678	−.249
390	84.59	1.1740	−51.65	−5225	107.5	867.7	1.382	.2951	−.228
400	86.16	1.1658	−48.91	−4143	108.9	842.5	1.375	.3229	−.206
410	87.78	1.1588	−46.21	−3047	110.2	818.8	1.369	.3510	−.184

420	89.46	1.1528	− 43.54	− 1938	111.6	796.4	1.363	.3793	− .162
430	91.18	1.1477	− 40.89	− 815	112.9	775.3	1.357	.4076	− .139
440	92.97	1.1436	− 38.28	321	114.2	755.6	1.351	.4358	− .116
450	94.80	1.1403	− 35.70	1470	115.5	737.1	1.345	.4639	− .094
475	99.64	1.1354	− 29.37	4397	118.6	696.1	1.331	.5323	− .037

Table 2—continued

P/MPa								50				
T_σ/K												
Liq Vap			V_σ	Z_σ	S_σ	H_σ	C_P	w_σ	γ_σ	$(f/P)_\sigma$	μ_σ
$\dfrac{T}{K}$			$\dfrac{V}{\text{cm}^3\,\text{mol}^{-1}}$	Z	$\dfrac{S}{\text{J K}^{-1}\text{mol}^{-1}}$	$\dfrac{H}{\text{J mol}^{-1}}$	$\dfrac{C_P}{\text{J K}^{-1}\text{mol}^{-1}}$	$\dfrac{w}{\text{m sec}^{-1}}$	γ	(f/P)	$\dfrac{\mu}{\text{K MPa}^{-1}}$
95			54.20	3.4309	−173.53	−31438				.6950E−08	
100			54.55	3.2803	−169.60	−31055				.2644E−07	
105			54.90	3.1445	−166.01	−30688				.8767E−07	
110			55.27	3.0214	−162.63	−30324	73.0	2172	1.525	.2584E−06	−.647
115			55.64	2.9093	−159.36	−29956	74.0	2128	1.526	.6881E−06	−.636
120			56.01	2.8068	−156.18	−29583	75.4	2084	1.521	.1676E−05	−.624
125			56.38	2.7126	−153.08	−29203	76.7	2041	1.514	.3777E−05	−.612
130			56.76	2.6259	−150.05	−28816	77.9	2002	1.507	.7943E−05	−.601
135			57.15	2.5456	−147.08	−28423	79.0	1966	1.502	.1571E−04	−.592
140			57.53	2.4713	−144.19	−28026	79.9	1933	1.497	.2944E−04	−.584
145			57.92	2.4021	−141.37	−27624	80.7	1901	1.493	.5253E−04	−.578
150			58.31	2.3377	−138.63	−27219	81.3	1872	1.491	.8973E−04	−.573
155			58.70	2.2776	−135.95	−26811	81.8	1843	1.489	.1474E−03	−.568
160			59.10	2.2213	−133.35	−26401	82.3	1817	1.488	.2336E−03	−.564
165			59.50	2.1685	−130.81	−25989	82.6	1791	1.488	.3585E−03	−.560

170	59.90	2.1189	−128.34	−25575	83.0	1765	1.487	.5344E−03	−.557
175	60.30	2.0723	−125.93	−25159	83.3	1741	1.487	.7759E−03	−.553
180	60.71	2.0284	−123.58	−24742	83.5	1716	1.487	.1100E−02	−.550
185	61.12	1.9869	−121.29	−24324	83.8	1692	1.487	.1524E−02	−.547
190	61.54	1.9478	−119.05	−23904	84.1	1669	1.487	.2071E−02	−.543
195	61.96	1.9108	−116.86	−23483	84.4	1645	1.486	.2761E−02	−.539
200	62.38	1.8757	−114.72	−23060	84.7	1622	1.486	.3620E−02	−.536
205	62.81	1.8425	−112.62	−22636	85.0	1599	1.485	.4671E−02	−.532
210	63.24	1.8110	−110.57	−22210	85.4	1576	1.484	.5940E−02	−.527
215	63.68	1.7811	−108.56	−21782	85.7	1553	1.483	.7453E−02	−.523
220	64.12	1.7527	−106.58	−21352	86.1	1531	1.482	.9236E−02	−.518
225	64.56	1.7256	−104.64	−20921	86.5	1508	1.480	.1131E−01	−.513
230	65.01	1.6999	−102.74	−20487	87.0	1486	1.478	.1371E−01	−.508
235	65.47	1.6754	−100.86	−20051	87.4	1463	1.476	.1645E−01	−.502
240	65.93	1.6521	−99.02	−19613	87.9	1441	1.473	.1956E−01	−.497
245	66.40	1.6298	−97.20	−19172	88.4	1419	1.471	.2304E−01	−.491
250	66.87	1.6086	−95.41	−18729	88.9	1398	1.468	.2693E−01	−.485
255	67.35	1.5883	−93.64	−18284	89.4	1376	1.465	.3124E−01	−.479
260	67.84	1.5690	−91.90	−17835	89.9	1355	1.462	.3598E−01	−.472
265	68.33	1.5505	−90.19	−17384	90.5	1334	1.459	.4116E−01	−.466
270	68.82	1.5329	−88.49	−16931	91.0	1313	1.456	.4680E−01	−.459
275	69.33	1.5161	−86.81	−16474	91.6	1292	1.452	.5289E−01	−.452
280	69.84	1.5000	−85.16	−16015	92.2	1272	1.449	.5943E−01	−.445
285	70.36	1.4846	−83.52	−15552	92.8	1252	1.445	.6645E−01	−.438
290	70.88	1.4699	−81.90	−15087	93.4	1232	1.441	.7392E−01	−.430

Table 2—continued

P/MPa					50				
T_σ/K									
Liq Vap	V_σ	Z_σ	S_σ	H_σ		w_σ	γ_σ	$(f/P)_\sigma$	μ_σ
$\dfrac{T}{K}$	$\dfrac{V}{cm^3\,mol^{-1}}$	Z	$\dfrac{S}{J\,K^{-1}\,mol^{-1}}$	$\dfrac{H}{J\,mol^{-1}}$	$\dfrac{C_P}{J\,K^{-1}\,mol^{-1}}$	$\dfrac{w}{m\,sec^{-1}}$	γ	(f/P)	$\dfrac{\mu}{K\,MPa^{-1}}$
295	71.41	1.4558	−80.30	−14618	94.0	1213	1.438	.8184E−01	−.423
300	71.95	1.4424	−78.71	−14146	94.7	1194	1.434	.9022E−01	−.415
305	72.50	1.4295	−77.14	−13671	95.3	1175	1.430	.9904E−01	−.408
310	73.06	1.4172	−75.59	−13193	95.9	1157	1.426	.1083	−.400
315	73.62	1.4055	−74.05	−12712	96.6	1138	1.423	.1180	−.392
320	74.19	1.3943	−72.52	−12227	97.2	1121	1.419	.1280	−.384
325	74.77	1.3836	−71.01	−11739	97.9	1103	1.415	.1385	−.376
330	75.36	1.3733	−69.51	−11248	98.6	1086	1.411	.1494	−.368
335	75.96	1.3635	−68.02	−10754	99.2	1069	1.407	.1606	−.359
340	76.56	1.3542	−66.55	−10256	99.9	1053	1.404	.1721	−.351
345	77.18	1.3453	−65.08	−9755	100.6	1037	1.400	.1840	−.342
350	77.80	1.3368	−63.63	−9250	101.2	1021	1.396	.1962	−.334
355	78.44	1.3287	−62.19	−8742	101.9	1006	1.393	.2087	−.325
360	79.08	1.3210	−60.76	−8231	102.6	991.2	1.389	.2214	−.317
365	79.73	1.3136	−59.34	−7716	103.3	976.6	1.385	.2344	−.308
370	80.39	1.3066	−57.93	−7198	103.9	962.4	1.382	.2476	−.299
380	81.75	1.2937	−55.14	−6152	105.3	935.0	1.375	.2746	−.281
390	83.14	1.2820	−52.39	−5093	106.6	909.0	1.368	.3022	−.263
400	84.57	1.2715	−49.67	−4020	107.9	884.4	1.361	.3304	−.244
410	86.05	1.2621	−46.99	−2934	109.3	861.1	1.355	.3588	−.226

420	87.57	1.2538	−44.34	−1835	110.6	839.1	1.348	.3874	−.207
430	89.13	1.2465	−41.73	−723	111.9	818.4	1.342	.4161	−.188
440	90.74	1.2401	−39.14	403	113.2	798.9	1.336	.4447	−.169
450	92.39	1.2346	−36.58	1541	114.4	780.5	1.330	.4731	−.150
475	96.70	1.2243	−30.31	4439	117.5	739.6	1.316	.5424	−.103

Table 2—*continued*

P/MPa		60									
T_σ/K											
Liq Vap		V_σ	Z_σ	S_σ	H_σ	C_P	w_σ	γ_σ	$(f/P)_\sigma$	μ_σ	
$\dfrac{T}{K}$		$\dfrac{V}{cm^3\,mol^{-1}}$	Z	$\dfrac{S}{J\,K^{-1}\,mol^{-1}}$	$\dfrac{H}{J\,mol^{-1}}$	$\dfrac{C_P}{J\,K^{-1}\,mol^{-1}}$	$\dfrac{w}{m\,sec^{-1}}$	γ	(f/P)	$\dfrac{\mu}{K\,MPa^{-1}}$	
95		54.01	4.1027	−174.20	−30961				.1149E−07		
100		54.34	3.9217	−170.29	−30579				.4241E−07		
105		54.69	3.7586	−166.72	−30214				.1369E−06		
110		55.04	3.6107	−163.35	−29851	72.8	2206	1.509	.3936E−06	−.650	
115		55.39	3.4760	−160.09	−29485	73.8	2162	1.511	.1025E−05	−.639	
120		55.75	3.3528	−156.92	−29112	75.2	2118	1.507	.2446E−05	−.626	
125		56.12	3.2397	−153.82	−28733	76.6	2076	1.502	.5408E−05	−.614	
130		56.48	3.1355	−150.80	−28347	77.8	2037	1.496	.1118E−04	−.603	
135		56.85	3.0391	−147.84	−27955	78.9	2001	1.490	.2176E−04	−.594	
140		57.22	2.9497	−144.95	−27558	79.8	1968	1.486	.4016E−04	−.586	
145		57.60	2.8666	−142.14	−27157	80.5	1937	1.483	.7068E−04	−.580	
150		57.98	2.7892	−139.40	−26753	81.2	1908	1.481	.1192E−03	−.575	
155		58.35	2.7169	−136.73	−26346	81.7	1880	1.479	.1934E−03	−.570	
160		58.74	2.6491	−134.13	−25937	82.1	1854	1.478	.3031E−03	−.566	
165		59.12	2.5856	−131.60	−25525	82.4	1829	1.477	.4603E−03	−.563	

170	59.51	2.5260	−129.13	−25112	82.8	1804	1.477	.6794E−03	−.560
175	59.89	2.4698	−126.73	−24698	83.0	1780	1.476	.9773E−03	−.557
180	60.29	2.4169	−124.38	−24282	83.3	1757	1.476	.1373E−02	−.554
185	60.68	2.3670	−122.10	−23864	83.6	1734	1.476	.1887E−02	−.551
190	61.08	2.3198	−119.86	−23446	83.9	1711	1.475	.2544E−02	−.548
195	61.48	2.2751	−117.68	−23026	84.1	1689	1.475	.3367E−02	−.545
200	61.88	2.2328	−115.55	−22605	84.4	1666	1.474	.4383E−02	−.541
205	62.29	2.1926	−113.46	−22182	84.7	1644	1.473	.5618E−02	−.538
210	62.70	2.1545	−111.42	−21757	85.0	1622	1.472	.7099E−02	−.534
215	63.11	2.1183	−109.41	−21331	85.4	1600	1.470	.8855E−02	−.530
220	63.53	2.0838	−107.44	−20904	85.7	1578	1.468	.1091E−01	−.526
225	63.95	2.0510	−105.51	−20474	86.1	1557	1.466	.1329E−01	−.521
230	64.37	2.0198	−103.62	−20042	86.5	1535	1.464	.1603E−01	−.517
235	64.80	1.9900	−101.75	−19609	87.0	1514	1.462	.1913E−01	−.512
240	65.24	1.9615	−99.91	−19173	87.4	1493	1.459	.2264E−01	−.507
245	65.67	1.9344	−98.11	−18735	87.9	1472	1.456	.2656E−01	−.502
250	66.12	1.9085	−96.33	−18294	88.3	1451	1.453	.3091E−01	−.497
255	66.56	1.8837	−94.57	−17851	88.8	1430	1.450	.3570E−01	−.492
260	67.01	1.8600	−92.84	−17406	89.3	1410	1.446	.4096E−01	−.486
265	67.47	1.8373	−91.14	−16958	89.9	1390	1.443	.4668E−01	−.480
270	67.93	1.8156	−89.45	−16507	90.4	1370	1.439	.5288E−01	−.475
275	68.40	1.7948	−87.79	−16054	91.0	1350	1.435	.5956E−01	−.469
280	68.87	1.7749	−86.14	−15597	91.5	1331	1.431	.6671E−01	−.463
285	69.34	1.7558	−84.52	−15138	92.1	1312	1.428	.7435E−01	−.456
290	69.83	1.7376	−82.91	−14676	92.7	1293	1.424	.8246E−01	−.450

Table 2—continued

P/MPa: **60**

T_σ/K

$\dfrac{T}{K}$	$\dfrac{V}{cm^3\,mol^{-1}}$	Z	$\dfrac{S}{J\,K^{-1}\,mol^{-1}}$	$\dfrac{H}{J\,mol^{-1}}$	$\dfrac{C_P}{J\,K^{-1}\,mol^{-1}}$	$\dfrac{w}{m\,sec^{-1}}$	γ	(f/P)	$\dfrac{\mu}{K\,MPa^{-1}}$
295	70.31	1.7200	−81.32	−14212	93.3	1274	1.420	.9104E−01	−.444
300	70.81	1.7032	−79.75	−13744	93.9	1256	1.416	.1001	−.438
305	71.30	1.6871	−78.19	−13273	94.5	1238	1.411	.1096	−.431
310	71.81	1.6716	−76.65	−12799	95.1	1221	1.407	.1195	−.425
315	72.32	1.6567	−75.13	−12322	95.7	1203	1.403	.1299	−.418
320	72.83	1.6425	−73.61	−11842	96.3	1186	1.399	.1407	−.411
325	73.35	1.6288	−72.12	−11358	97.0	1170	1.395	.1518	−.405
330	73.88	1.6157	−70.63	−10872	97.6	1153	1.391	.1634	−.398
335	74.42	1.6031	−69.16	−10382	98.2	1137	1.387	.1753	−.391
340	74.96	1.5910	−67.70	−9890	98.9	1122	1.383	.1875	−.384
345	75.50	1.5793	−66.25	−9394	99.5	1106	1.379	.2001	−.377
350	76.06	1.5682	−64.81	−8895	100.1	1091	1.375	.2130	−.371
355	76.62	1.5575	−63.39	−8392	100.8	1077	1.371	.2261	−.364
360	77.18	1.5472	−61.97	−7887	101.4	1063	1.367	.2395	−.357
365	77.76	1.5373	−60.57	−7378	102.1	1049	1.364	.2532	−.350
370	78.34	1.5279	−59.18	−6866	102.7	1035	1.360	.2670	−.343
380	79.52	1.5101	−56.42	−5832	104.0	1009	1.352	.2953	−.328
390	80.72	1.4937	−53.70	−4786	105.3	983.5	1.345	.3242	−.314
400	81.96	1.4786	−51.02	−3727	106.6	959.7	1.338	.3536	−.300
410	83.22	1.4648	−48.37	−2655	107.8	937.1	1.332	.3832	−.286

420	84.52	1.4521	−45.76	−1570	109.1	915.7	1.325	.4130	−.271
430	85.84	1.4406	−43.18	−473	110.3	895.5	1.319	.4428	−.257
440	87.19	1.4300	−40.63	637	111.6	876.3	1.313	.4725	−.243
450	88.57	1.4203	−38.11	1759	112.8	858.3	1.307	.5019	−.229
475	92.15	1.4000	−31.93	4616	115.8	817.5	1.293	.5740	−.194

Table 2—continued

P/MPa						**70**			
T_σ/K									
Liq	V_σ	Z_σ	S_σ	H_σ	C_P	w_σ	γ_σ	$(f/P)_\sigma$	μ_σ
Vap									
$\dfrac{T}{\text{K}}$	$\dfrac{V}{\text{cm}^3\,\text{mol}^{-1}}$	Z	$\dfrac{S}{\text{J K}^{-1}\,\text{mol}^{-1}}$	$\dfrac{H}{\text{J mol}^{-1}}$	$\dfrac{C_P}{\text{J K}^{-1}\,\text{mol}^{-1}}$	$\dfrac{w}{\text{m sec}^{-1}}$	γ	(f/P)	$\dfrac{\mu}{\text{K MPa}^{-1}}$
95	(53.83)	(4.7705)	(−174.85)	(−30483)				(.1949E−07)	
100	(54.15)	(4.5591)	(−170.95)	(−30103)				(.6981E−07)	
105	(54.48)	(4.3685)	(−167.40)	(−29740)				(.2192E−06)	
110	(54.82)	(4.1958)	(−164.04)	(−29378)	(72.5)	(2238)	(1.494)	(.6151E−06)	(−.653)
115	(55.16)	(4.0385)	(−160.79)	(−29013)	(73.6)	(2193)	(1.497)	(.1566E−05)	(−.641)
120	(55.51)	(3.8946)	(−157.63)	(−28641)	(75.0)	(2149)	(1.494)	(.3661E−05)	(−.628)
125	(55.86)	(3.7625)	(−154.54)	(−28263)	(76.4)	(2108)	(1.489)	(.7944E−05)	(−.616)
130	(56.22)	(3.6408)	(−151.52)	(−27878)	(77.7)	(2069)	(1.484)	(.1614E−04)	(−.605)
135	(56.57)	(3.5282)	(−148.57)	(−27486)	(78.8)	(2034)	(1.479)	(.3091E−04)	(−.595)
140	(56.93)	(3.4238)	(−145.69)	(−27090)	(79.7)	(2001)	(1.475)	(.5621E−04)	(−.588)
145	(57.30)	(3.3268)	(−142.88)	(−26690)	(80.4)	(1970)	(1.473)	(.9757E−04)	(−.582)
150	(57.66)	(3.2363)	(−140.14)	(−26286)	(81.0)	(1942)	(1.471)	(.1624E−03)	(−.576)
155	(58.03)	(3.1518)	(−137.47)	(−25880)	(81.5)	(1915)	(1.469)	(.2604E−03)	(−.572)
160	(58.39)	(3.0726)	(−134.88)	(−25471)	(82.0)	(1889)	(1.468)	(.4034E−03)	(−.569)
165	(58.76)	(2.9984)	(−132.35)	(−25060)	(82.3)	(1865)	(1.467)	(.6063E−03)	(−.565)

170	(59.13)	(2.9286)	(−129.89)	(−24648)	(82.6)	(1841)	(1.467)	(.8861E−03)	(−.562)
175	(59.51)	(2.8630)	(−127.49)	(−24234)	(82.9)	(1818)	(1.466)	(.1263E−02)	(−.560)
180	(59.89)	(2.8010)	(−125.15)	(−23819)	(83.1)	(1795)	(1.466)	(.1758E−02)	(−.557)
185	(60.26)	(2.7425)	(−122.87)	(−23403)	(83.4)	(1772)	(1.466)	(.2397E−02)	(−.554)
190	(60.64)	(2.6872)	(−120.64)	(−22986)	(83.6)	(1750)	(1.465)	(.3205E−02)	(−.551)
195	(61.03)	(2.6349)	(−118.47)	(−22567)	(83.9)	(1729)	(1.464)	(.4211E−02)	(−.549)
200	(61.41)	(2.5853)	(−116.34)	(−22146)	(84.2)	(1707)	(1.463)	(.5442E−02)	(−.546)
205	(61.80)	(2.5382)	(−114.26)	(−21725)	(84.5)	(1686)	(1.462)	(.6929E−02)	(−.542)
210	(62.19)	(2.4934)	(−112.22)	(−21302)	(84.8)	(1664)	(1.460)	(.8701E−02)	(−.539)
215	(62.59)	(2.4509)	(−110.22)	(−20877)	(85.1)	(1643)	(1.459)	(.1079E−01)	(−.536)
220	(62.98)	(2.4103)	(−108.26)	(−20451)	(85.4)	(1622)	(1.457)	(.1322E−01)	(−.532)
225	(63.38)	(2.3717)	(−106.34)	(−20023)	(85.8)	(1602)	(1.454)	(.1601E−01)	(−.528)
230	(63.79)	(2.3349)	(−104.45)	(−19593)	(86.2)	(1581)	(1.452)	(.1920E−01)	(−.524)
235	(64.19)	(2.2998)	(−102.59)	(−19161)	(86.6)	(1560)	(1.449)	(.2281E−01)	(−.520)
240	(64.60)	(2.2663)	(−100.76)	(−18727)	(87.0)	(1540)	(1.446)	(.2686E−01)	(−.516)
245	(65.02)	(2.2342)	(−98.96)	(−18290)	(87.5)	(1520)	(1.443)	(.3137E−01)	(−.511)
250	(65.43)	(2.2036)	(−97.19)	(−17852)	(87.9)	(1500)	(1.440)	(.3635E−01)	(−.506)
255	(65.85)	(2.1742)	(−95.44)	(−17411)	(88.4)	(1480)	(1.436)	(.4182E−01)	(−.502)
260	(66.28)	(2.1461)	(−93.72)	(−16968)	(88.9)	(1460)	(1.433)	(.4778E−01)	(−.497)
265	(66.70)	(2.1192)	(−92.02)	(−16522)	(89.4)	(1441)	(1.429)	(.5425E−01)	(−.492)
270	(67.13)	(2.0934)	(−90.35)	(−16074)	(89.9)	(1422)	(1.425)	(.6123E−01)	(−.487)
275	(67.57)	(2.0686)	(−88.69)	(−15623)	(90.5)	(1403)	(1.421)	(.6872E−01)	(−.481)
280	(68.01)	(2.0449)	(−87.06)	(−15169)	(91.0)	(1384)	(1.417)	(.7672E−01)	(−.476)
285	(68.45)	(2.0221)	(−85.44)	(−14712)	(91.6)	(1366)	(1.413)	(.8523E−01)	(−.471)
290	(68.90)	(2.0002)	(−83.84)	(−14253)	(92.1)	(1348)	(1.409)	(.9424E−01)	(−.465)

Table 2—continued

P/MPa

T_σ/K		70							
	V_σ	Z_σ	S_σ	H_σ	C_P	w_σ	γ_σ	$(f/P)_{t\sigma}$	μ_σ
Liq Vap $\dfrac{T}{K}$	$\dfrac{V}{\text{cm}^3\,\text{mol}^{-1}}$	Z	$\dfrac{S}{\text{J}\,\text{K}^{-1}\,\text{mol}^{-1}}$	$\dfrac{H}{\text{J}\,\text{mol}^{-1}}$	$\dfrac{C_P}{\text{J}\,\text{K}^{-1}\,\text{mol}^{-1}}$	$\dfrac{w}{\text{m}\,\text{sec}^{-1}}$	γ	(f/P)	$\dfrac{\mu}{\text{K}\,\text{MPa}^{-1}}$
295	(69.35)	(1.9792)	(−82.26)	(−13791)	(92.7)	(1330)	(1.404)	(.1037)	(−.460)
300	69.80	1.9590	−80.70	−13326	93.3	1313	1.400	.1137	−.454
305	70.26	1.9395	−79.15	−12858	93.9	1295	1.396	.1242	−.449
310	70.73	1.9208	−77.62	−12387	94.5	1279	1.392	.1351	−.443
315	71.19	1.9028	−76.11	−11913	95.1	1262	1.387	.1464	−.437
320	71.67	1.8855	−74.61	−11437	95.7	1246	1.383	.1582	−.432
325	72.14	1.8689	−73.12	−10957	96.3	1230	1.379	.1703	−.426
330	72.62	1.8528	−71.64	−10474	96.9	1214	1.375	.1829	−.420
335	73.11	1.8374	−70.18	−9988	97.5	1199	1.370	.1958	−.414
340	73.60	1.8225	−68.73	−9499	98.1	1183	1.366	.2090	−.409
345	74.10	1.8082	−67.29	−9007	98.7	1169	1.362	.2226	−.403
350	74.60	1.7944	−65.87	−8511	99.4	1154	1.358	.2365	−.397
355	75.10	1.7811	−64.46	−8013	100.0	1140	1.354	.2506	−.391
360	75.61	1.7683	−63.05	−7511	100.6	1126	1.350	.2650	−.385
365	76.12	1.7559	−61.66	−7007	101.2	1113	1.346	.2796	−.379
370	76.64	1.7440	−60.28	−6499	101.9	1099	1.342	.2944	−.373
380	77.70	1.7214	−57.55	−5474	103.1	1074	1.335	.3246	−.362
390	78.77	1.7004	−54.85	−4437	104.4	1050	1.327	.3553	−.350
400	79.86	1.6809	−52.19	−3387	105.6	1026	1.320	.3865	−.338
410	80.98	1.6628	−49.57	−2325	106.8	1004	1.313	.4179	−.327

420	82.11	1.6459	−46.98	−1250	108.1	983.4	1.307	.4493	−.315
430	83.26	1.6302	−44.42	−163	109.3	963.5	1.301	.4807	−.304
440	84.44	1.6157	−41.90	935	110.5	944.7	1.294	.5120	−.292
450	85.63	1.6021	−39.40	2046	111.7	926.8	1.289	.5429	−.281
475	88.71	1.5724	−33.28	4876	114.6	886.2	1.275	.6184	−.253

Table 2—continued

P/MPa						80			
T_σ/K									
Liq Vap	V_σ	Z_σ	S_σ	H_σ	C_P	w_σ	γ_σ	$(f/P)_\sigma$	μ_σ
$\dfrac{T}{K}$	$\dfrac{V}{\text{cm}^3\,\text{mol}^{-1}}$	Z	$\dfrac{S}{\text{J K}^{-1}\,\text{mol}^{-1}}$	$\dfrac{H}{\text{J mol}^{-1}}$	$\dfrac{C_P}{\text{J K}^{-1}\,\text{mol}^{-1}}$	$\dfrac{w}{\text{m sec}^{-1}}$	γ	(f/P)	$\dfrac{\mu}{\text{K MPa}^{-1}}$
95	(53.66)	(5.4346)	(−175.47)	(−30005)				(.3368E−07)	
100	(53.97)	(5.1927)	(−171.59)	(−29627)				(.1170E−06)	
105	(54.29)	(4.9746)	(−168.06)	(−29265)				(.3576E−06)	
110	(54.61)	(4.7770)	(−164.71)	(−28905)	(72.3)	(2267)	(1.478)	(.9790E−06)	(−.656)
115	(54.94)	(4.5970)	(−161.47)	(−28541)	(73.4)	(2223)	(1.482)	(.2437E−05)	(−.644)
120	(55.28)	(4.4324)	(−158.32)	(−28170)	(74.8)	(2179)	(1.480)	(.5582E−05)	(−.630)
125	(55.62)	(4.2813)	(−155.24)	(−27792)	(76.2)	(2138)	(1.476)	(.1188E−04)	(−.617)
130	(55.96)	(4.1421)	(−152.22)	(−27408)	(77.5)	(2099)	(1.472)	(.2372E−04)	(−.606)
135	(56.31)	(4.0133)	(−149.27)	(−27017)	(78.6)	(2064)	(1.468)	(.4472E−04)	(−.597)
140	(56.66)	(3.8939)	(−146.40)	(−26622)	(79.6)	(2032)	(1.465)	(.8012E−04)	(−.589)
145	(57.01)	(3.7829)	(−143.59)	(−26222)	(80.3)	(2002)	(1.462)	(.1371E−03)	(−.583)
150	(57.36)	(3.6794)	(−140.86)	(−25819)	(80.9)	(1974)	(1.461)	(.2254E−03)	(−.578)
155	(57.71)	(3.5826)	(−138.20)	(−25413)	(81.4)	(1948)	(1.459)	(.3569E−03)	(−.574)
160	(58.07)	(3.4921)	(−135.60)	(−25005)	(81.8)	(1922)	(1.458)	(.5469E−03)	(−.570)
165	(58.43)	(3.4071)	(−133.08)	(−24595)	(82.2)	(1898)	(1.458)	(.8132E−03)	(−.567)

170	(58.79)	(3.3272)	(−130.62)	(−24183)	(82.5)	(1875)	(1.457)	(.1177E−02)	(−.564)
175	(59.15)	(3.2520)	(−128.23)	(−23770)	(82.7)	(1852)	(1.457)	(.1661E−02)	(−.562)
180	(59.51)	(3.1810)	(−125.89)	(−23356)	(83.0)	(1830)	(1.456)	(.2292E−02)	(−.559)
185	(59.87)	(3.1140)	(−123.62)	(−22940)	(83.2)	(1809)	(1.456)	(.3099E−02)	(−.557)
190	(60.24)	(3.0506)	(−121.39)	(−22523)	(83.5)	(1787)	(1.455)	(.4112E−02)	(−.554)
195	(60.61)	(2.9906)	(−119.22)	(−22105)	(83.7)	(1766)	(1.454)	(.5361E−02)	(−.552)
200	(60.98)	(2.9336)	(−117.10)	(−21686)	(84.0)	(1745)	(1.453)	(.6880E−02)	(−.549)
205	(61.35)	(2.8796)	(−115.02)	(−21266)	(84.3)	(1725)	(1.452)	(.8701E−02)	(−.546)
210	(61.73)	(2.8282)	(−112.99)	(−20844)	(84.6)	(1704)	(1.450)	(.1086E−01)	(−.543)
215	(62.10)	(2.7793)	(−110.99)	(−20420)	(84.9)	(1684)	(1.448)	(.1338E−01)	(−.540)
220	(62.48)	(2.7327)	(−109.04)	(−19995)	(85.2)	(1663)	(1.446)	(.1629E−01)	(−.537)
225	(62.86)	(2.6883)	(−107.12)	(−19568)	(85.6)	(1643)	(1.443)	(.1963E−01)	(−.533)
230	(63.25)	(2.6459)	(−105.24)	(−19139)	(85.9)	(1623)	(1.441)	(.2342E−01)	(−.530)
235	(63.63)	(2.6054)	(−103.38)	(−18709)	(86.3)	(1604)	(1.438)	(.2769E−01)	(−.526)
240	(64.02)	(2.5668)	(−101.56)	(−18276)	(86.7)	(1584)	(1.435)	(.3244E−01)	(−.522)
245	(64.41)	(2.5298)	(−99.77)	(−17841)	(87.2)	(1564)	(1.431)	(.3771E−01)	(−.518)
250	(64.81)	(2.4944)	(−98.00)	(−17404)	(87.6)	(1545)	(1.428)	(.4351E−01)	(−.514)
255	(65.21)	(2.4604)	(−96.26)	(−16965)	(88.1)	(1526)	(1.424)	(.4984E−01)	(−.509)
260	(65.61)	(2.4279)	(−94.55)	(−16523)	(88.6)	(1507)	(1.420)	(.5672E−01)	(−.505)
265	(66.01)	(2.3968)	(−92.86)	(−16079)	(89.1)	(1489)	(1.416)	(.6415E−01)	(−.501)
270	(66.42)	(2.3668)	(−91.19)	(−15633)	(89.6)	(1470)	(1.412)	(.7214E−01)	(−.496)
275	(66.82)	(2.3381)	(−89.54)	(−15183)	(90.1)	(1452)	(1.408)	(.8067E−01)	(−.491)
280	(67.24)	(2.3105)	(−87.91)	(−14732)	(90.6)	(1434)	(1.404)	(.8976E−01)	(−.487)
285	(67.65)	(2.2840)	(−86.30)	(−14277)	(91.2)	(1416)	(1.400)	(.9938E−01)	(−.482)
290	(68.07)	(2.2585)	(−84.71)	(−13820)	(91.7)	(1399)	(1.396)	(.1095)	(−.477)

Table 2—*continued*

P/MPa					80				
T_σ/K									
Liq Vap									
$\dfrac{T}{\text{K}}$	V_σ $\dfrac{V}{\text{cm}^3\,\text{mol}^{-1}}$	Z_σ Z	S_σ $\dfrac{S}{\text{J K}^{-1}\,\text{mol}^{-1}}$	H_σ $\dfrac{H}{\text{J mol}^{-1}}$	C_P $\dfrac{C_P}{\text{J K}^{-1}\,\text{mol}^{-1}}$	w_σ $\dfrac{w}{\text{m sec}^{-1}}$	γ_σ γ	$(f/P)_\sigma$ (f/P)	μ_σ $\dfrac{\mu}{\text{K MPa}^{-1}}$
295	(68.49)	(2.2339)	(−83.14)	(−13360)	(92.3)	(1382)	(1.391)	(.1202)	(−.472)
300	68.91	2.2103	−81.58	−12897	92.8	1365	1.387	.1314	−.467
305	69.34	2.1876	−80.04	−12432	93.4	1348	1.382	.1431	−.462
310	69.77	2.1656	−78.52	−11963	94.0	1332	1.378	.1552	−.457
315	70.21	2.1445	−77.01	−11492	94.6	1316	1.374	.1678	−.452
320	70.64	2.1242	−75.52	−11017	95.2	1300	1.369	.1808	−.447
325	71.09	2.1045	−74.04	−10540	95.8	1284	1.365	.1943	−.442
330	71.53	2.0856	−72.57	−10059	96.4	1269	1.361	.2081	−.437
335	71.98	2.0673	−71.12	−9576	97.0	1254	1.356	.2223	−.432
340	72.43	2.0497	−69.68	−9090	97.6	1240	1.352	.2368	−.427
345	72.88	2.0327	−68.25	−8600	98.2	1225	1.348	.2516	−.422
350	73.34	2.0162	−66.83	−8108	98.8	1211	1.344	.2668	−.417
355	73.80	2.0003	−65.42	−7612	99.4	1197	1.340	.2822	−.411
360	74.27	1.9850	−64.03	−7114	100.0	1184	1.336	.2978	−.406
365	74.74	1.9702	−62.65	−6612	100.6	1171	1.332	.3137	−.401
370	75.21	1.9558	−61.27	−6108	101.2	1158	1.328	.3297	−.396
380	76.16	1.9285	−58.56	−5089	102.5	1133	1.320	.3623	−.386
390	77.13	1.9030	−55.88	−4058	103.7	1109	1.313	.3954	−.376
400	78.12	1.8791	−53.24	−3015	104.9	1087	1.306	.4288	−.366
410	79.12	1.8568	−50.63	−1960	106.1	1065	1.299	.4624	−.356

420	80.13	1.8358	−48.06	− 893	107.3	1044	1.292	.4959	−.347
430	81.16	1.8161	−45.52	186	108.5	1025	1.286	.5293	−.337
440	82.21	1.7977	−43.01	1278	109.7	1006	1.280	.5625	−.327
450	83.27	1.7804	−40.53	2381	110.9	988.4	1.274	.5953	−.318
475	85.98	1.7416	−34.46	5190	113.8	947.9	1.260	.6750	−.295

Table 2—*continued*

P/MPa						90			
T_σ/K									
Liq Vap									
$\dfrac{T}{\text{K}}$	V_σ $\dfrac{V}{\text{cm}^3\,\text{mol}^{-1}}$	Z_σ Z	S_σ $\dfrac{S}{\text{J K}^{-1}\,\text{mol}^{-1}}$	H_σ $\dfrac{H}{\text{J mol}^{-1}}$	C_P $\dfrac{C_P}{\text{J K}^{-1}\,\text{mol}^{-1}}$	w_σ $\dfrac{w}{\text{m sec}^{-1}}$	γ_σ γ	$(f/P)_\sigma$ (f/P)	μ_σ $\dfrac{\mu}{\text{K MPa}^{-1}}$
95	(53.49)	(6.0951)	(−176.07)	(−29526)				(.5898E−07)	
100	(53.79)	(5.8226)	(−172.21)	(−29150)				(.1989E−06)	
105	(54.10)	(5.5770)	(−168.69)	(−28789)				(.5913E−06)	
110	(54.41)	(5.3545)	(−165.36)	(−28431)	(72.0)	(2294)	(1.461)	(.1579E−05)	(−.659)
115	(54.73)	(5.1518)	(−162.13)	(−28068)	(73.2)	(2250)	(1.466)	(.3843E−05)	(−.646)
120	(55.06)	(4.9665)	(−158.99)	(−27698)	(74.7)	(2207)	(1.466)	(.8624E−05)	(−.632)
125	(55.39)	(4.7964)	(−155.91)	(−27321)	(76.1)	(2166)	(1.463)	(.1802E−04)	(−.619)
130	(55.72)	(4.6396)	(−152.90)	(−26938)	(77.4)	(2128)	(1.460)	(.3535E−04)	(−.608)
135	(56.05)	(4.4946)	(−149.96)	(−26548)	(78.5)	(2093)	(1.457)	(.6558E−04)	(−.598)
140	(56.39)	(4.3602)	(−147.08)	(−26153)	(79.4)	(2061)	(1.454)	(.1157E−03)	(−.591)
145	(56.73)	(4.2352)	(−144.28)	(−25754)	(80.2)	(2032)	(1.452)	(.1954E−03)	(−.584)
150	(57.07)	(4.1186)	(−141.55)	(−25351)	(80.8)	(2004)	(1.451)	(.3169E−03)	(−.579)
155	(57.42)	(4.0097)	(−138.89)	(−24946)	(81.3)	(1978)	(1.450)	(.4959E−03)	(−.575)
160	(57.76)	(3.9077)	(−136.31)	(−24538)	(81.7)	(1954)	(1.449)	(.7513E−03)	(−.572)
165	(58.11)	(3.8120)	(−133.78)	(−24128)	(82.1)	(1930)	(1.449)	(.1105E−02)	(−.569)

170	(58.45)	(3.7220)	(−131.33)	(−23717)	(82.4)	(1907)	(1.448)	(.1583E−02)	(−.566)
175	(58.80)	(3.6372)	(−128.94)	(−23305)	(82.6)	(1885)	(1.448)	(.2214E−02)	(−.564)
180	(59.15)	(3.5573)	(−126.61)	(−22891)	(82.9)	(1864)	(1.447)	(.3029E−02)	(−.561)
185	(59.50)	(3.4817)	(−124.33)	(−22476)	(83.1)	(1843)	(1.447)	(.4061E−02)	(−.559)
190	(59.86)	(3.4102)	(−122.11)	(−22060)	(83.3)	(1822)	(1.446)	(.5345E−02)	(−.557)
195	(60.21)	(3.3425)	(−119.95)	(−21643)	(83.6)	(1802)	(1.445)	(.6917E−02)	(−.554)
200	(60.57)	(3.2782)	(−117.83)	(−21224)	(83.8)	(1781)	(1.444)	(.8814E−02)	(−.552)
205	(60.93)	(3.2172)	(−115.75)	(−20804)	(84.1)	(1761)	(1.442)	(.1107E−01)	(−.549)
210	(61.29)	(3.1591)	(−113.72)	(−20383)	(84.4)	(1741)	(1.440)	(.1373E−01)	(−.547)
215	(61.65)	(3.1039)	(−111.73)	(−19960)	(84.7)	(1722)	(1.438)	(.1681E−01)	(−.544)
220	(62.01)	(3.0512)	(−109.78)	(−19536)	(85.0)	(1702)	(1.436)	(.2035E−01)	(−.541)
225	(62.38)	(3.0010)	(−107.87)	(−19110)	(85.4)	(1683)	(1.433)	(.2439E−01)	(−.538)
230	(62.75)	(2.9531)	(−105.99)	(−18682)	(85.7)	(1663)	(1.431)	(.2894E−01)	(−.534)
235	(63.12)	(2.9072)	(−104.14)	(−18253)	(86.1)	(1644)	(1.428)	(.3404E−01)	(−.531)
240	(63.49)	(2.8634)	(−102.32)	(−17821)	(86.5)	(1625)	(1.424)	(.3969E−01)	(−.527)
245	(63.86)	(2.8215)	(−100.54)	(−17388)	(86.9)	(1606)	(1.421)	(.4593E−01)	(−.523)
250	(64.24)	(2.7813)	(−98.78)	(−16952)	(87.4)	(1588)	(1.417)	(.5275E−01)	(−.520)
255	(64.61)	(2.7428)	(−97.04)	(−16514)	(87.8)	(1569)	(1.414)	(.6017E−01)	(−.516)
260	(64.99)	(2.7059)	(−95.33)	(−16074)	(88.3)	(1551)	(1.410)	(.6820E−01)	(−.512)
265	(65.38)	(2.6704)	(−93.64)	(−15631)	(88.8)	(1533)	(1.406)	(.7683E−01)	(−.508)
270	(65.76)	(2.6364)	(−91.98)	(−15186)	(89.3)	(1515)	(1.401)	(.8607E−01)	(−.503)
275	(66.15)	(2.6037)	(−90.34)	(−14738)	(89.8)	(1497)	(1.397)	(.9591E−01)	(−.499)
280	(66.54)	(2.5722)	(−88.71)	(−14288)	(90.3)	(1480)	(1.393)	(.1063)	(−.495)
285	(66.93)	(2.5420)	(−87.11)	(−13835)	(90.9)	(1463)	(1.389)	(.1173)	(−.490)
290	(67.32)	(2.5128)	(−85.53)	(−13379)	(91.4)	(1446)	(1.384)	(.1289)	(−.486)

Table 2—continued

P/MPa									90
Tσ/K									
Liq Vap									
$\dfrac{T}{K}$	$\dfrac{V}{\text{cm}^3\,\text{mol}^{-1}}$ (V_σ)	Z (Z_σ)	$\dfrac{S}{\text{J K}^{-1}\text{mol}^{-1}}$ (S_σ)	$\dfrac{H}{\text{J mol}^{-1}}$ (H_σ)	$\dfrac{C_P}{\text{J K}^{-1}\text{mol}^{-1}}$ ($C_{P\sigma}$)	$\dfrac{w}{\text{m sec}^{-1}}$ (w_σ)	γ (γ_σ)	(f/P) $(f/P)_\sigma$	$\dfrac{\mu}{\text{K MPa}^{-1}}$ (μ_σ)
295	(67.72)	(2.4848)	(−83.96)	(−12921)	(91.9)	(1429)	(1.380)	(.1411)	(−.481)
300	68.12	2.4578	−82.41	−12460	92.5	1413	1.375	.1537	−.477
305	68.52	2.4317	−80.88	−11996	93.1	1396	1.371	.1669	−.473
310	68.92	2.4066	−79.36	−11529	93.6	1381	1.366	.1805	−.468
315	69.33	2.3823	−77.85	−11059	94.2	1365	1.362	.1947	−.463
320	69.73	2.3589	−76.37	−10587	94.8	1350	1.358	.2092	−.459
325	70.15	2.3363	−74.89	−10111	95.4	1335	1.353	.2242	−.454
330	70.56	2.3145	−73.43	−9633	96.0	1320	1.349	.2396	−.450
335	70.98	2.2934	−71.98	−9152	96.6	1305	1.344	.2554	−.445
340	71.39	2.2730	−70.55	−8667	97.2	1291	1.340	.2714	−.441
345	71.82	2.2533	−69.13	−8180	97.8	1277	1.336	.2878	−.436
350	72.24	2.2342	−67.71	−7690	98.4	1264	1.332	.3045	−.432
355	72.67	2.2158	−66.31	−7196	99.0	1250	1.328	.3214	−.427
360	73.10	2.1979	−64.93	−6700	99.6	1237	1.324	.3386	−.423
365	73.53	2.1806	−63.55	−6201	100.2	1224	1.319	.3560	−.418
370	73.96	2.1639	−62.18	−5698	100.8	1212	1.316	.3735	−.414
380	74.84	2.1319	−59.48	−4684	102.0	1187	1.308	.4090	−.405
390	75.73	2.1019	−56.81	−3658	103.2	1164	1.300	.4449	−.396
400	76.63	2.0737	−54.19	−2621	104.4	1142	1.293	.4810	−.387
410	77.54	2.0472	−51.59	−1571	105.6	1120	1.286	.5171	−.379

420	78.46	2.0222	− 49.03	− 509	106.8	1100	1.280	.5532	− .370
430	79.39	1.9986	− 46.51	565	108.0	1081	1.273	.5890	− .362
440	80.34	1.9764	− 44.01	1651	109.2	1062	1.267	.6244	− .353
450	81.29	1.9555	− 41.55	2748	110.3	1045	1.261	.6593	− .345
475	83.73	1.9080	− 35.50	5542	113.2	1004	1.248	.7437	− .325

Table 2—continued

P/MPa: 100

T_σ/K — Liq, Vap

$\dfrac{T}{K}$	$\dfrac{V}{cm^3\,mol^{-1}}$	Z	$\dfrac{S}{J\,K^{-1}\,mol^{-1}}$	$\dfrac{H}{J\,mol^{-1}}$	$\dfrac{C_P}{J\,K^{-1}\,mol^{-1}}$	$\dfrac{w}{m\,sec^{-1}}$	γ	(f/P)	$\dfrac{\mu}{K\,MPa^{-1}}$
95	(53.33)	(6.7523)	(−176.64)	(−29046)				(.1044E − 06)	
100	(53.62)	(6.4492)	(−172.81)	(−28672)				(.3414E − 06)	
105	(53.92)	(6.1760)	(−169.30)	(−28313)				(.9880E − 06)	
110	(54.22)	(5.9285)	(−165.98)	(−27956)	(71.7)	(2320)	(1.445)	(.2574E − 05)	(− .662)
115	(54.53)	(5.7031)	(−162.77)	(−27595)	(73.0)	(2276)	(1.451)	(.6125E − 05)	(− .649)
120	(54.85)	(5.4971)	(−159.63)	(−27226)	(74.5)	(2233)	(1.452)	(.1346E − 04)	(− .634)
125	(55.16)	(5.3079)	(−156.56)	(−26850)	(75.9)	(2192)	(1.450)	(.2760E − 04)	(− .621)
130	(55.49)	(5.1335)	(−153.56)	(−26467)	(77.2)	(2155)	(1.448)	(.5321E − 04)	(− .609)
135	(55.81)	(4.9724)	(−150.62)	(−26078)	(78.4)	(2121)	(1.446)	(.9714E − 04)	(− .600)
140	(56.14)	(4.8229)	(−147.75)	(−25684)	(79.3)	(2089)	(1.444)	(.1689E − 03)	(− .592)
145	(56.47)	(4.6839)	(−144.95)	(−25285)	(80.1)	(2060)	(1.442)	(.2812E − 03)	(− .585)
150	(56.80)	(4.5543)	(−142.23)	(−24883)	(80.7)	(2033)	(1.441)	(.4503E − 03)	(− .580)
155	(57.13)	(4.4332)	(−139.57)	(−24478)	(81.2)	(2007)	(1.441)	(.6961E − 03)	(− .576)
160	(57.47)	(4.3198)	(−136.98)	(−24071)	(81.6)	(1983)	(1.440)	(.1043E − 02)	(− .573)
165	(57.80)	(4.2133)	(−134.47)	(−23662)	(82.0)	(1960)	(1.440)	(.1518E − 02)	(− .570)

170	(58.14)	(4.1132)	(−132.02)	(−23251)	(82.3)	(1938)	(1.440)	(.2153E−02)	(−.567)
175	(58.48)	(4.0189)	(−129.63)	(−22839)	(82.5)	(1917)	(1.439)	(.2982E−02)	(−.565)
180	(58.81)	(3.9300)	(−127.30)	(−22426)	(82.8)	(1896)	(1.439)	(.4043E−02)	(−.563)
185	(59.16)	(3.8459)	(−125.03)	(−22011)	(83.0)	(1875)	(1.438)	(.5375E−02)	(−.561)
190	(59.50)	(3.7663)	(−122.81)	(−21596)	(83.2)	(1855)	(1.437)	(.7018E−02)	(−.559)
195	(59.84)	(3.6909)	(−120.65)	(−21179)	(83.5)	(1835)	(1.436)	(.9015E−02)	(−.556)
200	(60.18)	(3.6193)	(−118.53)	(−20761)	(83.7)	(1815)	(1.435)	(.1141E−01)	(−.554)
205	(60.53)	(3.5513)	(−116.46)	(−20342)	(84.0)	(1796)	(1.433)	(.1423E−01)	(−.552)
210	(60.88)	(3.4866)	(−114.43)	(−19921)	(84.3)	(1777)	(1.432)	(.1753E−01)	(−.549)
215	(61.22)	(3.4250)	(−112.45)	(−19499)	(84.6)	(1757)	(1.429)	(.2133E−01)	(−.547)
220	(61.57)	(3.3663)	(−110.50)	(−19075)	(84.9)	(1738)	(1.427)	(.2568E−01)	(−.544)
225	(61.93)	(3.3102)	(−108.59)	(−18650)	(85.2)	(1720)	(1.424)	(.3060E−01)	(−.541)
230	(62.28)	(3.2567)	(−106.71)	(−18223)	(85.6)	(1701)	(1.421)	(.3612E−01)	(−.538)
235	(62.63)	(3.2055)	(−104.87)	(−17795)	(85.9)	(1682)	(1.418)	(.4226E−01)	(−.535)
240	(62.99)	(3.1566)	(−103.05)	(−17364)	(86.3)	(1664)	(1.415)	(.4905E−01)	(−.531)
245	(63.35)	(3.1097)	(−101.27)	(−16931)	(86.8)	(1645)	(1.411)	(.5648E−01)	(−.528)
250	(63.70)	(3.0648)	(−99.51)	(−16496)	(87.2)	(1627)	(1.408)	(.6458E−01)	(−.524)
255	(64.06)	(3.0217)	(−97.78)	(−16059)	(87.6)	(1609)	(1.404)	(.7336E−01)	(−.521)
260	(64.43)	(2.9804)	(−96.07)	(−15620)	(88.1)	(1592)	(1.400)	(.8280E−01)	(−.517)
265	(64.79)	(2.9406)	(−94.39)	(−15178)	(88.6)	(1574)	(1.396)	(.9291E−01)	(−.513)
270	(65.16)	(2.9025)	(−92.73)	(−14734)	(89.1)	(1557)	(1.392)	(.1037)	(−.509)
275	(65.52)	(2.8658)	(−91.09)	(−14287)	(89.6)	(1540)	(1.387)	(.1151)	(−.505)
280	(65.89)	(2.8305)	(−89.47)	(−13838)	(90.1)	(1523)	(1.383)	(.1272)	(−.501)
285	(66.27)	(2.7965)	(−87.87)	(−13387)	(90.6)	(1506)	(1.379)	(.1399)	(−.497)
290	(66.64)	(2.7637)	(−86.29)	(−12932)	(91.1)	(1490)	(1.374)	(.1532)	(−.493)

Table 2—*continued*

T/K	$\dfrac{V_\sigma}{\text{cm}^3\,\text{mol}^{-1}}$ $\dfrac{V}{}$	Z_σ Z	$\dfrac{S_\sigma}{\text{J K}^{-1}\,\text{mol}^{-1}}$ $\dfrac{S}{}$	$\dfrac{H_\sigma}{\text{J mol}^{-1}}$ $\dfrac{H}{}$	$\dfrac{C_P}{\text{J K}^{-1}\,\text{mol}^{-1}}$ $\dfrac{C_P}{}$	$\dfrac{w_\sigma}{\text{m sec}^{-1}}$ $\dfrac{w}{}$	γ_σ γ	$(f/P)_\sigma$ (f/P)	$\dfrac{\mu_\sigma}{\text{K MPa}^{-1}}$ $\dfrac{\mu}{}$
295	(67.01)	(2.7322)	(−84.73)	(−12475)	(91.7)	(1473)	(1.370)	(.1671)	(−.489)
300	67.39	2.7017	−83.19	−12015	92.2	1457	1.365	.1815	−.485
305	67.77	2.6724	−81.66	−11553	92.8	1442	1.361	.1965	−.481
310	68.15	2.6440	−80.14	−11087	93.4	1426	1.356	.2120	−.477
315	68.53	2.6167	−78.64	−10619	93.9	1411	1.352	.2280	−.472
320	68.92	2.5902	−77.16	−10148	94.5	1396	1.347	.2444	−.468
325	69.30	2.5647	−75.69	−9674	95.1	1381	1.343	.2612	−.464
330	69.69	2.5400	−74.23	−9197	95.7	1367	1.338	.2785	−.460
335	70.08	2.5161	−72.79	−8717	96.3	1353	1.334	.2960	−.456
340	70.47	2.4929	−71.36	−8234	96.9	1339	1.330	.3140	−.452
345	70.87	2.4705	−69.94	−7748	97.4	1325	1.325	.3322	−.447
350	71.26	2.4489	−68.54	−7260	98.0	1312	1.321	.3507	−.443
355	71.66	2.4279	−67.14	−6768	98.6	1299	1.317	.3694	−.439
360	72.06	2.4075	−65.76	−6273	99.2	1286	1.313	.3883	−.435
365	72.46	2.3878	−64.38	−5776	99.8	1274	1.309	.4075	−.431
370	72.87	2.3686	−63.02	−5275	100.4	1261	1.305	.4267	−.427
380	73.68	2.3321	−60.33	−4265	101.6	1237	1.297	.4656	−.419
390	74.50	2.2976	−57.67	−3242	102.8	1215	1.290	.5047	−.411
400	75.33	2.2652	−55.05	−2208	104.0	1193	1.282	.5440	−.403
410	76.17	2.2345	−52.47	−1162	105.2	1172	1.276	.5831	−.396

P/MPa   100

T_σ/K   Liq   Vap

420	77.02	2.2056	− 49.92	− 104	106.4	1152	1.269	.6220	−.388
430	77.87	2.1782	− 47.40	966	107.6	1132	1.263	.6604	−.381
440	78.74	2.1523	− 44.92	2047	108.7	1114	1.256	.6984	−.373
450	79.61	2.1277	− 42.46	3140	109.9	1097	1.251	.7356	−.366
475	81.82	2.0718	− 36.44	5923	112.8	1056	1.237	.8254	−.348

Table 2—continued

P/MPa									
					150				
T_σ/K									
Liq Vap	V_σ	Z_σ	S_σ	H_σ	C_P	w_σ	γ_σ	$(f/P)_\sigma$	μ_σ
$\dfrac{T}{K}$	$\dfrac{V}{\text{cm}^3\,\text{mol}^{-1}}$	Z	$\dfrac{S}{\text{J K}^{-1}\,\text{mol}^{-1}}$	$\dfrac{H}{\text{J mol}^{-1}}$	$\dfrac{C_P}{\text{J K}^{-1}\,\text{mol}^{-1}}$	$\dfrac{w}{\text{m sec}^{-1}}$	γ	(f/P)	$\dfrac{\mu}{\text{K MPa}^{-1}}$
100	(52.86)	(9.5371)	(−175.47)	(−26277)				(.5591E−05)	
105	(53.11)	(9.1253)	(−172.05)	(−25927)				(.1411E−04)	
110	(53.37)	(8.7524)	(−168.80)	(−25578)	(70.3)	(2432)	(1.368)	(.3247E−04)	(−.678)
115	(53.63)	(8.4131)	(−165.65)	(−25223)	(71.7)	(2389)	(1.379)	(.6900E−04)	(−.663)
120	(53.90)	(8.1029)	(−162.56)	(−24860)	(73.4)	(2347)	(1.385)	(.1368E−03)	(−.646)
125	(54.17)	(7.8183)	(−159.54)	(−24489)	(75.0)	(2308)	(1.388)	(.2551E−03)	(−.630)
130	(54.45)	(7.5562)	(−156.57)	(−24111)	(76.5)	(2273)	(1.390)	(.4507E−03)	(−.617)
135	(54.73)	(7.3139)	(−153.66)	(−23725)	(77.7)	(2240)	(1.392)	(.7590E−03)	(−.606)
140	(55.01)	(7.0894)	(−150.81)	(−23334)	(78.7)	(2211)	(1.394)	(.1225E−02)	(−.597)
145	(55.30)	(6.8805)	(−148.03)	(−22938)	(79.6)	(2183)	(1.395)	(.1902E−02)	(−.590)
150	(55.59)	(6.6859)	(−145.32)	(−22538)	(80.3)	(2158)	(1.396)	(.2854E−02)	(−.584)
155	(55.88)	(6.5039)	(−142.68)	(−22135)	(80.8)	(2135)	(1.398)	(.4153E−02)	(−.580)
160	(56.17)	(6.3335)	(−140.11)	(−21730)	(81.3)	(2113)	(1.399)	(.5877E−02)	(−.576)
165	(56.46)	(6.1736)	(−137.60)	(−21323)	(81.7)	(2092)	(1.400)	(.8110E−02)	(−.573)
170	(56.75)	(6.0231)	(−135.16)	(−20914)	(82.0)	(2072)	(1.401)	(.1094E−01)	(−.571)

175	(57.05)	(5.8813)	(−132.78)	(−20503)	(82.2)	(2053)	(1.401)	(.1446E−01)	(−.569)
180	(57.34)	(5.7474)	(−130.46)	(−20091)	(82.5)	(2034)	(1.401)	(.1875E−01)	(−.567)
185	(57.64)	(5.6208)	(−128.19)	(−19678)	(82.7)	(2016)	(1.401)	(.2389E−01)	(−.565)
190	(57.93)	(5.5010)	(−125.98)	(−19264)	(82.9)	(1998)	(1.401)	(.2998E−01)	(−.563)
195	(58.23)	(5.3873)	(−123.83)	(−18849)	(83.2)	(1980)	(1.400)	(.3708E−01)	(−.561)
200	(58.53)	(5.2793)	(−121.72)	(−18433)	(83.4)	(1963)	(1.399)	(.4526E−01)	(−.560)
205	(58.82)	(5.1766)	(−119.66)	(−18015)	(83.7)	(1946)	(1.397)	(.5457E−01)	(−.558)
210	(59.12)	(5.0789)	(−117.64)	(−17596)	(83.9)	(1929)	(1.395)	(.6506E−01)	(−.556)
215	(59.42)	(4.9857)	(−115.66)	(−17175)	(84.2)	(1912)	(1.393)	(.7677E−01)	(−.554)
220	(59.71)	(4.8968)	(−113.72)	(−16754)	(84.5)	(1895)	(1.391)	(.8973E−01)	(−.552)
225	(60.01)	(4.8118)	(−111.81)	(−16330)	(84.9)	(1879)	(1.388)	(.1040)	(−.549)
230	(60.31)	(4.7306)	(−109.95)	(−15905)	(85.2)	(1862)	(1.385)	(.1194)	(−.547)
235	(60.61)	(4.6528)	(−108.11)	(−15478)	(85.6)	(1846)	(1.382)	(.1362)	(−.544)
240	(60.91)	(4.5783)	(−106.30)	(−15049)	(85.9)	(1829)	(1.378)	(.1542)	(−.542)
245	(61.20)	(4.5069)	(−104.53)	(−14619)	(86.3)	(1813)	(1.375)	(.1734)	(−.539)
250	(61.50)	(4.4383)	(−102.78)	(−14186)	(86.8)	(1797)	(1.371)	(.1939)	(−.537)
255	(61.80)	(4.3725)	(−101.06)	(−13751)	(87.2)	(1781)	(1.367)	(.2154)	(−.534)
260	(62.10)	(4.3092)	(−99.36)	(−13314)	(87.6)	(1766)	(1.363)	(.2381)	(−.531)
265	(62.40)	(4.2483)	(−97.69)	(−12875)	(88.1)	(1750)	(1.358)	(.2619)	(−.528)
270	(62.70)	(4.1897)	(−96.04)	(−12433)	(88.6)	(1735)	(1.354)	(.2867)	(−.525)

Table 2—continued

	V_σ	Z_σ	S_σ	H_σ	C_P	w_σ	γ_σ	$(f/P)_\sigma$	μ_σ
P/MPa					150				
T_σ/K									
Liq Vap $\dfrac{T}{K}$	V $\dfrac{V}{\text{cm}^3\,\text{mol}^{-1}}$	Z	S $\dfrac{S}{\text{J K}^{-1}\,\text{mol}^{-1}}$	H $\dfrac{H}{\text{J mol}^{-1}}$	C_P $\dfrac{C_P}{\text{J K}^{-1}\,\text{mol}^{-1}}$	w $\dfrac{w}{\text{m sec}^{-1}}$	γ	(f/P)	μ $\dfrac{\mu}{\text{K MPa}^{-1}}$
275	(63.00)	(4.1332)	(−94.41)	(−11989)	(89.1)	(1719)	(1.350)	(.3124)	(−.522)
280	(63.30)	(4.0788)	(−92.80)	(−11542)	(89.6)	(1704)	(1.345)	(.3390)	(−.519)
285	(63.60)	(4.0263)	(−91.21)	(−11093)	(90.1)	(1689)	(1.341)	(.3665)	(−.516)
290	(63.90)	(3.9756)	(−89.64)	(−10641)	(90.6)	(1675)	(1.336)	(.3947)	(−.512)
295	(64.21)	(3.9266)	(−88.08)	(−10187)	(91.1)	(1660)	(1.332)	(.4236)	(−.509)
300	64.51	3.8793	−86.55	−9730	91.7	1646	1.327	.4531	−.506
305	64.81	3.8336	−85.03	−9270	92.2	1632	1.322	.4832	−.503
310	65.11	3.7894	−83.52	−8808	92.8	1618	1.318	.5138	−.500
315	65.42	3.7466	−82.03	−8343	93.3	1604	1.313	.5448	−.496
320	65.72	3.7051	−80.56	−7875	93.9	1591	1.309	.5762	−.493
325	66.02	3.6649	−79.10	−7404	94.5	1578	1.304	.6078	−.490
330	66.33	3.6260	−77.65	−6930	95.0	1564	1.300	.6398	−.487
335	66.63	3.5882	−76.22	−6453	95.6	1552	1.295	.6719	−.484
340	66.93	3.5516	−74.80	−5974	96.2	1539	1.291	.7041	−.481
345	67.24	3.5161	−73.39	−5491	96.8	1527	1.287	.7363	−.478
350	67.54	3.4815	−71.99	−5006	97.4	1514	1.283	.7686	−.475
355	67.85	3.4480	−70.61	−4518	97.9	1502	1.278	.8009	−.472
360	68.15	3.4154	−69.23	−4027	98.5	1491	1.274	.8331	−.469
365	68.46	3.3837	−67.87	−3533	99.1	1479	1.270	.8652	−.466
370	68.76	3.3529	−66.52	−3036	99.7	1468	1.266	.8971	−.463

380	69.38	3.2937	− 63.84	− 2033	100.9	1446	1.259	.9603	− .457
390	69.99	3.2377	− 61.21	− 1018	102.0	1424	1.251	1.022	− .451
400	70.60	3.1844	− 58.61	8	103.2	1404	1.244	1.083	− .446
410	71.22	3.1339	− 56.05	1046	104.4	1384	1.237	1.142	− .440
420	71.84	3.0858	− 53.52	2096	105.5	1365	1.230	1.200	− .435
430	72.45	3.0399	− 51.02	3157	106.7	1346	1.224	1.256	− .430
440	73.07	2.9962	− 48.55	4230	107.8	1329	1.218	1.309	− .425
450	73.69	2.9545	− 46.12	5314	109.0	1312	1.212	1.361	− .420
475	75.25	2.8581	− 40.15	8074	111.8	1272	1.199	1.481	− .408

Table 2—continued

P/MPa: 200

T$_\sigma$/K

Liq
Vap

$\dfrac{T}{\mathrm{K}}$	V_σ	$\dfrac{V}{\mathrm{cm^3\,mol^{-1}}}$	Z_σ	Z	S_σ	$\dfrac{S}{\mathrm{J\,K^{-1}\,mol^{-1}}}$	H_σ	$\dfrac{H}{\mathrm{J\,mol^{-1}}}$	C_P	$\dfrac{C_P}{\mathrm{J\,K^{-1}\,mol^{-1}}}$	w_σ	$\dfrac{w}{\mathrm{m\,sec^{-1}}}$	γ_σ	γ	$(f/P)_\sigma$	(f/P)	μ_σ	$\dfrac{\mu}{\mathrm{K\,MPa^{-1}}}$
105		(52.43)		(12.011)		(−174.36)		(−23531)								(.2172E−03)		
110		(52.64)		(11.512)		(−171.19)		(−23191)		(68.6)		(2526)		(1.298)		(.4413E−03)		(−.698)
115		(52.86)		(11.058)		(−168.11)		(−22844)		(70.2)		(2484)		(1.313)		(.8375E−03)		(−.679)
120		(53.09)		(10.643)		(−165.08)		(−22488)		(72.0)		(2444)		(1.323)		(.1497E−02)		(−.659)
125		(53.33)		(10.263)		(−162.11)		(−22124)		(73.8)		(2406)		(1.331)		(.2539E−02)		(−.641)
130		(53.57)		(9.9128)		(−159.18)		(−21750)		(75.4)		(2372)		(1.338)		(.4110E−02)		(−.626)
135		(53.82)		(9.5895)		(−156.30)		(−21370)		(76.8)		(2341)		(1.343)		(.6384E−02)		(−.613)
140		(54.07)		(9.2898)		(−153.49)		(−20982)		(78.0)		(2313)		(1.348)		(.9557E−02)		(−.603)
145		(54.32)		(9.0114)		(−150.73)		(−20590)		(79.0)		(2287)		(1.352)		(.1384E−01)		(−.595)
150		(54.57)		(8.7518)		(−148.04)		(−20193)		(79.8)		(2264)		(1.356)		(.1947E−01)		(−.588)
155		(54.83)		(8.5094)		(−145.41)		(−19792)		(80.4)		(2242)		(1.360)		(.2666E−01)		(−.583)
160		(55.09)		(8.2823)		(−142.85)		(−19389)		(80.9)		(2221)		(1.363)		(.3565E−01)		(−.578)
165		(55.35)		(8.0691)		(−140.36)		(−18983)		(81.4)		(2202)		(1.365)		(.4664E−01)		(−.575)
170		(55.61)		(7.8687)		(−137.92)		(−18575)		(81.7)		(2184)		(1.368)		(.5984E−01)		(−.572)
175		(55.87)		(7.6798)		(−135.55)		(−18166)		(82.0)		(2166)		(1.369)		(.7542E−01)		(−.569)

180	(56.13)	(7.5014)	(−133.23)	(−17755)	(82.3)	(2149)	(1.370)	(.9353E − 01)	(−.567)
185	(56.39)	(7.3328)	(−130.97)	(−17343)	(82.6)	(2133)	(1.371)	(.1143)	(−.565)
190	(56.66)	(7.1730)	(−128.77)	(−16929)	(82.8)	(2116)	(1.371)	(.1378)	(−.563)
195	(56.92)	(7.0215)	(−126.61)	(−16514)	(83.1)	(2101)	(1.371)	(.1641)	(−.562)
200	(57.18)	(6.8776)	(−124.50)	(−16098)	(83.4)	(2085)	(1.370)	(.1932)	(−.560)
205	(57.45)	(6.7406)	(−122.44)	(−15681)	(83.6)	(2069)	(1.369)	(.2251)	(−.558)
210	(57.71)	(6.6102)	(−120.42)	(−15262)	(83.9)	(2054)	(1.368)	(.2598)	(−.556)
215	(57.97)	(6.4859)	(−118.45)	(−14841)	(84.2)	(2039)	(1.366)	(.2971)	(−.554)
220	(58.23)	(6.3671)	(−116.51)	(−14420)	(84.5)	(2024)	(1.364)	(.3371)	(−.552)
225	(58.49)	(6.2537)	(−114.60)	(−13996)	(84.9)	(2009)	(1.361)	(.3797)	(−.550)
230	(58.76)	(6.1451)	(−112.73)	(−13571)	(85.2)	(1994)	(1.358)	(.4246)	(−.548)
235	(59.02)	(6.0411)	(−110.90)	(−13144)	(85.6)	(1979)	(1.355)	(.4717)	(−.546)
240	(59.28)	(5.9415)	(−109.09)	(−12715)	(86.0)	(1964)	(1.352)	(.5209)	(−.544)
245	(59.54)	(5.8458)	(−107.32)	(−12284)	(86.4)	(1949)	(1.348)	(.5720)	(−.541)
250	(59.80)	(5.7540)	(−105.57)	(−11852)	(86.8)	(1935)	(1.345)	(.6249)	(−.539)
255	(60.06)	(5.6657)	(−103.84)	(−11416)	(87.2)	(1920)	(1.341)	(.6794)	(−.536)
260	(60.32)	(5.5808)	(−102.15)	(−10979)	(87.7)	(1906)	(1.337)	(.7352)	(−.534)
265	(60.58)	(5.4991)	(−100.47)	(−10540)	(88.1)	(1892)	(1.332)	(.7923)	(−.531)
270	(60.84)	(5.4204)	(−98.82)	(−10098)	(88.6)	(1878)	(1.328)	(.8504)	(−.529)
275	(61.10)	(5.3445)	(−97.19)	(−9653)	(89.1)	(1864)	(1.324)	(.9093)	(−.526)

Table 2—continued

P/MPa					200				
T_σ/K									
Liq Vap	V_σ	Z_σ	S_σ	H_σ		w_σ	γ_σ	$(f/P)_\sigma$	μ_σ
$\dfrac{T}{K}$	$\dfrac{V}{\text{cm}^3\,\text{mol}^{-1}}$	Z	$\dfrac{S}{\text{J K}^{-1}\,\text{mol}^{-1}}$	$\dfrac{H}{\text{J mol}^{-1}}$	$\dfrac{C_P}{\text{J K}^{-1}\,\text{mol}^{-1}}$	$\dfrac{w}{\text{m sec}^{-1}}$	γ	(f/P)	$\dfrac{\mu}{\text{K MPa}^{-1}}$
280	(61.36)	(5.2713)	(−95.58)	(−9207)	(89.6)	(1850)	(1.319)	(.9690)	(−.523)
285	(61.62)	(5.2006)	(−93.99)	(−8757)	(90.1)	(1836)	(1.315)	(1.029)	(−.520)
290	(61.87)	(5.1323)	(−92.42)	(−8305)	(90.7)	(1823)	(1.310)	(1.090)	(−.518)
295	(62.13)	(5.0663)	(−90.86)	(−7851)	(91.2)	(1809)	(1.306)	(1.150)	(−.515)
300	62.39	5.0025	−89.32	−7393	91.7	1796	1.301	1.211	−.512
305	62.65	4.9408	−87.80	−6933	92.3	1783	1.297	1.272	−.509
310	62.90	4.8810	−86.30	−6471	92.8	1770	1.292	1.333	−.506
315	63.16	4.8231	−84.81	−6005	93.4	1758	1.288	1.393	−.504
320	63.41	4.7669	−83.33	−5537	94.0	1745	1.283	1.453	−.501
325	63.67	4.7125	−81.87	−5066	94.5	1733	1.279	1.512	−.498
330	63.92	4.6597	−80.42	−4592	95.1	1720	1.275	1.571	−.495
335	64.18	4.6084	−78.99	−4115	95.7	1709	1.270	1.628	−.493
340	64.43	4.5586	−77.57	−3635	96.3	1697	1.266	1.686	−.490
345	64.69	4.5102	−76.16	−3152	96.8	1685	1.262	1.742	−.487
350	64.94	4.4632	−74.76	−2666	97.4	1674	1.257	1.797	−.485
355	65.19	4.4175	−73.38	−2178	98.0	1662	1.253	1.852	−.482
360	65.45	4.3730	−72.00	−1686	98.6	1651	1.249	1.905	−.479
365	65.70	4.3297	−70.64	−1192	99.2	1640	1.245	1.957	−.477
370	65.95	4.2876	−69.28	−695	99.8	1630	1.241	2.008	−.474
380	66.45	4.2066	−66.61	309	100.9	1609	1.234	2.107	−.469

390	66.95	4.1296	− 63.97	1324	102.1	1588	1.226	2.201	− .465
400	67.45	4.0564	− 61.37	2351	103.3	1569	1.219	2.290	− .460
410	67.95	3.9866	− 58.80	3390	104.5	1550	1.212	2.374	− .455
420	68.45	3.9201	− 56.27	4441	105.6	1531	1.206	2.453	− .451
430	68.94	3.8567	− 53.77	5503	106.8	1513	1.199	2.527	− .447
440	69.43	3.7960	− 51.31	6576	107.9	1496	1.193	2.597	− .443
450	69.93	3.7379	− 48.87	7661	109.1	1480	1.187	2.661	− .439
475	71.15	3.6031	− 42.90	10423	111.9	1440	1.174	2.802	− .429

Table 2—continued

P/MPa					250				
T_σ/K									
Liq Vap									
V_σ	Z_σ	S_σ	H_σ	C_P	w_σ	γ_σ	$(f/P)_\sigma$	μ_σ	
$\dfrac{V}{\mathrm{cm^3\,mol^{-1}}}$	Z	$\dfrac{S}{\mathrm{J\,K^{-1}\,mol^{-1}}}$	$\dfrac{H}{\mathrm{J\,mol^{-1}}}$	$\dfrac{C_P}{\mathrm{J\,K^{-1}\,mol^{-1}}}$	$\dfrac{w}{\mathrm{m\,sec^{-1}}}$	γ	(f/P)	$\dfrac{\mu}{\mathrm{K\,MPa^{-1}}}$	
$\dfrac{T}{\mathrm{K}}$									
105	(51.84)	(14.844)	(−176.28)	(−21126)				(.3439E−02)	
110	(52.02)	(14.219)	(−173.20)	(−20796)	(66.7)	(2612)	(1.238)	(.6167E−02)	(−.719)
115	(52.21)	(13.650)	(−170.20)	(−20458)	(68.4)	(2570)	(1.255)	(.1045E−01)	(−.698)
120	(52.40)	(13.130)	(−167.25)	(−20111)	(70.4)	(2531)	(1.269)	(.1683E−01)	(−.676)
125	(52.61)	(12.654)	(−164.33)	(−19754)	(72.4)	(2494)	(1.280)	(.2596E−01)	(−.655)
130	(52.82)	(12.216)	(−161.46)	(−19387)	(74.2)	(2461)	(1.290)	(.3850E−01)	(−.637)
135	(53.03)	(11.812)	(−158.63)	(−19012)	(75.7)	(2431)	(1.299)	(.5515E−01)	(−.623)
140	(53.25)	(11.437)	(−155.85)	(−18630)	(77.1)	(2404)	(1.307)	(.7660E−01)	(−.610)
145	(53.47)	(11.089)	(−153.12)	(−18242)	(78.2)	(2380)	(1.314)	(.1035)	(−.601)
150	(53.70)	(10.765)	(−150.46)	(−17849)	(79.1)	(2357)	(1.320)	(.1364)	(−.592)
155	(53.93)	(10.462)	(−147.85)	(−17451)	(79.8)	(2336)	(1.326)	(.1758)	(−.586)
160	(54.16)	(10.179)	(−145.31)	(−17051)	(80.5)	(2317)	(1.331)	(.2221)	(−.581)
165	(54.40)	(9.9127)	(−142.82)	(−16647)	(81.0)	(2299)	(1.335)	(.2756)	(−.576)
170	(54.63)	(9.6626)	(−140.40)	(−16241)	(81.4)	(2282)	(1.339)	(.3363)	(−.573)
175	(54.87)	(9.4270)	(−138.03)	(−15833)	(81.8)	(2266)	(1.342)	(.4043)	(−.570)

180	(55.10)	(9.2046)	(− 135.72)	(− 15423)	(82.2)	(2250)	(1.344)	(.4796)	(− .567)
185	(55.34)	(8.9943)	(− 133.46)	(− 15011)	(82.5)	(2235)	(1.346)	(.5618)	(− .565)
190	(55.58)	(8.7952)	(− 131.26)	(− 14598)	(82.8)	(2220)	(1.347)	(.6508)	(− .562)
195	(55.81)	(8.6063)	(− 129.11)	(− 14183)	(83.1)	(2205)	(1.347)	(.7462)	(− .560)
200	(56.05)	(8.4268)	(− 127.00)	(− 13767)	(83.4)	(2191)	(1.347)	(.8476)	(− .558)
205	(56.29)	(8.2561)	(− 124.94)	(− 13349)	(83.7)	(2176)	(1.347)	(.9544)	(− .556)
210	(56.53)	(8.0935)	(− 122.92)	(− 12930)	(84.0)	(2162)	(1.346)	(1.066)	(− .554)
215	(56.76)	(7.9384)	(− 120.94)	(− 12509)	(84.3)	(2148)	(1.344)	(1.182)	(− .552)
220	(57.00)	(7.7903)	(− 118.99)	(− 12087)	(84.7)	(2134)	(1.343)	(1.302)	(− .550)
225	(57.23)	(7.6487)	(− 117.09)	(− 11663)	(85.0)	(2120)	(1.340)	(1.425)	(− .548)
230	(57.47)	(7.5132)	(− 115.21)	(− 11237)	(85.4)	(2107)	(1.338)	(1.551)	(− .546)
235	(57.71)	(7.3835)	(− 113.37)	(− 10809)	(85.8)	(2093)	(1.335)	(1.679)	(− .544)
240	(57.94)	(7.2591)	(− 111.56)	(− 10379)	(86.2)	(2079)	(1.332)	(1.809)	(− .542)
245	(58.17)	(7.1397)	(− 109.78)	(− 9947)	(86.6)	(2066)	(1.328)	(1.940)	(− .539)
250	(58.41)	(7.0250)	(− 108.03)	(− 9513)	(87.0)	(2052)	(1.325)	(2.071)	(− .537)
255	(58.64)	(6.9147)	(− 106.30)	(− 9077)	(87.5)	(2039)	(1.321)	(2.202)	(− .535)
260	(58.87)	(6.8086)	(− 104.60)	(− 8638)	(87.9)	(2025)	(1.317)	(2.333)	(− .532)
265	(59.10)	(6.7064)	(− 102.92)	(− 8197)	(88.4)	(2012)	(1.313)	(2.463)	(− .530)
270	(59.34)	(6.6080)	(− 101.26)	(− 7754)	(88.9)	(1999)	(1.309)	(2.593)	(− .527)
275	(59.57)	(6.5130)	(− 99.63)	(− 7308)	(89.4)	(1986)	(1.305)	(2.720)	(− .525)

301

Table 2—continued

P/MPa 250

T_σ/K (Liq/Vap)	V_σ	Z_σ	S_σ	H_σ	C_P	w_σ	γ_σ	$(f/P)_\sigma$	μ_σ
$\dfrac{T}{K}$	$\dfrac{V}{cm^3\,mol^{-1}}$	Z	$\dfrac{S}{J\,K^{-1}\,mol^{-1}}$	$\dfrac{H}{J\,mol^{-1}}$	$\dfrac{C_P}{J\,K^{-1}\,mol^{-1}}$	$\dfrac{w}{m\,sec^{-1}}$	γ	(f/P)	$\dfrac{\mu}{K\,MPa^{-1}}$
280	(59.80)	(6.4214)	(−98.01)	(−6860)	(89.9)	(1973)	(1.300)	(2.846)	(−.522)
285	(60.03)	(6.3329)	(−96.41)	(−6409)	(90.4)	(1960)	(1.296)	(2.970)	(−.520)
290	(60.25)	(6.2474)	(−94.84)	(−5955)	(91.0)	(1948)	(1.292)	(3.091)	(−.517)
295	(60.48)	(6.1647)	(−93.28)	(−5499)	(91.5)	(1935)	(1.287)	(3.210)	(−.514)
300	60.71	6.0847	−91.73	−5040	92.1	1923	1.283	3.326	−.512
305	60.93	6.0073	−90.21	−4579	92.6	1910	1.278	3.439	−.509
310	61.16	5.9323	−88.70	−4114	93.2	1898	1.274	3.549	−.507
315	61.39	5.8596	−87.20	−3647	93.7	1886	1.269	3.656	−.504
320	61.61	5.7891	−85.72	−3177	94.3	1874	1.265	3.760	−.501
325	61.83	5.7207	−84.25	−2704	94.9	1863	1.260	3.860	−.499
330	62.06	5.6543	−82.80	−2228	95.5	1851	1.256	3.957	−.496
335	62.28	5.5898	−81.36	−1749	96.1	1840	1.252	4.051	−.494
340	62.50	5.5272	−79.93	−1267	96.6	1829	1.247	4.141	−.491
345	62.72	5.4663	−78.52	−783	97.2	1818	1.243	4.227	−.489
350	62.94	5.4071	−77.11	−295	97.8	1807	1.239	4.311	−.486
355	63.16	5.3495	−75.72	196	98.4	1796	1.235	4.390	−.484
360	63.38	5.2935	−74.34	689	99.0	1785	1.231	4.466	−.482
365	63.59	5.2389	−72.97	1186	99.6	1775	1.227	4.539	−.479
370	63.81	5.1858	−71.61	1685	100.2	1765	1.223	4.609	−.477
380	64.24	5.0835	−68.93	2693	101.4	1744	1.215	4.738	−.472

390	64.67	4.9862	−66.28	3713	102.6	1725	1.208	4.854	− .468
400	65.10	4.8936	−63.67	4744	103.7	1706	1.201	4.959	− .464
410	65.52	4.8053	−61.09	5788	104.9	1688	1.194	5.051	− .460
420	65.94	4.7209	−58.55	6843	106.1	1670	1.188	5.133	− .456
430	66.36	4.6404	−56.04	7910	107.3	1652	1.181	5.204	− .452
440	66.78	4.5632	−53.56	8988	108.4	1636	1.175	5.265	− .448
450	67.19	4.4894	−51.11	10078	109.5	1619	1.169	5.317	− .445
475	68.21	4.3175	−45.11	12852	112.4	1580	1.156	5.412	− .436

Table 2—continued

P/MPa									
T_σ/K					300				
Liq	V_σ	Z_σ	S_σ	H_σ	C_P	w_σ	γ_σ	$(f/P)_\sigma$	μ_σ
Vap									
$\dfrac{T}{K}$	$\dfrac{V}{\text{cm}^3\,\text{mol}^{-1}}$	Z	$\dfrac{S}{\text{J K}^{-1}\,\text{mol}^{-1}}$	$\dfrac{H}{\text{J mol}^{-1}}$	$\dfrac{C_P}{\text{J K}^{-1}\,\text{mol}^{-1}}$	$\dfrac{w}{\text{m sec}^{-1}}$	γ	(f/P)	$\dfrac{\mu}{\text{K MPa}^{-1}}$
110	(51.47)	(16.883)	(−174.89)	(−18395)	(64.6)	(2693)	(1.187)	(.8696E−01)	(−.744)
115	(51.63)	(16.199)	(−171.98)	(−18067)	(66.5)	(2652)	(1.205)	(.1314)	(−.720)
120	(51.80)	(15.574)	(−169.11)	(−17729)	(68.7)	(2613)	(1.221)	(.1909)	(−.694)
125	(51.97)	(15.002)	(−166.26)	(−17381)	(70.8)	(2577)	(1.235)	(.2676)	(−.671)
130	(52.15)	(14.475)	(−163.45)	(−17022)	(72.7)	(2545)	(1.247)	(.3636)	(−.651)
135	(52.34)	(13.990)	(−160.67)	(−16654)	(74.4)	(2515)	(1.258)	(.4803)	(−.634)
140	(52.54)	(13.540)	(−157.94)	(−16278)	(75.9)	(2489)	(1.269)	(.6190)	(−.620)
145	(52.73)	(13.123)	(−155.25)	(−15896)	(77.1)	(2465)	(1.278)	(.7801)	(−.608)
150	(52.94)	(12.734)	(−152.62)	(−15507)	(78.2)	(2443)	(1.287)	(.9636)	(−.599)
155	(53.14)	(12.371)	(−150.04)	(−15114)	(79.1)	(2424)	(1.295)	(1.169)	(−.591)
160	(53.35)	(12.032)	(−147.51)	(−14717)	(79.8)	(2405)	(1.302)	(1.396)	(−.584)
165	(53.56)	(11.713)	(−145.05)	(−14316)	(80.5)	(2388)	(1.308)	(1.642)	(−.579)
170	(53.78)	(11.414)	(−142.64)	(−13912)	(81.0)	(2372)	(1.313)	(1.906)	(−.574)
175	(53.99)	(11.132)	(−140.28)	(−13506)	(81.5)	(2356)	(1.317)	(2.186)	(−.570)
180	(54.20)	(10.866)	(−137.98)	(−13097)	(81.9)	(2341)	(1.321)	(2.480)	(−.567)

185	(54.42)	(10.614)	(−135.73)	(−12686)	(82.3)	(2327)	(1.324)	(2.786)	(−.564)
190	(54.64)	(10.376)	(−133.53)	(−12274)	(82.7)	(2313)	(1.326)	(3.102)	(−.561)
195	(54.85)	(10.150)	(−131.38)	(−11859)	(83.1)	(2299)	(1.327)	(3.425)	(−.559)
200	(55.07)	(9.9355)	(−129.27)	(−11443)	(83.4)	(2286)	(1.328)	(3.753)	(−.556)
205	(55.29)	(9.7314)	(−127.20)	(−11025)	(83.8)	(2272)	(1.328)	(4.085)	(−.554)
210	(55.51)	(9.5370)	(−125.18)	(−10606)	(84.1)	(2259)	(1.328)	(4.417)	(−.552)
215	(55.72)	(9.3516)	(−123.20)	(−10184)	(84.5)	(2246)	(1.327)	(4.749)	(−.549)
220	(55.94)	(9.1746)	(−121.25)	(−9761)	(84.8)	(2233)	(1.325)	(5.078)	(−.547)
225	(56.16)	(9.0053)	(−119.34)	(−9336)	(85.2)	(2220)	(1.323)	(5.404)	(−.545)
230	(56.37)	(8.8434)	(−117.46)	(−8909)	(85.6)	(2207)	(1.321)	(5.724)	(−.543)
235	(56.59)	(8.6882)	(−115.62)	(−8479)	(86.0)	(2194)	(1.319)	(6.038)	(−.540)
240	(56.80)	(8.5395)	(−113.80)	(−8048)	(86.5)	(2181)	(1.316)	(6.345)	(−.538)
245	(57.01)	(8.3967)	(−112.01)	(−7615)	(86.9)	(2168)	(1.313)	(6.643)	(−.536)
250	(57.23)	(8.2595)	(−110.25)	(−7179)	(87.4)	(2156)	(1.309)	(6.932)	(−.533)
255	(57.44)	(8.1276)	(−108.52)	(−6741)	(87.8)	(2143)	(1.306)	(7.211)	(−.531)
260	(57.65)	(8.0006)	(−106.81)	(−6301)	(88.3)	(2130)	(1.302)	(7.479)	(−.528)
265	(57.86)	(7.8784)	(−105.12)	(−5858)	(88.8)	(2118)	(1.298)	(7.737)	(−.526)
270	(58.07)	(7.7606)	(−103.46)	(−5413)	(89.3)	(2106)	(1.294)	(7.984)	(−.524)
275	(58.28)	(7.6469)	(−101.81)	(−4965)	(89.8)	(2093)	(1.290)	(8.219)	(−.521)
280	(58.49)	(7.5372)	(−100.19)	(−4515)	(90.3)	(2081)	(1.286)	(8.443)	(−.518)

Table 2—*continued*

P/MPa																			
T_σ/K										**300**									
Liq Vap																			
$\dfrac{T}{K}$	V_σ	$\dfrac{V}{\text{cm}^3\,\text{mol}^{-1}}$	Z_σ	Z	S_σ	$\dfrac{S}{\text{J K}^{-1}\,\text{mol}^{-1}}$	H_σ	$\dfrac{H}{\text{J mol}^{-1}}$	C_P	$\dfrac{C_P}{\text{J K}^{-1}\,\text{mol}^{-1}}$	w_σ	$\dfrac{w}{\text{m sec}^{-1}}$	γ_σ	γ	$(f/P)_b$	(f/P)	μ_σ	$\dfrac{\mu}{\text{K MPa}^{-1}}$
285	(58.70)		(7.4312)		(−98.59)		(−4062)		(90.9)		(2069)		(1.281)		(8.656)		(−.516)	
290	(58.90)		(7.3288)		(−97.00)		(−3606)		(91.4)		(2057)		(1.277)		(8.857)		(−.513)	
295	(59.11)		(7.2298)		(−95.43)		(−3147)		(92.0)		(2045)		(1.273)		(9.047)		(−.511)	
300	59.31		7.1339		−93.88		−2686		92.5		2033		1.268		9.226		−.508	
305	59.52		7.0411		−92.35		−2222		93.1		2021		1.264		9.393		−.506	
310	59.72		6.9512		−90.83		−1755		93.7		2010		1.259		9.549		−.503	
315	59.92		6.8641		−89.33		−1285		94.3		1998		1.255		9.695		−.501	
320	60.13		6.7795		−87.84		−812		94.8		1987		1.251		9.831		−.498	
325	60.33		6.6975		−86.36		−337		95.4		1976		1.246		9.956		−.496	
330	60.53		6.6178		−84.90		142		96.0		1965		1.242		10.07		−.493	
335	60.72		6.5405		−83.45		623		96.6		1954		1.238		10.18		−.491	
340	60.92		6.4653		−82.02		1108		97.2		1943		1.233		10.27		−.489	
345	61.12		6.3922		−80.60		1595		97.8		1932		1.229		10.36		−.486	
350	61.31		6.3211		−79.18		2086		98.4		1922		1.225		10.44		−.484	
355	61.51		6.2519		−77.78		2579		99.0		1912		1.221		10.51		−.482	
360	61.70		6.1845		−76.40		3076		99.6		1901		1.217		10.58		−.479	
365	61.90		6.1189		−75.02		3575		100.2		1891		1.213		10.63		−.477	
370	62.09		6.0550		−73.65		4078		100.8		1881		1.209		10.68		−.475	
380	62.47		5.9319		−70.95		5092		102.0		1862		1.201		10.76		−.471	
390	62.85		5.8149		−68.28		6118		103.2		1843		1.194		10.81		−.467	

400	63.23	5.7033	− 65.65	7156	104.4	1824	1.187	10.84	−.463
410	63.60	5.5969	− 63.06	8206	105.6	1807	1.180	10.85	−.459
420	63.97	5.4952	− 60.50	9267	106.8	1789	1.174	10.84	−.455
430	64.33	5.3980	− 57.98	10341	107.9	1772	1.167	10.81	−.452
440	64.69	5.3049	− 55.48	11426	109.1	1756	1.161	10.77	−.448
450	65.05	5.2157	− 53.02	12523	110.3	1740	1.155	10.72	−.445
475	65.93	5.0079	− 46.98	15315	113.1	1701	1.142	10.54	−.437

Table 2—continued

P/MPa					350				
T_σ/K									
Liq Vap	V_σ	Z_σ	S_σ	H_σ	C_P	w_σ	γ_σ	$(f/P)_\sigma$	μ_σ
$\dfrac{T}{K}$	$\dfrac{V}{cm^3\,mol^{-1}}$	Z	$\dfrac{S}{J\,K^{-1}\,mol^{-1}}$	$\dfrac{H}{J\,mol^{-1}}$	$\dfrac{C_P}{J\,K^{-1}\,mol^{-1}}$	$\dfrac{w}{m\,sec^{-1}}$	γ	(f/P)	$\dfrac{\mu}{K\,MPa^{-1}}$
115	(51.12)	(18.711)	(−173.48)	(−15672)	(64.4)	(2731)	(1.162)	(1.653)	(−.744)
120	(51.26)	(17.981)	(−170.69)	(−15344)	(66.7)	(2692)	(1.179)	(2.163)	(−.716)
125	(51.41)	(17.312)	(−167.93)	(−15005)	(68.9)	(2657)	(1.194)	(2.757)	(−.690)
130	(51.57)	(16.698)	(−165.18)	(−14655)	(71.0)	(2625)	(1.209)	(3.430)	(−.667)
135	(51.73)	(16.131)	(−162.46)	(−14295)	(72.9)	(2596)	(1.222)	(4.180)	(−.648)
140	(51.90)	(15.606)	(−159.78)	(−13927)	(74.5)	(2570)	(1.235)	(4.998)	(−.631)
145	(52.08)	(15.119)	(−157.15)	(−13551)	(75.9)	(2546)	(1.246)	(5.875)	(−.618)
150	(52.26)	(14.666)	(−154.55)	(−13168)	(77.1)	(2525)	(1.257)	(6.802)	(−.606)
155	(52.44)	(14.243)	(−152.01)	(−12780)	(78.1)	(2506)	(1.266)	(7.769)	(−.597)
160	(52.63)	(13.848)	(−149.51)	(−12387)	(79.0)	(2488)	(1.275)	(8.764)	(−.589)
165	(52.82)	(13.477)	(−147.07)	(−11990)	(79.8)	(2471)	(1.283)	(9.778)	(−.582)
170	(53.02)	(13.129)	(−144.68)	(−11589)	(80.5)	(2456)	(1.289)	(10.80)	(−.577)
175	(53.21)	(12.801)	(−142.33)	(−11185)	(81.0)	(2441)	(1.295)	(11.82)	(−.572)
180	(53.41)	(12.491)	(−140.04)	(−10779)	(81.6)	(2426)	(1.300)	(12.83)	(−.567)
185	(53.61)	(12.199)	(−137.80)	(−10370)	(82.1)	(2413)	(1.304)	(13.82)	(−.564)

190	(53.81)	(11.922)	(−135.61)	(−9958)	(82.5)	(2399)	(1.307)	(14.79)	(−.560)
195	(54.01)	(11.659)	(−133.46)	(−9545)	(82.9)	(2386)	(1.309)	(15.73)	(−.557)
200	(54.21)	(11.410)	(−131.35)	(−9129)	(83.4)	(2373)	(1.311)	(16.63)	(−.554)
205	(54.41)	(11.173)	(−129.29)	(−8711)	(83.8)	(2361)	(1.312)	(17.49)	(−.551)
210	(54.61)	(10.947)	(−127.27)	(−8291)	(84.2)	(2348)	(1.312)	(18.32)	(−.549)
215	(54.81)	(10.732)	(−125.28)	(−7869)	(84.6)	(2336)	(1.312)	(19.09)	(−.546)
220	(55.01)	(10.526)	(−123.33)	(−7445)	(85.0)	(2323)	(1.311)	(19.83)	(−.544)
225	(55.21)	(10.329)	(−121.42)	(−7019)	(85.4)	(2311)	(1.309)	(20.51)	(−.541)
230	(55.41)	(10.141)	(−119.53)	(−6591)	(85.9)	(2299)	(1.307)	(21.15)	(−.539)
235	(55.61)	(9.9612)	(−117.68)	(−6160)	(86.3)	(2286)	(1.305)	(21.74)	(−.536)
240	(55.81)	(9.7885)	(−115.86)	(−5727)	(86.8)	(2274)	(1.303)	(22.29)	(−.534)
245	(56.00)	(9.6226)	(−114.06)	(−5292)	(87.3)	(2262)	(1.300)	(22.79)	(−.531)
250	(56.20)	(9.4633)	(−112.30)	(−4855)	(87.7)	(2250)	(1.296)	(23.24)	(−.529)
255	(56.40)	(9.3101)	(−110.55)	(−4415)	(88.2)	(2237)	(1.293)	(23.65)	(−.526)
260	(56.59)	(9.1627)	(−108.84)	(−3972)	(88.7)	(2225)	(1.289)	(24.02)	(−.524)
265	(56.79)	(9.0207)	(−107.14)	(−3527)	(89.2)	(2213)	(1.286)	(24.35)	(−.521)
270	(56.98)	(8.8838)	(−105.47)	(−3080)	(89.8)	(2202)	(1.282)	(24.64)	(−.519)
275	(57.17)	(8.7518)	(−103.82)	(−2630)	(90.3)	(2190)	(1.278)	(24.89)	(−.516)
280	(57.36)	(8.6243)	(−102.18)	(−2177)	(90.8)	(2178)	(1.274)	(25.11)	(−.514)
285	(57.56)	(8.5012)	(−100.57)	(−1721)	(91.4)	(2166)	(1.269)	(25.29)	(−.511)

Table 2—continued

P/MPa																			
T_σ/K																			
									350										
Liq Vap	V_σ		Z_σ		S_σ		H_σ		C_P		w_σ		γ_σ		$(f/P)_\sigma$		μ_σ		
$\dfrac{T}{K}$	$\dfrac{V}{\text{cm}^3\,\text{mol}^{-1}}$		Z		$\dfrac{S}{\text{J K}^{-1}\,\text{mol}^{-1}}$		$\dfrac{H}{\text{J mol}^{-1}}$		$\dfrac{C_P}{\text{J K}^{-1}\,\text{mol}^{-1}}$		$\dfrac{w}{\text{m sec}^{-1}}$		γ		(f/P)		$\dfrac{\mu}{\text{K MPa}^{-1}}$		
290	(57.75)		(8.3822)		(−98.98)		(−1263)		(92.0)		(2155)		(1.265)		(25.44)		(−.509)	
295	(57.93)		(8.2671)		(−97.40)		(−802)		(92.5)		(2143)		(1.261)		(25.56)		(−.506)	
300	58.12		8.1557		−95.84		−338		93.1		2132		1.256		25.65		−.503	
305	58.31		8.0478		−94.30		129		93.7		2121		1.252		25.72		−.501	
310	58.50		7.9432		−92.77		599		94.3		2109		1.248		25.76		−.499	
315	58.68		7.8419		−91.26		1072		94.8		2098		1.243		25.77		−.496	
320	58.86		7.7436		−89.76		1548		95.4		2087		1.239		25.77		−.494	
325	59.05		7.6481		−88.27		2026		96.0		2077		1.235		25.74		−.491	
330	59.23		7.5555		−86.80		2508		96.6		2066		1.230		25.70		−.489	
335	59.41		7.4655		−85.35		2993		97.2		2055		1.226		25.64		−.487	
340	59.59		7.3780		−83.90		3480		97.8		2045		1.222		25.56		−.484	
345	59.77		7.2929		−82.47		3971		98.5		2035		1.218		25.46		−.482	
350	59.95		7.2101		−81.05		4465		99.1		2024		1.214		25.36		−.480	
355	60.12		7.1296		−79.64		4962		99.7		2014		1.210		25.24		−.477	
360	60.30		7.0511		−78.24		5462		100.3		2004		1.206		25.11		−.475	
365	60.48		6.9747		−76.85		5965		100.9		1995		1.202		24.96		−.473	
370	60.65		6.9003		−75.47		6471		101.5		1985		1.198		24.81		−.471	
380	60.99		6.7569		−72.75		7492		102.7		1966		1.190		24.48		−.467	
390	61.34		6.6204		−70.07		8525		103.9		1948		1.183		24.13		−.463	
400	61.67		6.4904		−67.42		9571		105.2		1929		1.176		23.74		−.459	

410	62.01	6.3663	−64.81	10628	106.4	1912	1.169	23.34	−.455
420	62.33	6.2477	−62.23	11698	107.6	1895	1.162	22.93	−.452
430	62.66	6.1342	−59.69	12779	108.7	1878	1.156	22.51	−.448
440	62.98	6.0255	−57.17	13873	109.9	1862	1.150	22.08	−.445
450	63.30	5.9213	−54.69	14978	111.1	1846	1.144	21.64	−.442
475	64.07	5.6785	−48.61	17791	114.0	1808	1.130	20.56	−.435

Table 2—continued

P/MPa						400			
T_σ/K									
	V_σ	Z_σ	S_σ	H_σ	C_P	w_σ	γ_σ	$(f/P)_\sigma$	μ_σ
Liq Vap									
$\dfrac{T}{K}$	$\dfrac{V}{cm^3\,mol^{-1}}$	Z	$\dfrac{S}{J\,K^{-1}\,mol^{-1}}$	$\dfrac{H}{J\,mol^{-1}}$	$\dfrac{C_P}{J\,K^{-1}\,mol^{-1}}$	$\dfrac{w}{m\,sec^{-1}}$	γ	(f/P)	$\dfrac{\mu}{K\,MPa^{-1}}$
115	(50.66)	(21.192)	(−174.74)	(−13272)	(62.3)	(2810)	(1.125)	(20.70)	(−.771)
120	(50.78)	(20.357)	(−172.04)	(−12955)	(64.6)	(2771)	(1.143)	(24.40)	(−.740)
125	(50.90)	(19.592)	(−169.36)	(−12626)	(66.9)	(2735)	(1.159)	(28.26)	(−.711)
130	(51.04)	(18.889)	(−166.69)	(−12286)	(69.1)	(2703)	(1.175)	(32.21)	(−.685)
135	(51.18)	(18.241)	(−164.04)	(−11935)	(71.1)	(2674)	(1.189)	(36.19)	(−.664)
140	(51.34)	(17.641)	(−161.42)	(−11575)	(72.9)	(2648)	(1.203)	(40.14)	(−.645)
145	(51.49)	(17.085)	(−158.84)	(−11207)	(74.5)	(2625)	(1.216)	(44.02)	(−.629)
150	(51.65)	(16.567)	(−156.29)	(−10831)	(75.8)	(2604)	(1.229)	(47.78)	(−.616)
155	(51.82)	(16.084)	(−153.78)	(−10449)	(77.0)	(2585)	(1.240)	(51.37)	(−.605)
160	(51.99)	(15.633)	(−151.32)	(−10061)	(78.0)	(2567)	(1.250)	(54.77)	(−.596)
165	(52.16)	(15.210)	(−148.91)	(−9669)	(78.9)	(2551)	(1.259)	(57.95)	(−.588)
170	(52.34)	(14.813)	(−146.54)	(−9272)	(79.7)	(2536)	(1.267)	(60.90)	(−.581)
175	(52.52)	(14.439)	(−144.22)	(−8872)	(80.4)	(2521)	(1.274)	(63.61)	(−.575)
180	(52.70)	(14.086)	(−141.94)	(−8468)	(81.1)	(2507)	(1.280)	(66.06)	(−.569)
185	(52.89)	(13.753)	(−139.71)	(−8061)	(81.7)	(2494)	(1.286)	(68.26)	(−.564)

190	(53.07)	(13.438)	(−137.53)	(−7651)	(82.2)	(2481)	(1.290)	(70.21)	(−.560)
195	(53.25)	(13.139)	(−135.38)	(−7239)	(82.7)	(2468)	(1.293)	(71.92)	(−.556)
200	(53.44)	(12.855)	(−133.28)	(−6824)	(83.2)	(2456)	(1.295)	(73.39)	(−.553)
205	(53.63)	(12.585)	(−131.22)	(−6407)	(83.7)	(2444)	(1.297)	(74.64)	(−.549)
210	(53.81)	(12.328)	(−129.20)	(−5987)	(84.2)	(2431)	(1.298)	(75.67)	(−.546)
215	(54.00)	(12.083)	(−127.21)	(−5564)	(84.7)	(2419)	(1.298)	(76.51)	(−.543)
220	(54.19)	(11.849)	(−125.26)	(−5140)	(85.2)	(2407)	(1.298)	(77.14)	(−.540)
225	(54.37)	(11.626)	(−123.34)	(−4713)	(85.6)	(2395)	(1.297)	(77.61)	(−.537)
230	(54.56)	(11.412)	(−121.45)	(−4284)	(86.1)	(2384)	(1.295)	(77.91)	(−.534)
235	(54.74)	(11.207)	(−119.60)	(−3852)	(86.6)	(2372)	(1.293)	(78.06)	(−.532)
240	(54.93)	(11.010)	(−117.77)	(−3418)	(87.1)	(2360)	(1.291)	(78.07)	(−.529)
245	(55.11)	(10.822)	(−115.97)	(−2981)	(87.6)	(2348)	(1.289)	(77.95)	(−.526)
250	(55.29)	(10.641)	(−114.19)	(−2542)	(88.1)	(2336)	(1.286)	(77.72)	(−.524)
255	(55.48)	(10.467)	(−112.44)	(−2100)	(88.6)	(2325)	(1.282)	(77.39)	(−.521)
260	(55.66)	(10.299)	(−110.72)	(−1655)	(89.2)	(2313)	(1.279)	(76.96)	(−.518)
265	(55.84)	(10.137)	(−109.01)	(−1208)	(89.7)	(2301)	(1.275)	(76.45)	(−.516)
270	(56.02)	(9.9819)	(−107.33)	(−758)	(90.3)	(2290)	(1.272)	(75.86)	(−.513)
275	(56.20)	(9.8317)	(−105.67)	(−305)	(90.8)	(2278)	(1.268)	(75.21)	(−.511)
280	(56.38)	(9.6868)	(−104.03)	(150)	(91.4)	(2267)	(1.264)	(74.50)	(−.508)
285	(56.55)	(9.5468)	(−102.41)	(608)	(91.9)	(2255)	(1.260)	(73.73)	(−.505)

Table 2—continued

	V_σ	Z_σ	S_σ	H_σ	C_P	w_σ	γ_σ	$(f/P)_\sigma$	μ_σ
P/MPa				400					
T_σ/K									
Liq	$\dfrac{V}{\text{cm}^3\,\text{mol}^{-1}}$	Z	$\dfrac{S}{\text{J K}^{-1}\,\text{mol}^{-1}}$	$\dfrac{H}{\text{J mol}^{-1}}$	$\dfrac{C_P}{\text{J K}^{-1}\,\text{mol}^{-1}}$	$\dfrac{w}{\text{m sec}^{-1}}$	γ	(f/P)	$\dfrac{\mu}{\text{K MPa}^{-1}}$
Vap									
$\dfrac{T}{\text{K}}$									
290	(56.73)	(9.4114)	(−100.80)	(1069)	(92.5)	(2244)	(1.255)	(72.93)	(−.503)
295	(56.91)	(9.2805)	(−99.21)	(1533)	(93.1)	(2233)	(1.251)	(72.08)	(−.500)
300	57.08	9.1538	−97.65	2000	93.7	2222	1.247	71.20	−.498
305	57.25	9.0311	−96.09	2470	94.3	2211	1.243	70.29	−.495
310	57.43	8.9122	−94.55	2943	94.9	2200	1.238	69.36	−.493
315	57.60	8.7969	−93.03	3419	95.5	2189	1.234	68.41	−.491
320	57.77	8.6850	−91.52	3898	96.1	2179	1.230	67.44	−.488
325	57.94	8.5765	−90.03	4380	96.7	2168	1.225	66.47	−.486
330	58.11	8.4710	−88.55	4865	97.3	2158	1.221	65.48	−.483
335	58.27	8.3686	−87.08	5354	97.9	2148	1.217	64.49	−.481
340	58.44	8.2690	−85.62	5845	98.6	2137	1.213	63.50	−.479
345	58.60	8.1722	−84.18	6339	99.2	2127	1.209	62.50	−.477
350	58.77	8.0779	−82.75	6837	99.8	2117	1.204	61.51	−.474
355	58.93	7.9862	−81.33	7337	100.4	2108	1.200	60.52	−.472
360	59.09	7.8969	−79.92	7841	101.0	2098	1.196	59.54	−.470
365	59.25	7.8099	−78.52	8347	101.7	2088	1.193	58.56	.468
370	59.41	7.7251	−77.13	8857	102.3	2079	1.189	57.59	.466
380	59.73	7.5619	−74.39	9886	103.5	2060	1.181	55.67	.462
390	60.04	7.4064	−71.69	10927	104.7	2042	1.174	53.80	.458
400	60.35	7.2582	−69.02	11981	106.0	2024	1.167	51.98	.454

410	60.65	7.1168	− 66.39	13047	107.2	2007	1.160	50.21	−.451
420	60.95	6.9816	− 63.79	14125	108.4	1990	1.153	48.49	−.447
430	61.24	6.8522	− 61.22	15215	109.6	1974	1.147	46.83	−.444
440	61.54	6.7282	− 58.69	16317	110.8	1958	1.141	45.23	−.441
450	61.82	6.6093	− 56.19	17431	112.0	1942	1.135	43.68	−.438
475	62.52	6.3322	− 50.05	20268	114.9	1904	1.121	40.07	−.431

315

Table 2—continued

P/MPa					450				
T_σ/K									
Liq Vap									
$\dfrac{T}{K}$	V_σ $\dfrac{V}{cm^3\,mol^{-1}}$	Z_σ Z	S_σ $\dfrac{S}{J\,K^{-1}\,mol^{-1}}$	H_σ $\dfrac{H}{J\,mol^{-1}}$	C_P $\dfrac{C_P}{J\,K^{-1}\,mol^{-1}}$	w_σ $\dfrac{w}{m\,sec^{-1}}$	γ_σ γ	$(f/P)_\sigma$ (f/P)	μ_σ $\dfrac{\mu}{K\,MPa^{-1}}$
120	(50.34)	(22.706)	(−173.18)	(−10564)	(62.4)	(2849)	(1.111)	(273.3)	(−.767)
125	(50.45)	(21.845)	(−170.58)	(−10246)	(64.8)	(2813)	(1.128)	(287.6)	(−.735)
130	(50.57)	(21.053)	(−167.99)	(−9916)	(67.1)	(2781)	(1.144)	(300.2)	(−.706)
135	(50.69)	(20.323)	(−165.42)	(−9575)	(69.2)	(2752)	(1.160)	(311.0)	(−.682)
140	(50.82)	(19.649)	(−162.87)	(−9224)	(71.1)	(2726)	(1.175)	(320.1)	(−.661)
145	(50.96)	(19.023)	(−160.34)	(−8863)	(72.8)	(2702)	(1.189)	(327.4)	(−.643)
150	(51.11)	(18.441)	(−157.84)	(−8495)	(74.3)	(2681)	(1.203)	(333.1)	(−.628)
155	(51.26)	(17.898)	(−155.38)	(−8120)	(75.7)	(2662)	(1.215)	(337.2)	(−.615)
160	(51.41)	(17.391)	(−152.96)	(−7739)	(76.8)	(2644)	(1.227)	(339.8)	(−.604)
165	(51.57)	(16.916)	(−150.58)	(−7352)	(77.9)	(2628)	(1.237)	(341.1)	(−.594)
170	(51.73)	(16.470)	(−148.24)	(−6960)	(78.8)	(2613)	(1.246)	(341.1)	(−.586)
175	(51.90)	(16.050)	(−145.95)	(−6564)	(79.7)	(2599)	(1.255)	(340.0)	(−.579)
180	(52.06)	(15.655)	(−143.69)	(−6164)	(80.4)	(2585)	(1.262)	(337.9)	(−.572)
185	(52.23)	(15.281)	(−141.48)	(−5760)	(81.1)	(2572)	(1.268)	(334.9)	(−.566)
190	(52.40)	(14.927)	(−139.31)	(−5353)	(81.8)	(2559)	(1.273)	(331.2)	(−.561)

195	(52.57)	(14.592)	(−137.17)	(−4942)	(82.4)	(2547)	(1.278)	(326.8)	(−.556)
200	(52.75)	(14.274)	(−135.08)	(−4528)	(83.0)	(2534)	(1.281)	(321.9)	(−.552)
205	(52.92)	(13.972)	(−133.02)	(−4112)	(83.6)	(2522)	(1.283)	(316.6)	(−.548)
210	(53.09)	(13.684)	(−131.00)	(−3693)	(84.1)	(2511)	(1.285)	(310.8)	(−.544)
215	(53.27)	(13.410)	(−129.01)	(−3271)	(84.7)	(2499)	(1.286)	(304.8)	(−.541)
220	(53.44)	(13.148)	(−127.06)	(−2846)	(85.2)	(2487)	(1.286)	(298.5)	(−.537)
225	(53.62)	(12.897)	(−125.14)	(−2418)	(85.8)	(2475)	(1.285)	(292.0)	(−.534)
230	(53.79)	(12.658)	(−123.25)	(−1988)	(86.3)	(2464)	(1.284)	(285.4)	(−.531)
235	(53.96)	(12.429)	(−121.39)	(−1555)	(86.8)	(2452)	(1.283)	(278.8)	(−.528)
240	(54.14)	(12.209)	(−119.55)	(−1120)	(87.4)	(2440)	(1.281)	(272.1)	(−.525)
245	(54.31)	(11.998)	(−117.75)	(−682)	(87.9)	(2429)	(1.279)	(265.4)	(−.522)
250	(54.48)	(11.795)	(−115.96)	(−241)	(88.5)	(2417)	(1.276)	(258.7)	(−.519)
255	(54.65)	(11.600)	(−114.21)	(203)	(89.0)	(2406)	(1.273)	(252.0)	(−.516)
260	(54.82)	(11.413)	(−112.47)	(650)	(89.6)	(2395)	(1.270)	(245.4)	(−.513)
265	(54.99)	(11.232)	(−110.76)	(1099)	(90.2)	(2383)	(1.267)	(239.0)	(−.510)
270	(55.16)	(11.058)	(−109.07)	(1551)	(90.7)	(2372)	(1.263)	(232.6)	(−.508)
275	(55.33)	(10.890)	(−107.40)	(2007)	(91.3)	(2361)	(1.259)	(226.3)	(−.505)
280	(55.50)	(10.728)	(−105.75)	(2465)	(91.9)	(2349)	(1.255)	(220.1)	(−.502)
285	(55.66)	(10.571)	(−104.12)	(2926)	(92.5)	(2338)	(1.251)	(214.1)	(−.500)
290	(55.83)	(10.420)	(−102.50)	(3390)	(93.1)	(2327)	(1.247)	(208.2)	(−.497)

Table 2—continued

P/MPa									
T_σ/K				**450**					
Liq Vap	V_σ	Z_σ	S_σ	H_σ	C_P	w_σ	γ_σ	$(f/P)_\sigma$	μ_σ
$\dfrac{T}{K}$	$\dfrac{V}{\text{cm}^3\,\text{mol}^{-1}}$	Z	$\dfrac{S}{\text{J K}^{-1}\,\text{mol}^{-1}}$	$\dfrac{H}{\text{J mol}^{-1}}$	$\dfrac{C_P}{\text{J K}^{-1}\,\text{mol}^{-1}}$	$\dfrac{w}{\text{m sec}^{-1}}$	γ	(f/P)	$\dfrac{\mu}{\text{K MPa}^{-1}}$
295	(55.99)	(10.273)	(−100.91)	(3857)	(93.7)	(2316)	(1.243)	(202.5)	(−.494)
300	56.16	10.131	−99.33	4327	94.3	2306	1.239	196.9	−.492
305	56.32	9.9941	−97.76	4800	94.9	2295	1.235	191.4	−.489
310	56.48	9.8610	−96.21	5276	95.5	2284	1.230	186.1	−.487
315	56.64	9.7320	−94.68	5755	96.2	2274	1.226	180.9	−.485
320	56.80	9.6068	−93.16	6238	96.8	2263	1.222	175.9	−.482
325	56.96	9.4853	−91.66	6723	97.4	2253	1.217	171.0	−.480
330	57.11	9.3673	−90.16	7212	98.0	2243	1.213	166.3	−.477
335	57.27	9.2527	−88.68	7703	98.7	2232	1.209	161.7	−.475
340	57.42	9.1412	−87.22	8198	99.3	2222	1.205	157.2	−.473
345	57.58	9.0328	−85.76	8696	99.9	2213	1.201	152.9	−.471
350	57.73	8.9273	−84.32	9198	100.6	2203	1.197	148.7	−.468
355	57.88	8.8246	−82.89	9702	101.2	2193	1.193	144.7	−.466
360	58.03	8.7247	−81.47	10209	101.8	2184	1.189	140.7	−.464
365	58.18	8.6273	−80.06	10720	102.4	2174	1.185	136.9	−.462
370	58.33	8.5323	−78.66	11234	103.1	2165	1.181	133.2	−.460
380	58.62	8.3494	−75.90	12271	104.3	2146	1.173	126.2	−.456
390	58.91	8.1754	−73.17	13321	105.6	2129	1.166	119.6	−.452
400	59.19	8.0094	−70.48	14383	106.8	2111	1.159	113.5	−.449
410	59.47	7.8509	−67.83	15458	108.1	2094	1.152	107.7	−.445

450 MPa

420	59.75	7.6994	−65.21	16545	109.3	2077	1.145	102.3	−.442
430	60.02	7.5544	−62.62	17644	110.5	2061	1.139	97.17	−.439
440	60.28	7.4154	−60.07	18755	111.8	2045	1.133	92.41	−.436
450	60.55	7.2821	−57.54	19879	113.0	2030	1.127	87.94	−.433
475	61.18	6.9713	−51.36	22740	115.9	1992	1.113	77.93	−.426

319

Table 2—continued

P/MPa									
T_σ/K									
Liq Vap	V_σ	Z_σ	S_σ	H_σ	C_P	w_σ	γ_σ	$(f/P)_\sigma$	μ_σ
					500				
$\dfrac{T}{K}$	$\dfrac{V}{\text{cm}^3\,\text{mol}^{-1}}$	Z	$\dfrac{S}{\text{J K}^{-1}\text{mol}^{-1}}$	$\dfrac{H}{\text{J mol}^{-1}}$	$\dfrac{C_P}{\text{J K}^{-1}\text{mol}^{-1}}$	$\dfrac{w}{\text{m sec}^{-1}}$	γ	(f/P)	$\dfrac{\mu}{\text{K MPa}^{-1}}$
120	(49.95)	(25.032)	(−174.12)	(−8170)	(60.1)	(2928)	(1.085)	(3035)	(−.796)
125	(50.04)	(24.074)	(−171.62)	(−7863)	(62.6)	(2891)	(1.101)	(2903)	(−.761)
130	(50.14)	(23.194)	(−169.11)	(−7544)	(65.0)	(2858)	(1.117)	(2774)	(−.730)
135	(50.25)	(22.383)	(−166.62)	(−7213)	(67.2)	(2828)	(1.133)	(2651)	(−.703)
140	(50.36)	(21.633)	(−164.14)	(−6872)	(69.2)	(2802)	(1.149)	(2531)	(−.679)
145	(50.48)	(20.938)	(−161.67)	(−6521)	(71.1)	(2778)	(1.164)	(2415)	(−.659)
150	(50.61)	(20.291)	(−159.24)	(−6161)	(72.7)	(2757)	(1.178)	(2303)	(−.642)
155	(50.75)	(19.689)	(−156.83)	(−5794)	(74.2)	(2738)	(1.192)	(2195)	(−.627)
160	(50.89)	(19.126)	(−154.45)	(−5420)	(75.5)	(2720)	(1.204)	(2091)	(−.614)
165	(51.03)	(18.599)	(−152.11)	(−5039)	(76.7)	(2704)	(1.216)	(1991)	(−.603)
170	(51.18)	(18.104)	(−149.81)	(−4653)	(77.8)	(2688)	(1.227)	(1895)	(−.593)
175	(51.33)	(17.639)	(−147.54)	(−4262)	(78.7)	(2674)	(1.236)	(1802)	(−.584)
180	(51.48)	(17.200)	(−145.31)	(−3866)	(79.6)	(2660)	(1.244)	(1715)	(−.576)
185	(51.64)	(16.786)	(−143.11)	(−3466)	(80.5)	(2647)	(1.252)	(1630)	(−.569)
190	(51.80)	(16.394)	(−140.96)	(−3061)	(81.2)	(2635)	(1.258)	(1550)	(−.563)

195	(51.96)	(16.023)	(−138.84)	(−2653)	(82.0)	(2622)	(1.263)	(1474)	(−.557)
200	(52.12)	(15.671)	(−136.75)	(−2242)	(82.7)	(2610)	(1.267)	(1402)	(−.552)
205	(52.28)	(15.336)	(−134.70)	(−1827)	(83.3)	(2598)	(1.270)	(1333)	(−.547)
210	(52.44)	(15.018)	(−132.69)	(−1408)	(84.0)	(2587)	(1.273)	(1267)	(−.543)
215	(52.61)	(14.714)	(−130.70)	(−987)	(84.6)	(2575)	(1.274)	(1206)	(−.539)
220	(52.77)	(14.425)	(−128.75)	(−562)	(85.2)	(2563)	(1.275)	(1147)	(−.535)
225	(52.93)	(14.148)	(−126.83)	(−135)	(85.8)	(2552)	(1.275)	(1091)	(−.531)
230	(53.10)	(13.883)	(−124.94)	(296)	(86.4)	(2540)	(1.274)	(1039)	(−.527)
235	(53.26)	(13.630)	(−123.07)	(730)	(87.0)	(2529)	(1.273)	(989.2)	(−.524)
240	(53.42)	(13.387)	(−121.23)	(1166)	(87.6)	(2517)	(1.272)	(942.1)	(−.520)
245	(53.59)	(13.153)	(−119.42)	(1606)	(88.2)	(2506)	(1.270)	(897.7)	(−.517)
250	(53.75)	(12.929)	(−117.63)	(2048)	(88.8)	(2494)	(1.268)	(855.6)	(−.514)
255	(53.91)	(12.714)	(−115.87)	(2494)	(89.4)	(2483)	(1.265)	(815.9)	(−.511)
260	(54.07)	(12.507)	(−114.13)	(2942)	(90.0)	(2472)	(1.262)	(778.2)	(−.508)
265	(54.23)	(12.307)	(−112.41)	(3394)	(90.6)	(2461)	(1.259)	(742.6)	(−.505)
270	(54.39)	(12.115)	(−110.71)	(3848)	(91.2)	(2449)	(1.255)	(708.9)	(−.502)
275	(54.55)	(11.929)	(−109.03)	(4306)	(91.8)	(2438)	(1.252)	(677.1)	(−.499)
280	(54.71)	(11.750)	(−107.37)	(4766)	(92.4)	(2427)	(1.248)	(646.9)	(−.497)
285	(54.86)	(11.577)	(−105.72)	(5230)	(93.1)	(2416)	(1.244)	(618.4)	(−.494)
290	(55.02)	(11.409)	(−104.10)	(5697)	(93.7)	(2405)	(1.240)	(591.4)	(−.491)

Table 2—continued

P/MPa									
T_σ/K					500				
Liq Vap	V_σ	Z_σ	S_σ	H_σ	$C_{P\sigma}$	w_σ	γ_σ	$(f/P)_\sigma$	μ_σ
$\dfrac{T}{K}$	$\dfrac{V}{cm^3\,mol^{-1}}$	Z	$\dfrac{S}{J\,K^{-1}\,mol^{-1}}$	$\dfrac{H}{J\,mol^{-1}}$	$\dfrac{C_P}{J\,K^{-1}\,mol^{-1}}$	$\dfrac{w}{m\,sec^{-1}}$	γ	(f/P)	$\dfrac{\mu}{K\,MPa^{-1}}$
295	(55.17)	(11.248)	(−102.49)	(6167)	(94.3)	(2395)	(1.236)	(565.8)	(−.489)
300	55.33	11.091	−100.90	6640	94.9	2384	1.232	541.5	−.486
305	55.48	10.939	−99.33	7116	95.6	2373	1.228	518.6	−.483
310	55.63	10.792	−97.77	7596	96.2	2363	1.223	496.8	−.481
315	55.78	10.650	−96.23	8078	96.8	2352	1.219	476.1	−.478
320	55.93	10.511	−94.70	8564	97.5	2342	1.215	456.5	−.476
325	56.08	10.377	−93.18	9053	98.1	2332	1.211	437.9	−.474
330	56.23	10.247	−91.68	9545	98.8	2322	1.206	420.3	−.471
335	56.37	10.120	−90.19	10041	99.4	2312	1.202	403.5	−.469
340	56.52	9.9968	−88.71	10539	100.0	2302	1.198	387.6	−.467
345	56.66	9.8770	−87.24	11041	100.7	2292	1.194	372.4	−.464
350	56.81	9.7604	−85.79	11546	101.3	2282	1.190	358.0	−.462
355	56.95	9.6469	−84.35	12055	102.0	2273	1.186	344.3	−.460
360	57.09	9.5364	−82.92	12566	102.6	2263	1.182	331.3	−.458
365	57.23	9.4288	−81.50	13081	103.3	2254	1.178	318.8	−.456
370	57.37	9.3238	−80.09	13599	103.9	2245	1.174	307.0	−.454
380	57.64	9.1217	−77.30	14644	105.2	2227	1.167	285.0	−.450
390	57.91	8.9293	−74.55	15703	106.5	2209	1.159	264.9	−.446
400	58.17	8.7457	−71.84	16774	107.7	2192	1.152	246.7	−.443
410	58.43	8.5705	−69.16	17857	109.0	2175	1.145	230.1	−.439

420	58.69	8.4030	− 66.52	18954	110.3	2158	1.139	214.8	− .436
430	58.94	8.2426	− 63.91	20062	111.5	2142	1.132	200.9	− .433
440	59.18	8.0889	− 61.34	21184	112.7	2126	1.126	188.1	− .430
450	59.43	7.9415	− 58.79	22317	114.0	2111	1.120	176.4	− .427
475	60.01	7.5975	− 52.55	25203	117.0	2073	1.106	151.0	− .420

Table 2—continued

P/MPa					600				
T_σ/K									
Liq									
Vap									
	V_σ	Z_σ	S_σ	H_σ	C_P	w_σ	γ_σ	$(f/P)_\sigma$	μ_σ
$\dfrac{T}{K}$	$\dfrac{V}{cm^3\,mol^{-1}}$	Z	$\dfrac{S}{J\,K^{-1}\,mol^{-1}}$	$\dfrac{H}{J\,mol^{-1}}$	$\dfrac{C_P}{J\,K^{-1}\,mol^{-1}}$	$\dfrac{w}{m\,sec^{-1}}$	γ	(f/P)	$\dfrac{\mu}{K\,MPa^{-1}}$
125	(49.32)	(28.474)	(−173.20)	(−3094)	(58.0)	(3049)	(1.058)	(.2878E+06)	(−.822)
130	(49.39)	(27.418)	(−170.88)	(−2797)	(60.6)	(3013)	(1.073)	(.2307E+06)	(−.784)
135	(49.47)	(26.444)	(−168.55)	(−2489)	(62.9)	(2981)	(1.088)	(.1874E+06)	(−.751)
140	(49.55)	(25.544)	(−166.22)	(−2168)	(65.1)	(2953)	(1.104)	(.1541E+06)	(−.722)
145	(49.65)	(24.709)	(−163.89)	(−1838)	(67.2)	(2928)	(1.120)	(.1279E+06)	(−.697)
150	(49.75)	(23.934)	(−161.58)	(−1497)	(69.1)	(2906)	(1.135)	(.1072E+06)	(−.675)
155	(49.86)	(23.212)	(−159.29)	(−1147)	(70.8)	(2886)	(1.150)	(.9057E+05)	(−.656)
160	(49.97)	(22.538)	(−157.02)	(−789)	(72.4)	(2868)	(1.164)	(.7709E+05)	(−.639)
165	(50.09)	(21.906)	(−154.77)	(−423)	(73.9)	(2851)	(1.177)	(.6606E+05)	(−.624)
170	(50.21)	(21.314)	(−152.54)	(−51)	(75.2)	(2835)	(1.189)	(.5695E+05)	(−.611)
175	(50.34)	(20.757)	(−150.34)	(329)	(76.5)	(2820)	(1.201)	(.4938E+05)	(−.599)
180	(50.47)	(20.233)	(−148.17)	(714)	(77.6)	(2807)	(1.211)	(.4303E+05)	(−.589)
185	(50.60)	(19.738)	(−146.03)	(1105)	(78.7)	(2793)	(1.220)	(.3767E+05)	(−.579)
190	(50.74)	(19.271)	(−143.92)	(1501)	(79.8)	(2780)	(1.228)	(.3313E+05)	(−.571)
195	(50.88)	(18.828)	(−141.83)	(1902)	(80.7)	(2768)	(1.235)	(.2926E+05)	(−.563)

200	(51.02)	(18.408)	(−139.78)	(2308)	(81.6)	(2756)	(1.241)	(.2593E + 05)	(−.556)
205	(51.16)	(18.009)	(−137.75)	(2719)	(82.5)	(2744)	(1.245)	(.2307E + 05)	(−.549)
210	(51.30)	(17.630)	(−135.75)	(3133)	(83.4)	(2732)	(1.249)	(.2059E + 05)	(−.543)
215	(51.45)	(17.268)	(−133.78)	(3552)	(84.2)	(2720)	(1.252)	(.1843E + 05)	(−.537)
220	(51.59)	(16.923)	(−131.84)	(3975)	(85.0)	(2708)	(1.254)	(.1655E + 05)	(−.532)
225	(51.74)	(16.594)	(−129.92)	(4402)	(85.7)	(2697)	(1.255)	(.1491E + 05)	(−.527)
230	(51.88)	(16.279)	(−128.03)	(4832)	(86.5)	(2685)	(1.256)	(.1346E + 05)	(−.522)
235	(52.03)	(15.977)	(−126.16)	(5266)	(87.2)	(2673)	(1.256)	(.1219E + 05)	(−.518)
240	(52.18)	(15.689)	(−124.32)	(5704)	(87.9)	(2662)	(1.255)	(.1106E + 05)	(−.514)
245	(52.32)	(15.411)	(−122.50)	(6145)	(88.6)	(2651)	(1.254)	(.1006E + 05)	(−.510)
250	(52.47)	(15.145)	(−120.70)	(6590)	(89.3)	(2639)	(1.252)	(9169)	(−.506)
255	(52.61)	(14.889)	(−118.92)	(7039)	(90.0)	(2628)	(1.250)	(8376)	(−.502)
260	(52.76)	(14.643)	(−117.17)	(7490)	(90.7)	(2616)	(1.248)	(7667)	(−.499)
265	(52.90)	(14.406)	(−115.43)	(7946)	(91.4)	(2605)	(1.245)	(7031)	(−.495)
270	(53.04)	(14.178)	(−113.72)	(8404)	(92.1)	(2594)	(1.242)	(6460)	(−.492)
275	(53.19)	(13.957)	(−112.02)	(8866)	(92.8)	(2583)	(1.239)	(5946)	(−.489)
280	(53.33)	(13.745)	(−110.35)	(9332)	(93.4)	(2572)	(1.235)	(5483)	(−.486)
285	(53.47)	(13.539)	(−108.69)	(9801)	(94.1)	(2561)	(1.231)	(5064)	(−.483)
290	(53.61)	(13.341)	(−107.04)	(10273)	(94.8)	(2550)	(1.228)	(4684)	(−.480)
295	(53.75)	(13.149)	(−105.42)	(10749)	(95.5)	(2539)	(1.224)	(4339)	(−.477)

Table 2—continued

P/MPa											
T_σ/K						600					
Liq Vap		V_σ	Z_σ	S_σ	H_σ		w_σ	γ_σ	$(f/P)_\sigma$	μ_σ	
$\dfrac{T}{K}$		$\dfrac{V}{cm^3\,mol^{-1}}$	Z	$\dfrac{S}{J\,K^{-1}\,mol^{-1}}$	$\dfrac{H}{J\,mol^{-1}}$	$\dfrac{C_P}{J\,K^{-1}\,mol^{-1}}$	$\dfrac{w}{m\,sec^{-1}}$	γ	(f/P)	$\dfrac{\mu}{K\,MPa^{-1}}$	
300		53.89	12.963	−103.81	11228	96.2	2529	1.220	4026	−.474	
305		54.03	12.783	−102.21	11710	96.8	2518	1.216	3741	−.472	
310		54.16	12.608	−100.63	12196	97.5	2508	1.212	3480	−.469	
315		54.30	12.439	−99.07	12685	98.2	2497	1.208	3242	−.467	
320		54.43	12.275	−97.52	13178	98.9	2487	1.203	3025	−.464	
325		54.57	12.116	−95.98	13674	99.5	2477	1.199	2825	−.462	
330		54.70	11.961	−94.45	14173	100.2	2467	1.195	2642	−.459	
335		54.83	11.811	−92.94	14676	100.9	2457	1.191	2473	−.457	
340		54.96	11.665	−91.44	15182	101.6	2447	1.187	2318	−.454	
345		55.09	11.523	−89.95	15692	102.2	2437	1.183	2175	−.452	
350		55.21	11.384	−88.48	16205	102.9	2428	1.179	2043	−.450	
355		55.34	11.250	−87.01	16721	103.6	2418	1.175	1921	−.448	
360		55.47	11.119	−85.56	17241	104.3	2409	1.171	1808	−.446	
365		55.59	10.991	−84.12	17764	104.9	2400	1.167	1703	−.444	
370		55.71	10.866	−82.68	18290	105.6	2391	1.163	1606	−.442	

380	55.96	10.627	−79.85	19353	106.9	2373	1.155	1432	−.438
390	56.20	10.398	−77.06	20429	108.3	2355	1.148	1282	−.434
400	56.43	10.180	−74.30	21518	109.6	2338	1.141	1150	−.431
410	56.66	9.9724	−71.58	22620	110.9	2321	1.134	1036	−.427
420	56.88	9.7735	−68.89	23735	112.2	2305	1.128	935.8	−.424
430	57.10	9.5831	−66.23	24863	113.5	2288	1.121	847.6	−.421
440	57.32	9.4005	−63.61	26004	114.7	2273	1.115	769.9	−.418
450	57.53	9.2254	−61.02	27158	116.0	2257	1.109	701.0	−.415
475	58.03	8.8168	−54.66	30097	119.1	2220	1.095	560.4	−.409

Table 2—continued

P/MPa									700
T_σ/K									
	V_σ	Z_σ	S_σ	H_σ	C_P	w_σ	γ_σ	$(f/P)_\sigma$	μ_σ
$\dfrac{T}{K}$ (Liq / Vap)	$\dfrac{V}{cm^3\,mol^{-1}}$	Z	$\dfrac{S}{J\,K^{-1}\,mol^{-1}}$	$\dfrac{H}{J\,mol^{-1}}$	$\dfrac{C_P}{J\,K^{-1}\,mol^{-1}}$	$\dfrac{w}{m\,sec^{-1}}$	γ	(f/P)	$\dfrac{\mu}{K\,MPa^{-1}}$
130	(48.76)	(31.577)	(−172.09)	(1951)	(55.9)	(3170)	(1.039)	(.1852E+08)	(−.849)
135	(48.81)	(30.441)	(−169.94)	(2237)	(58.4)	(3135)	(1.053)	(.1279E+08)	(−.809)
140	(48.87)	(29.391)	(−167.77)	(2535)	(60.8)	(3104)	(1.068)	(.9048E+07)	(−.774)
145	(48.94)	(28.418)	(−165.60)	(2844)	(63.0)	(3077)	(1.083)	(.6538E+07)	(−.743)
150	(49.02)	(27.514)	(−163.42)	(3165)	(65.1)	(3053)	(1.098)	(.4815E+07)	(−.716)
155	(49.10)	(26.672)	(−161.26)	(3495)	(67.0)	(3032)	(1.113)	(.3607E+07)	(−.692)
160	(49.19)	(25.886)	(−159.10)	(3835)	(68.9)	(3012)	(1.128)	(.2744E+07)	(−.671)
165	(49.29)	(25.151)	(−156.96)	(4183)	(70.6)	(2994)	(1.142)	(.2117E+07)	(−.652)
170	(49.39)	(24.461)	(−154.82)	(4540)	(72.2)	(2978)	(1.155)	(.1654E+07)	(−.635)
175	(49.50)	(23.813)	(−152.71)	(4905)	(73.7)	(2963)	(1.168)	(.1307E+07)	(−.620)
180	(49.61)	(23.204)	(−150.61)	(5277)	(75.2)	(2948)	(1.179)	(.1044E+07)	(−.606)
185	(49.72)	(22.628)	(−148.53)	(5657)	(76.5)	(2934)	(1.190)	(.8416E+06)	(−.594)
190	(49.84)	(22.085)	(−146.48)	(6042)	(77.8)	(2921)	(1.199)	(.6847E+06)	(−.583)
195	(49.96)	(21.570)	(−144.44)	(6434)	(79.0)	(2908)	(1.208)	(.5617E+06)	(−.572)
200	(50.08)	(21.083)	(−142.43)	(6832)	(80.2)	(2895)	(1.215)	(.4643E+06)	(−.563)

205	(50.21)	(20.620)	(−140.43)	(7236)	(81.3)	(2883)	(1.221)	(.3865E + 06)	(− .554)
210	(50.33)	(20.180)	(−138.46)	(7645)	(82.3)	(2871)	(1.226)	(.3239E + 06)	(− .546)
215	(50.46)	(19.760)	(−136.51)	(8059)	(83.3)	(2858)	(1.231)	(.2731E + 06)	(− .539)
220	(50.59)	(19.361)	(−134.58)	(8478)	(84.3)	(2846)	(1.234)	(.2316E + 06)	(− .532)
225	(50.72)	(18.979)	(−132.68)	(8902)	(85.3)	(2834)	(1.236)	(.1975E + 06)	(− .526)
230	(50.85)	(18.615)	(−130.80)	(9331)	(86.2)	(2822)	(1.238)	(.1692E + 06)	(− .520)
235	(50.98)	(18.266)	(−128.93)	(9764)	(87.1)	(2810)	(1.239)	(.1457E + 06)	(− .515)
240	(51.12)	(17.931)	(−127.09)	(10201)	(87.9)	(2799)	(1.239)	(.1260E + 06)	(− .509)
245	(51.25)	(17.611)	(−125.27)	(10643)	(88.8)	(2787)	(1.239)	(.1095E + 06)	(− .505)
250	(51.38)	(17.303)	(−123.47)	(11089)	(89.6)	(2775)	(1.238)	(.9549E + 05)	(− .500)
255	(51.51)	(17.007)	(−121.68)	(11539)	(90.4)	(2763)	(1.236)	(.8361E + 05)	(− .496)
260	(51.64)	(16.722)	(−119.92)	(11993)	(91.2)	(2752)	(1.234)	(.7347E + 05)	(− .491)
265	(51.77)	(16.448)	(−118.18)	(12451)	(92.0)	(2740)	(1.232)	(.6478E + 05)	(− .487)
270	(51.90)	(16.185)	(−116.45)	(12913)	(92.8)	(2729)	(1.230)	(.5731E + 05)	(− .484)
275	(52.03)	(15.930)	(−114.74)	(13379)	(93.5)	(2717)	(1.227)	(.5085E + 05)	(− .480)
280	(52.16)	(15.685)	(−113.05)	(13848)	(94.3)	(2706)	(1.224)	(.4526E + 05)	(− .477)
285	(52.29)	(15.447)	(−111.37)	(14322)	(95.1)	(2695)	(1.220)	(.4040E + 05)	(− .473)
290	(52.42)	(15.218)	(−109.71)	(14799)	(95.8)	(2684)	(1.217)	(.3616E + 05)	(− .470)
295	(52.55)	(14.996)	(−108.07)	(15280)	(96.5)	(2673)	(1.213)	(.3245E + 05)	(− .467)
300	52.67	14.782	− 106.44	15764	97.3	2662	1.210	.2920E + 05	− .464

Table 2—continued

P/MPa										
					700					
T_σ/K										
Liq										
Vap	V_σ	Z_σ	S_σ	H_σ	C_P	w_σ	γ_σ	$(f/P)_\sigma$	μ_σ	
$\dfrac{T}{K}$	$\dfrac{V}{\mathrm{cm^3\,mol^{-1}}}$	Z	$\dfrac{S}{\mathrm{J\,K^{-1}\,mol^{-1}}}$	$\dfrac{H}{\mathrm{J\,mol^{-1}}}$	$\dfrac{C_P}{\mathrm{J\,K^{-1}\,mol^{-1}}}$	$\dfrac{w}{\mathrm{m\,sec^{-1}}}$	γ	(f/P)	$\dfrac{\mu}{\mathrm{K\,MPa^{-1}}}$	
305	52.80	14.574	−104.83	16253	98.0	2651	1.206	.2633E+05	−.461	
310	52.92	14.373	−103.23	16745	98.8	2641	1.202	.2380E+05	−.458	
315	53.05	14.178	−101.64	17240	99.5	2630	1.198	.2156E+05	−.455	
320	53.17	13.988	−100.07	17739	100.2	2620	1.194	.1957E+05	−.453	
325	53.29	13.805	−98.51	18242	100.9	2610	1.190	.1780E+05	−.450	
330	53.41	13.626	−96.96	18749	101.6	2600	1.186	.1623E+05	−.448	
335	53.53	13.453	−95.43	19259	102.4	2589	1.182	.1482E+05	−.445	
340	53.65	13.284	−93.91	19772	103.1	2580	1.178	.1356E+05	−.443	
345	53.76	13.120	−92.40	20289	103.8	2570	1.174	.1242E+05	−.441	
350	53.88	12.961	−90.90	20810	104.5	2560	1.170	.1141E+05	−.438	
355	54.00	12.805	−89.41	21334	105.2	2551	1.166	.1049E+05	−.436	
360	54.11	12.654	−87.94	21862	105.9	2541	1.162	9658	−.434	
365	54.22	12.507	−86.47	22393	106.6	2532	1.158	8909	−.432	
370	54.33	12.363	−85.01	22928	107.3	2523	1.154	8229	−.430	
380	54.55	12.087	−82.14	24008	108.7	2504	1.147	7052	−.426	

390	54.77	11.823	−79.29	25102	110.1	2487	1.139	6076	−.422
400	54.98	11.572	−76.49	26209	111.4	2469	1.132	5261	−.419
410	55.18	11.332	−73.72	27330	112.8	2453	1.125	4577	−.415
420	55.39	11.102	−70.99	28464	114.1	2436	1.119	4000	−.412
430	55.58	10.883	−68.29	29612	115.4	2420	1.112	3511	−.409
440	55.77	10.672	−65.62	30773	116.8	2404	1.106	3094	−.406
450	55.96	10.470	−62.98	31948	118.1	2389	1.100	2736	−.403
475	56.41	9.9984	−56.51	34940	121.3	2351	1.086	2044	−.397

Table 2—continued

P/MPa	800								
T_σ/K									
	V_σ	Z_σ	S_σ	H_σ	C_P	w_σ	γ_σ	$(f/P)_\sigma$	μ_σ
Liq Vap T/K	$\dfrac{V}{\text{cm}^3\,\text{mol}^{-1}}$	Z	$\dfrac{S}{\text{J K}^{-1}\,\text{mol}^{-1}}$	$\dfrac{H}{\text{J mol}^{-1}}$	$\dfrac{C_P}{\text{J K}^{-1}\,\text{mol}^{-1}}$	$\dfrac{w}{\text{m sec}^{-1}}$	γ	(f/P)	$\dfrac{\mu}{\text{K MPa}^{-1}}$
135	(48.25)	(34.388)	(−170.88)	(6962)	(53.8)	(3291)	(1.027)	(.8441E+09)	(−.878)
140	(48.29)	(33.188)	(−168.88)	(7237)	(56.3)	(3256)	(1.039)	(.5139E+09)	(−.836)
145	(48.34)	(32.077)	(−166.86)	(7524)	(58.6)	(3226)	(1.053)	(.3232E+09)	(−.798)
150	(48.40)	(31.044)	(−164.83)	(7823)	(60.9)	(3199)	(1.067)	(.2092E+09)	(−.765)
155	(48.46)	(30.083)	(−162.80)	(8133)	(63.0)	(3176)	(1.082)	(.1390E+09)	(−.736)
160	(48.53)	(29.185)	(−160.77)	(8453)	(65.0)	(3154)	(1.096)	(.9449E+08)	(−.710)
165	(48.61)	(28.346)	(−158.74)	(8783)	(67.0)	(3135)	(1.110)	(.6563E+08)	(−.686)
170	(48.69)	(27.559)	(−156.71)	(9123)	(68.8)	(3118)	(1.124)	(.4646E+08)	(−.665)
175	(48.78)	(26.820)	(−154.69)	(9472)	(70.6)	(3101)	(1.137)	(.3347E+08)	(−.646)
180	(48.87)	(26.124)	(−152.68)	(9829)	(72.3)	(3086)	(1.150)	(.2450E+08)	(−.629)
185	(48.97)	(25.468)	(−150.68)	(10194)	(73.9)	(3071)	(1.161)	(.1820E+08)	(−.613)
190	(49.07)	(24.849)	(−148.68)	(10568)	(75.4)	(3057)	(1.172)	(.1370E+08)	(−.599)
195	(49.17)	(24.263)	(−146.71)	(10948)	(76.9)	(3044)	(1.182)	(.1045E+08)	(−.586)
200	(49.28)	(23.708)	(−144.74)	(11336)	(78.3)	(3030)	(1.190)	(.8055E+07)	(−.574)
205	(49.39)	(23.181)	(−142.79)	(11731)	(79.6)	(3017)	(1.198)	(.6277E+07)	(−.563)

210	(49.50)	(22.680)	(−140.86)	(12132)	(80.9)	(3005)	(1.204)	(.4940E + 07)	(− .553)
215	(49.61)	(22.204)	(−138.94)	(12540)	(82.1)	(2992)	(1.209)	(.3924E + 07)	(− .544)
220	(49.73)	(21.749)	(−137.04)	(12954)	(83.3)	(2979)	(1.214)	(.3144E + 07)	(− .536)
225	(49.84)	(21.316)	(−135.15)	(13373)	(84.5)	(2967)	(1.217)	(.2538E + 07)	(− .528)
230	(49.96)	(20.901)	(−133.28)	(13798)	(85.6)	(2954)	(1.220)	(.2065E + 07)	(− .521)
235	(50.08)	(20.505)	(−131.43)	(14229)	(86.6)	(2942)	(1.222)	(.1692E + 07)	(− .514)
240	(50.20)	(20.126)	(−129.60)	(14664)	(87.6)	(2929)	(1.223)	(.1396E + 07)	(− .507)
245	(50.32)	(19.762)	(−127.78)	(15105)	(88.6)	(2917)	(1.224)	(.1158E + 07)	(− .502)
250	(50.44)	(19.413)	(−125.98)	(15551)	(89.6)	(2905)	(1.224)	(.9670E + 06)	(− .496)
255	(50.56)	(19.077)	(−124.20)	(16001)	(90.6)	(2893)	(1.223)	(.8117E + 06)	(− .491)
260	(50.68)	(18.755)	(−122.43)	(16456)	(91.5)	(2881)	(1.222)	(.6850E + 06)	(− .486)
265	(50.80)	(18.444)	(−120.68)	(16916)	(92.4)	(2869)	(1.220)	(.5809E + 06)	(− .481)
270	(50.92)	(18.145)	(−118.94)	(17380)	(93.3)	(2857)	(1.218)	(.4950E + 06)	(− .477)
275	(51.04)	(17.857)	(−117.22)	(17849)	(94.1)	(2845)	(1.216)	(.4237E + 06)	(− .473)
280	(51.16)	(17.579)	(−115.52)	(18321)	(95.0)	(2833)	(1.213)	(.3641E + 06)	(− .469)
285	(51.27)	(17.310)	(−113.83)	(18799)	(95.8)	(2822)	(1.210)	(.3143E + 06)	(− .465)
290	(51.39)	(17.051)	(−112.16)	(19280)	(96.7)	(2810)	(1.207)	(.2723E + 06)	(− .461)
295	(51.51)	(16.800)	(−110.50)	(19765)	(97.5)	(2799)	(1.204)	(.2367E + 06)	(− .458)
300	51.62	16.557	−108.85	20255	98.3	2788	1.200	.2066E + 06	− .455
305	51.74	16.322	−107.22	20748	99.1	2777	1.197	.1809E + 06	− .451

Table 2—continued

P/MPa					800				
T_σ/K									
Liq Vap	V_σ	Z_σ	S_σ	H_σ	C_P	w_σ	γ_σ	$(f/P)_b$	μ_σ
$\dfrac{T}{K}$	$\dfrac{V}{cm^3\,mol^{-1}}$	Z	$\dfrac{S}{J\,K^{-1}\,mol^{-1}}$	$\dfrac{H}{J\,mol^{-1}}$	$\dfrac{C_P}{J\,K^{-1}\,mol^{-1}}$	$\dfrac{w}{m\,sec^{-1}}$	γ	(f/P)	$\dfrac{\mu}{K\,MPa^{-1}}$
310	51.85	16.094	−105.60	21246	99.9	2766	1.193	.1589E+06	−.448
315	51.97	15.874	−104.00	21747	100.7	2755	1.189	.1400E+06	−.445
320	52.08	15.659	−102.41	22252	101.5	2744	1.185	.1237E+06	−.443
325	52.19	15.452	−100.83	22762	102.2	2734	1.182	.1096E+06	−.440
330	52.30	15.250	−99.26	23275	103.0	2723	1.178	.9742E+05	−.437
335	52.41	15.054	−97.71	23792	103.8	2713	1.174	.8680E+05	−.435
340	52.52	14.863	−96.16	24312	104.5	2703	1.170	.7753E+05	−.432
345	52.63	14.678	−94.63	24837	105.3	2693	1.166	.6942E+05	−.430
350	52.73	14.497	−93.11	25365	106.0	2683	1.162	.6230E+05	−.427
355	52.84	14.322	−91.60	25897	106.8	2673	1.158	.5603E+05	−.425
360	52.94	14.151	−90.10	26432	107.5	2663	1.154	.5051E+05	−.423
365	53.05	13.984	−88.62	26972	108.2	2654	1.150	.4562E+05	−.421
370	53.15	13.822	−87.14	27515	109.0	2644	1.147	.4129E+05	−.419
380	53.35	13.509	−84.21	28611	110.4	2626	1.139	.3402E+05	−.415
390	53.55	13.211	−81.33	29723	111.8	2608	1.132	.2823E+05	−.411

400	53.74	12.927	−78.48	30848	113.3	2591	1.125	.2359E+05	−.407
410	53.93	12.656	−75.66	31988	114.7	2573	1.118	.1984E+05	−.404
420	54.11	12.397	−72.88	33142	116.1	2557	1.111	.1678E+05	−.401
430	54.29	12.148	−70.14	34309	117.4	2540	1.105	.1427E+05	−.398
440	54.46	11.910	−67.42	35490	118.8	2524	1.099	.1221E+05	−.395
450	54.63	11.682	−64.74	36685	120.2	2509	1.093	.1049E+05	−.392
475	55.04	11.149	−58.15	39731	123.5	2471	1.079	7331	−.386

Table 2—continued

				P/MPa	900				
				T_σ/K					
	V_σ	Z_σ	S_σ	H_σ	C_P	w_σ	γ_σ	$(f/P)_\sigma$	μ_σ
Liq Vap	$\dfrac{V}{\text{cm}^3\,\text{mol}^{-1}}$	Z	$\dfrac{S}{\text{J K}^{-1}\,\text{mol}^{-1}}$	$\dfrac{H}{\text{J mol}^{-1}}$	$\dfrac{C_P}{\text{J K}^{-1}\,\text{mol}^{-1}}$	$\dfrac{w}{\text{m sec}^{-1}}$	γ	(f/P)	$\dfrac{\mu}{\text{K MPa}^{-1}}$
$\dfrac{T}{\text{K}}$									
140	(47.78)	(36.944)	(−169.61)	(11937)	(51.7)	(3410)	(1.019)	(.2829E + 11)	(−.909)
145	(47.81)	(35.694)	(−167.75)	(12202)	(54.1)	(3375)	(1.030)	(.1549E + 11)	(−.864)
150	(47.85)	(34.533)	(−165.88)	(12479)	(56.5)	(3345)	(1.042)	(.8810E + 10)	(−.823)
155	(47.90)	(33.452)	(−163.98)	(12767)	(58.8)	(3318)	(1.055)	(.5190E + 10)	(−.787)
160	(47.96)	(32.444)	(−162.08)	(13067)	(61.0)	(3295)	(1.069)	(.3155E + 10)	(−.755)
165	(48.02)	(31.500)	(−160.17)	(13378)	(63.2)	(3274)	(1.083)	(.1973E + 10)	(−.727)
170	(48.08)	(30.616)	(−158.25)	(13699)	(65.2)	(3255)	(1.096)	(.1266E + 10)	(−.701)
175	(48.15)	(29.786)	(−156.33)	(14030)	(67.2)	(3237)	(1.110)	(.8316E + 09)	(−.678)
180	(48.23)	(29.005)	(−154.41)	(14371)	(69.1)	(3221)	(1.123)	(.5581E + 09)	(−.656)
185	(48.31)	(28.268)	(−152.49)	(14721)	(71.0)	(3205)	(1.135)	(.3820E + 09)	(−.637)
190	(48.40)	(27.573)	(−150.58)	(15080)	(72.7)	(3190)	(1.146)	(.2662E + 09)	(−.620)
195	(48.49)	(26.916)	(−148.67)	(15448)	(74.4)	(3176)	(1.156)	(.1886E + 09)	(−.604)
200	(48.58)	(26.293)	(−146.76)	(15824)	(76.1)	(3162)	(1.166)	(.1357E + 09)	(−.589)
205	(48.68)	(25.702)	(−144.86)	(16209)	(77.6)	(3148)	(1.174)	(.9903E + 08)	(−.576)
210	(48.77)	(25.141)	(−142.97)	(16601)	(79.1)	(3135)	(1.182)	(.7322E + 08)	(−.564)

215	(48.87)	(24.607)	(− 141.10)	(17000)	(80.6)	(3121)	(1.188)	(.5480E + 08) (− .552)
220	(48.98)	(24.098)	(− 139.23)	(17407)	(82.0)	(3108)	(1.194)	(.4148E + 08) (− .542)
225	(49.08)	(23.612)	(− 137.37)	(17820)	(83.3)	(3095)	(1.198)	(.3173E + 08) (− .532)
230	(49.19)	(23.149)	(− 135.52)	(18240)	(84.6)	(3081)	(1.202)	(.2452E + 08) (− .524)
235	(49.29)	(22.705)	(− 133.69)	(18666)	(85.9)	(3068)	(1.205)	(.1912E + 08) (− .515)
240	(49.40)	(22.281)	(− 131.87)	(19098)	(87.1)	(3055)	(1.207)	(.1504E + 08) (− .508)
245	(49.51)	(21.874)	(− 130.06)	(19537)	(88.2)	(3043)	(1.208)	(.1193E + 08) (− .501)
250	(49.62)	(21.484)	(− 128.27)	(19981)	(89.4)	(3030)	(1.209)	(.9536E + 07) (− .494)
255	(49.73)	(21.109)	(− 126.49)	(20430)	(90.5)	(3017)	(1.209)	(.7677E + 07) (− .488)
260	(49.84)	(20.748)	(− 124.72)	(20885)	(91.5)	(3004)	(1.209)	(.6223E + 07) (− .482)
265	(49.95)	(20.402)	(− 122.97)	(21345)	(92.6)	(2992)	(1.208)	(.5078E + 07) (− .477)
270	(50.06)	(20.068)	(− 121.23)	(21811)	(93.6)	(2979)	(1.206)	(.4168E + 07) (− .472)
275	(50.16)	(19.746)	(− 119.50)	(22281)	(94.5)	(2967)	(1.205)	(.3442E + 07) (− .467)
280	(50.27)	(19.436)	(− 117.79)	(22756)	(95.5)	(2955)	(1.202)	(.2858E + 07) (− .463)
285	(50.38)	(19.136)	(− 116.09)	(23236)	(96.4)	(2942)	(1.200)	(.2385E + 07) (− .458)
290	(50.49)	(18.846)	(− 114.41)	(23720)	(97.4)	(2931)	(1.197)	(.2001E + 07) (− .454)
295	(50.60)	(18.567)	(− 112.73)	(24210)	(98.3)	(2919)	(1.194)	(.1686E + 07) (− .450)
300	50.71	18.296	− 111.07	24703	99.2	2907	1.191	.1428E + 07 − .447
305	50.81	18.034	− 109.43	25201	100.0	2895	1.188	.1214E + 07 − .443
310	50.92	17.780	− 107.79	25703	100.9	2884	1.185	.1037E + 07 − .440

Table 2—continued

P/MPa				900					
T_σ/K									
Liq Vap	V_σ	Z_σ	S_σ	H_σ	C_P	w_σ	γ_σ	$(f/P)_\sigma$	μ_σ
$\dfrac{T}{\text{K}}$	$\dfrac{V}{\text{cm}^3\,\text{mol}^{-1}}$	Z	$\dfrac{S}{\text{J K}^{-1}\text{mol}^{-1}}$	$\dfrac{H}{\text{J mol}^{-1}}$	$\dfrac{C_P}{\text{J K}^{-1}\text{mol}^{-1}}$	$\dfrac{w}{\text{m sec}^{-1}}$	γ	(f/P)	$\dfrac{\mu}{\text{K MPa}^{-1}}$
315	51.02	17.534	−106.17	26210	101.8	2873	1.181	.8886E+06	−.437
320	51.13	17.295	−104.56	26721	102.6	2862	1.177	.7646E+06	−.434
325	51.23	17.063	−102.97	27236	103.4	2851	1.174	.6603E+06	−.431
330	51.33	16.838	−101.38	27755	104.3	2840	1.170	.5722E+06	−.428
335	51.44	16.620	−99.81	28279	105.1	2829	1.166	.4975E+06	−.425
340	51.54	16.408	−98.25	28806	105.9	2819	1.162	.4340E+06	−.422
345	51.64	16.201	−96.69	29337	106.7	2809	1.159	.3797E+06	−.420
350	51.73	16.000	−95.15	29873	107.5	2798	1.155	.3332E+06	−.417
355	51.83	15.805	−93.62	30412	108.2	2788	1.151	.2932E+06	−.415
360	51.93	15.614	−92.10	30955	109.0	2778	1.147	.2587E+06	−.413
365	52.02	15.429	−90.59	31502	109.8	2769	1.144	.2289E+06	−.411
370	52.12	15.248	−89.10	32053	110.6	2759	1.140	.2031E+06	−.408
380	52.31	14.900	−86.13	33166	112.1	2740	1.132	.1609E+06	−.404
390	52.49	14.568	−83.20	34295	113.6	2722	1.125	.1287E+06	−.400
400	52.66	14.252	−80.30	35438	115.1	2704	1.118	.1038E+06	−.397

410	52.84	13.950	−77.44	36596	116.5	2686	1.111	.8439E + 05	−.393
420	53.01	13.661	−74.62	37769	118.0	2669	1.105	.6912E + 05	−.390
430	53.17	13.385	−71.82	38956	119.4	2653	1.098	.5701E + 05	−.387
440	53.33	13.120	−69.06	40157	120.8	2636	1.092	.4733E + 05	−.384
450	53.48	12.865	−66.33	41373	122.3	2620	1.086	.3954E + 05	−.381
475	53.85	12.272	−59.63	44472	125.7	2582	1.072	.2585E + 05	−.375

Table 2—continued

P/MPa: 1000

T_σ/K

Liq
Vap

$\frac{T}{K}$	$\frac{V}{cm^3\,mol^{-1}}$ V_σ	Z Z_σ	$\frac{S}{J\,K^{-1}\,mol^{-1}}$ S_σ	$\frac{H}{J\,mol^{-1}}$ H_σ	$\frac{C_P}{J\,K^{-1}\,mol^{-1}}$ C_P	$\frac{w}{m\,sec^{-1}}$ w_σ	γ γ_σ	(f/P) $(f/P)_\sigma$	$\frac{\mu}{K\,MPa^{-1}}$ μ_σ
145	(47.35)	(39.277)	(−168.33)	(16876)	(49.6)	(3527)	(1.013)	(.7212E+12)	(−.941)
150	(47.38)	(37.989)	(−166.60)	(17131)	(52.1)	(3491)	(1.022)	(.3607E+12)	(−.892)
155	(47.41)	(36.789)	(−164.85)	(17398)	(54.6)	(3461)	(1.034)	(.1884E+12)	(−.848)
160	(47.45)	(35.669)	(−163.08)	(17676)	(56.9)	(3434)	(1.046)	(.1024E+12)	(−.809)
165	(47.50)	(34.622)	(−161.30)	(17967)	(59.2)	(3411)	(1.059)	(.5767E+11)	(−.774)
170	(47.55)	(33.640)	(−159.49)	(18269)	(61.5)	(3389)	(1.072)	(.3355E+11)	(−.743)
175	(47.61)	(32.719)	(−157.68)	(18582)	(63.7)	(3370)	(1.085)	(.2010E+11)	(−.715)
180	(47.67)	(31.852)	(−155.86)	(18905)	(65.8)	(3352)	(1.098)	(.1237E+11)	(−.689)
185	(47.74)	(31.035)	(−154.03)	(19239)	(67.8)	(3336)	(1.110)	(.7800E+10)	(−.666)
190	(47.81)	(30.264)	(−152.19)	(19583)	(69.8)	(3320)	(1.122)	(.5032E+10)	(−.645)
195	(47.89)	(29.535)	(−150.35)	(19937)	(71.7)	(3304)	(1.133)	(.3315E+10)	(−.625)
200	(47.97)	(28.845)	(−148.51)	(20300)	(73.6)	(3290)	(1.143)	(.2225E+10)	(−.608)
205	(48.05)	(28.190)	(−146.68)	(20673)	(75.4)	(3275)	(1.152)	(.1521E+10)	(−.592)
210	(48.13)	(27.568)	(−144.84)	(21054)	(77.1)	(3261)	(1.160)	(.1057E+10)	(−.577)
215	(48.22)	(26.976)	(−143.00)	(21444)	(78.8)	(3247)	(1.168)	(.7453E+09)	(−.563)

220	(48.31)	(26.413)	(−141.18)	(21842)	(80.4)	(3233)	(1.174)	(.5332E+09)	(−.551)
225	(48.41)	(25.876)	(−139.35)	(22247)	(81.9)	(3219)	(1.180)	(.3865E+09)	(−.539)
230	(48.50)	(25.362)	(−137.54)	(22661)	(83.4)	(3205)	(1.184)	(.2837E+09)	(−.529)
235	(48.60)	(24.872)	(−135.73)	(23081)	(84.8)	(3191)	(1.188)	(.2106E+09)	(−.519)
240	(48.69)	(24.403)	(−133.93)	(23509)	(86.2)	(3178)	(1.191)	(.1581E+09)	(−.510)
245	(48.79)	(23.953)	(−132.13)	(23943)	(87.6)	(3164)	(1.193)	(.1198E+09)	(−.502)
250	(48.89)	(23.521)	(−130.35)	(24384)	(88.8)	(3151)	(1.194)	(.9174E+08)	(−.494)
255	(48.99)	(23.107)	(−128.58)	(24832)	(90.1)	(3137)	(1.195)	(.7085E+08)	(−.487)
260	(49.09)	(22.709)	(−126.82)	(25285)	(91.3)	(3124)	(1.196)	(.5519E+08)	(−.480)
265	(49.19)	(22.326)	(−125.07)	(25745)	(92.5)	(3111)	(1.195)	(.4333E+08)	(−.474)
270	(49.29)	(21.958)	(−123.33)	(26210)	(93.6)	(3097)	(1.195)	(.3428E+08)	(−.468)
275	(49.39)	(21.603)	(−121.60)	(26681)	(94.7)	(3085)	(1.193)	(.2731E+08)	(−.463)
280	(49.49)	(21.260)	(−119.88)	(27157)	(95.8)	(3072)	(1.192)	(.2191E+08)	(−.458)
285	(49.59)	(20.929)	(−118.18)	(27639)	(96.9)	(3059)	(1.190)	(.1769E+08)	(−.453)
290	(49.69)	(20.610)	(−116.49)	(28126)	(97.9)	(3046)	(1.188)	(.1437E+08)	(−.448)
295	(49.79)	(20.302)	(−114.80)	(28618)	(98.9)	(3034)	(1.185)	(.1174E+08)	(−.444)
300	49.89	20.003	−113.13	29115	99.9	3022	1.182	.9650E+07	−.440
305	49.99	19.714	−111.48	29616	100.8	3010	1.179	.7971E+07	−.436
310	50.09	19.434	−109.83	30123	101.8	2998	1.176	.6618E+07	−.432
315	50.19	19.163	−108.19	30634	102.7	2986	1.173	.5520E+07	−.429

Table 2—continued

		1000							
P/MPa									
T_σ/K		V_σ	Z_σ	S_σ	H_σ	C_P	w_σ	γ_σ	$(f/P)_b$	μ_σ

$\dfrac{T}{K}$	$\dfrac{V}{cm^3\,mol^{-1}}$	Z	$\dfrac{S}{J\,K^{-1}\,mol^{-1}}$	$\dfrac{H}{J\,mol^{-1}}$	$\dfrac{C_P}{J\,K^{-1}\,mol^{-1}}$	$\dfrac{w}{m\,sec^{-1}}$	γ	(f/P)	$\dfrac{\mu}{K\,MPa^{-1}}$
320	50.29	18.900	−106.57	31150	103.6	2974	1.170	.4626E+07	−.426
325	50.38	18.645	−104.95	31670	104.5	2963	1.166	.3894E+07	−.422
330	50.48	18.397	−103.35	32195	105.4	2952	1.163	.3292E+07	−.419
335	50.57	18.157	−101.76	32724	106.3	2941	1.159	.2794E+07	−.416
340	50.67	17.923	−100.18	33258	107.1	2930	1.156	.2381E+07	−.414
345	50.76	17.695	−98.61	33795	108.0	2919	1.152	.2036E+07	−.411
350	50.85	17.474	−97.05	34337	108.8	2908	1.148	.1747E+07	−.408
355	50.94	17.259	−95.50	34884	109.7	2898	1.145	.1505E+07	−.406
360	51.03	17.049	−93.96	35434	110.5	2888	1.141	.1300E+07	−.404
365	51.12	16.845	−92.43	35989	111.3	2878	1.137	.1127E+07	−.401
370	51.21	16.646	−90.91	36547	112.1	2868	1.134	.9797E+06	−.399
380	51.38	16.263	−87.90	37676	113.7	2848	1.126	.7471E+06	−.395
390	51.55	15.898	−84.93	38821	115.3	2829	1.119	.5759E+06	−.391
400	51.71	15.550	−81.99	39982	116.8	2811	1.112	.4486E+06	−.387
410	51.88	15.218	−79.08	41158	118.4	2793	1.106	.3527E+06	−.384

420	52.03	14.900	−76.21	42349	119.9	2775	1.099	.2798E+06	−.380
430	52.18	14.596	−73.38	43555	121.4	2758	1.093	.2238E+06	−.377
440	52.33	14.304	−70.57	44776	122.8	2742	1.087	.1805E+06	−.374
450	52.47	14.024	−67.79	46012	124.3	2725	1.081	.1466E+06	−.371
475	52.81	13.371	−60.97	49165	127.9	2686	1.067	.8976E+05	−.365

Table 3

THE VARIATION OF PRESSURE, MOLAR ENTROPY, MOLAR INTERNAL ENERGY AND MOLAR ISOCHORIC HEAT CAPACITY WITH DENSITY AND TEMPERATURE IN THE SINGLE-PHASE REGION

Notes:

1. Expressions such as $1.0\,E-9$ are to be read as 1.0×10^{-9}.
2. Interpolation into the region bounded by the isochores $5\ \text{mol dm}^{-3}$ and $6\ \text{mol dm}^{-3}$ and the saturation curve and the 370 K isotherm can only be approximate.
3. Numbers in parentheses are interpolations into regions unsupported by experiment.
4. In the units used in this table,

$$\text{on p. 346--9,}\quad \frac{P}{\text{mPa}} \div \frac{\rho}{\text{mol dm}^{-3}} = 10^{-6}\,\frac{P/\rho}{\text{J mol}^{-1}},$$

$$\text{on p. 350--3,}\quad \frac{P}{\text{Pa}} \div \frac{\rho}{\text{mol dm}^{-3}} = 10^{-3}\,\frac{P/\rho}{\text{J mol}^{-1}},$$

$$\text{on p. 354--7,}\quad \frac{P}{\text{kPa}} \div \frac{\rho}{\text{mol dm}^{-3}} = \frac{P/\rho}{\text{J mol}^{-1}},$$

on the remaining pages.

$$\frac{P}{\text{MPa}} \div \frac{\rho}{\text{mol dm}^{-3}} = 10^{3}\,\frac{P/\rho}{\text{J mol}^{-1}}.$$

5. In this table the molar entropy is given a value of zero at 298.15 K and 1 atm (0.101 325 MPa) in the ideal gas state.

$$S^{\text{id}}(298.15\ \text{K},\ 1\ \text{atm}) - S^{\text{ic}}(0\ \text{K}) = 266.62\ \text{J K}^{-1}\ \text{mol}^{-1}.$$

6. In this table the molar internal energy has a value of $-2479\ \text{J mol}^{-1}$ at 298.15 K in the ideal gas state.

$$U^{\text{id}}(298.15\ \text{K}) - U^{\text{id}}(0\ \text{K}) = 11\,067\ \text{J mol}^{-1}.$$

$$U^{\text{id}}(298.15\ \text{K}) - U^{\text{ic}}(0\ \text{K}) = 38\,224\ \text{J mol}^{-1}.$$

Table 3. DENSITY-TEMPERATURE CO-ORDINATES

ρ/mol dm^{-3}	1.0 E-9				10.0 E-9				100.0 E-9			
T_σ/K					93.961				102.068			
	P_σ	S_σ	U_σ	C_V	P_σ 7.8122	S_σ 80.79	U_σ −11056	C_V 30.0	P_σ 84.862	S_σ 64.18	U_σ −10808	C_V 31.0
$\dfrac{T}{K}$	$\dfrac{P}{\text{mPa}}$	$\dfrac{S}{\text{J K}^{-1}\text{mol}^{-1}}$	$\dfrac{U}{\text{J mol}^{-1}}$	$\dfrac{C_V}{\text{J K}^{-1}\text{mol}^{-1}}$	$\dfrac{P}{\text{mPa}}$	$\dfrac{S}{\text{J K}^{-1}\text{mol}^{-1}}$	$\dfrac{U}{\text{J mol}^{-1}}$	$\dfrac{C_V}{\text{J K}^{-1}\text{mol}^{-1}}$	$\dfrac{P}{\text{mPa}}$	$\dfrac{S}{\text{J K}^{-1}\text{mol}^{-1}}$	$\dfrac{U}{\text{J mol}^{-1}}$	$\dfrac{C_V}{\text{J K}^{-1}\text{mol}^{-1}}$
90	.74829	98.66	−11173	29.5								
95	.78986	100.27	−11024	30.1	7.8986	81.13	−11024	30.1				
100	.83143	101.83	−10872	30.8	8.3143	82.69	−10872	30.8				
105	.87301	103.35	−10717	31.4	8.7301	84.20	−10717	31.4	87.301	65.06	−10717	31.4
110	.91458	104.82	−10558	31.9	9.1458	85.68	−10558	31.9	91.458	66.53	−10558	31.9
115	.95615	106.25	−10397	32.5	9.5615	87.11	−10397	32.5	95.615	67.96	−10397	32.5
120	.99772	107.65	−10233	33.0	9.9772	88.50	−10233	33.0	99.772	69.36	−10233	33.0
125	1.0393	109.01	−10067	33.6	10.393	89.86	−10067	33.6	103.93	70.72	−10067	33.6
130	1.0809	110.33	−9898	34.1	10.809	91.19	−9898	34.1	108.09	72.04	−9898	34.1
135	1.1224	111.63	−9726	34.6	11.224	92.48	−9726	34.6	112.24	73.34	−9726	34.6
140	1.1640	112.89	−9552	35.1	11.640	93.75	−9552	35.1	116.40	74.61	−9552	35.1
145	1.2056	114.13	−9376	35.6	12.056	94.99	−9376	35.6	120.56	75.85	−9376	35.6
150	1.2472	115.35	−9196	36.1	12.472	96.20	−9196	36.1	124.72	77.06	−9196	36.1
155	1.2887	116.54	−9015	36.6	12.887	97.40	−9015	36.6	128.87	78.25	−9015	36.6
160	1.3303	117.71	−8831	37.1	13.303	98.57	−8831	37.1	133.03	79.42	−8831	37.1

165	1.3719	118.86	−8644	37.6	13.719	99.72	−8644	37.6	137.19	80.57	−8644	37.6
170	1.4134	119.99	−8454	38.2	14.134	100.85	−8454	38.2	141.34	81.70	−8454	38.2
175	1.4550	121.11	−8262	38.8	14.550	101.96	−8262	38.8	145.50	82.82	−8262	38.8
180	1.4966	122.21	−8066	39.3	14.966	103.06	−8066	39.3	149.66	83.92	−8066	39.3
185	1.5382	123.29	−7868	39.9	15.382	104.15	−7868	39.9	153.82	85.00	−7868	39.9
190	1.5797	124.36	−7667	40.5	15.797	105.22	−7667	40.5	157.97	86.08	−7667	40.5
195	1.6213	125.43	−7463	41.1	16.213	106.28	−7463	41.1	162.13	87.14	−7463	41.1
200	1.6629	126.47	−7256	41.8	16.629	107.33	−7256	41.8	166.29	88.19	−7256	41.8
205	1.7044	127.51	−7046	42.4	17.044	108.37	−7046	42.4	170.44	89.22	−7046	42.4
210	1.7460	128.54	−6832	43.0	17.460	109.40	−6832	43.0	174.60	90.25	−6832	43.0
215	1.7876	129.56	−6615	43.7	17.876	110.42	−6615	43.7	178.76	91.27	−6615	43.7
220	1.8292	130.58	−6395	44.4	18.292	111.43	−6395	44.4	182.92	92.29	−6395	44.4
225	1.8707	131.58	−6171	45.1	18.707	112.44	−6171	45.1	187.07	93.29	−6171	45.1
230	1.9123	132.58	−5944	45.8	19.123	113.44	−5944	45.8	191.23	94.29	−5944	45.8
235	1.9539	133.57	−5713	46.5	19.539	114.43	−5713	46.5	195.39	95.28	−5713	46.5
240	1.9954	134.56	−5479	47.2	19.954	115.41	−5479	47.2	199.54	96.27	−5479	47.2
245	2.0370	135.54	−5241	47.9	20.370	116.39	−5241	47.9	203.70	97.25	−5241	47.9
250	2.0786	136.52	−5000	48.7	20.786	117.37	−5000	48.7	207.86	98.23	−5000	48.7
255	2.1202	137.49	−4754	49.4	21.202	118.34	−4754	49.4	212.02	99.20	−4754	49.4
260	2.1617	138.45	−4505	50.2	21.617	119.31	−4505	50.2	216.17	100.17	−4505	50.2
265	2.2033	139.42	−4253	50.9	22.033	120.27	−4253	50.9	220.33	101.13	−4253	50.9
270	2.2449	140.38	−3996	51.7	22.449	121.23	−3996	51.7	224.49	102.09	−3996	51.7
275	2.2864	141.33	−3735	52.5	22.864	122.19	−3735	52.5	228.64	103.04	−3735	52.5
280	2.3280	142.28	−3471	53.3	23.280	123.14	−3471	53.3	232.80	104.00	−3471	53.3
285	2.3696	143.23	−3203	54.0	23.696	124.09	−3203	54.0	236.96	104.95	−3203	54.0

Table 3—continued

$\dfrac{T}{K}$	1.0 E-9 $\dfrac{P}{\text{mPa}}$	$\dfrac{S}{\text{J K}^{-1}\text{mol}^{-1}}$	$\dfrac{U}{\text{J mol}^{-1}}$	$\dfrac{C_v}{\text{J K}^{-1}\text{mol}^{-1}}$	10.0 E-9 (93.961) $P_σ$ 7.8122 $\dfrac{P}{\text{mPa}}$	$S_σ$ 80.79 $\dfrac{S}{\text{J K}^{-1}\text{mol}^{-1}}$	$U_σ$ −11056 $\dfrac{U}{\text{J mol}^{-1}}$	C_v 30.0 $\dfrac{C_v}{\text{J K}^{-1}\text{mol}^{-1}}$	100.0 E-9 (102.068) $P_σ$ 84.862 $\dfrac{P}{\text{mPa}}$	$S_σ$ 64.18 $\dfrac{S}{\text{J K}^{-1}\text{mol}^{-1}}$	$U_σ$ −10808 $\dfrac{U}{\text{J mol}^{-1}}$	C_v 31.0 $\dfrac{C_v}{\text{J K}^{-1}\text{mol}^{-1}}$
290	2.4112	144.18	−2931	54.8	24.112	125.04	−2931	54.8	241.12	105.89	−2931	54.8
295	2.4527	145.12	−2655	55.6	24.527	125.98	−2655	55.6	245.27	106.84	−2655	55.6
300	2.4943	146.07	−2375	56.4	24.943	126.92	−2375	56.4	249.43	107.78	−2375	56.4
305	2.5359	147.00	−2091	57.2	25.359	127.86	−2091	57.2	253.59	108.71	−2091	57.2
310	2.5774	147.94	−1803	58.0	25.774	128.80	−1803	58.0	257.74	109.65	−1803	58.0
315	2.6190	148.87	−1511	58.8	26.190	129.73	−1511	58.8	261.90	110.58	−1511	58.8
320	2.6606	149.80	−1216	59.5	26.606	130.66	−1216	59.5	266.06	111.52	−1216	59.5
325	2.7022	150.73	−916	60.3	27.022	131.59	−916	60.3	270.22	112.45	−916	60.3
330	2.7437	151.66	−612	61.1	27.437	132.52	−612	61.1	274.37	113.37	−612	61.1
335	2.7853	152.59	−304	61.9	27.853	133.44	−304	61.9	278.53	114.30	−304	61.9
340	2.8269	153.51	7	62.7	28.269	134.37	7	62.7	282.69	115.22	7	62.7
345	2.8684	154.43	323	63.5	28.684	135.29	323	63.5	286.84	116.14	323	63.5
350	2.9100	155.35	642	64.3	29.100	136.21	642	64.3	291.00	117.06	642	64.3
355	2.9516	156.27	966	65.1	29.516	137.12	966	65.1	295.16	117.98	966	65.1
360	2.9932	157.19	1293	65.9	29.932	138.04	1293	65.9	299.32	118.90	1293	65.9
365	3.0347	158.10	1625	66.7	30.347	138.96	1625	66.7	303.47	119.81	1625	66.7
370	3.0763	159.01	1960	67.5	30.763	139.87	1960	67.5	307.63	120.72	1960	67.5
380	3.1594	160.83	2643	69.0	31.594	141.69	2643	69.0	315.94	122.54	2643	69.0
390	3.2426	162.65	3341	70.6	32.426	143.50	3341	70.6	324.26	124.36	3341	70.6
400	3.3257	164.45	4054	72.1	33.257	145.31	4054	72.1	332.57	126.16	4054	72.1

$ρ/\text{mol dm}^{-3}$

410	3.4089	166.25	4783	73.7	34.089	147.11	4783	73.7	340.89	127.96	4783	73.7
420	3.4920	168.04	5527	75.2	34.920	148.90	5527	75.2	349.20	129.76	5527	75.2
430	3.5752	169.83	6287	76.7	35.752	150.69	6287	76.7	357.52	131.54	6287	76.7
440	3.6583	171.61	7061	78.2	36.583	152.47	7061	78.2	365.83	133.32	7061	78.2
450	3.7415	173.38	7850	79.6	37.415	154.24	7850	79.6	374.15	135.10	7850	79.6
475	3.9493	177.79	9886	83.2	39.493	158.64	9886	83.2	394.93	139.50	9886	83.2
500	4.1572	182.15	12011	86.8	41.572	163.00	12011	86.8	415.72	143.86	12011	86.8
525	4.3650	186.46	14223	90.1	43.650	167.32	14223	90.1	436.50	148.17	14223	90.1
550	4.5729	190.73	16518	93.4	45.729	171.59	16518	93.4	457.29	152.44	16518	93.4
575	4.7807	194.96	18893	96.6	47.807	175.81	18893	96.6	478.07	156.67	18893	96.6

Table 3—continued

$\rho/\text{mol dm}^{-3}$	1.0 E-6				10.0 E-6				100.0 E-6			
T/K	111.893				124.095				139.775			
	P_σ 0.93032	S_σ 47.93	U_σ −10498	C_v 32.2	P_σ 10.317	S_σ 32.19	U_σ −10097	C_v 33.5	P_σ 116.19	S_σ 17.11	U_σ −9561	C_v 35.1
$\dfrac{T}{\text{K}}$	$\dfrac{P}{\text{Pa}}$	$\dfrac{S}{\text{J K}^{-1}\text{mol}^{-1}}$	$\dfrac{U}{\text{J mol}^{-1}}$	$\dfrac{C_v}{\text{J K}^{-1}\text{mol}^{-1}}$	$\dfrac{P}{\text{Pa}}$	$\dfrac{S}{\text{J K}^{-1}\text{mol}^{-1}}$	$\dfrac{U}{\text{J mol}^{-1}}$	$\dfrac{C_v}{\text{J K}^{-1}\text{mol}^{-1}}$	$\dfrac{P}{\text{Pa}}$	$\dfrac{S}{\text{J K}^{-1}\text{mol}^{-1}}$	$\dfrac{U}{\text{J mol}^{-1}}$	$\dfrac{C_v}{\text{J K}^{-1}\text{mol}^{-1}}$
115	.95615	48.82	−10397	32.5								
120	.99772	50.21	−10233	33.0								
125	1.0393	51.57	−10067	33.6	10.393	32.43	−10067	33.6				
130	1.0809	52.90	−9898	34.1	10.808	33.76	−9898	34.1				
135	1.1224	54.19	−9726	34.6	11.224	35.05	−9726	34.6				
140	1.1640	55.46	−9552	35.1	11.640	36.32	−9552	35.1	116.38	17.17	−9553	35.1
145	1.2056	56.70	−9376	35.6	12.056	37.56	−9376	35.6	120.54	18.41	−9376	35.6
150	1.2471	57.92	−9196	36.1	12.471	38.77	−9197	36.1	124.70	19.62	−9197	36.1
155	1.2887	59.11	−9015	36.6	12.887	39.96	−9015	36.6	128.85	20.82	−9015	36.6
160	1.3303	60.28	−8831	37.1	13.303	41.13	−8831	37.1	133.01	21.99	−8831	37.1
165	1.3719	61.43	−8644	37.6	13.718	42.28	−8644	37.6	137.17	23.14	−8644	37.7
170	1.4134	62.56	−8454	38.2	14.134	43.41	−8454	38.2	141.33	24.27	−8454	38.2
175	1.4550	63.67	−8262	38.8	14.550	44.53	−8262	38.8	145.49	25.38	−8262	38.8
180	1.4966	64.77	−8066	39.3	14.966	45.63	−8066	39.3	149.64	26.48	−8067	39.3
185	1.5382	65.86	−7868	39.9	15.381	46.71	−7868	39.9	153.80	27.57	−7869	39.9

190	1.5797	66.93	−7667	40.5	15.797	47.79	−7667	157.96	28.64	−7668	40.5
195	1.6213	67.99	−7463	41.1	16.213	48.85	−7463	162.12	29.70	−7464	41.1
200	1.6629	69.04	−7256	41.8	16.629	49.90	−7256	166.27	30.75	−7256	41.8
205	1.7044	70.08	−7046	42.4	17.044	50.93	−7046	170.43	31.79	−7046	42.4
210	1.7460	71.11	−6832	43.0	17.460	51.96	−6832	174.59	32.82	−6832	43.0
215	1.7876	72.13	−6615	43.7	17.876	52.99	−6615	178.75	33.84	−6615	43.7
220	1.8292	73.14	−6395	44.4	18.291	54.00	−6395	182.90	34.85	−6395	44.4
225	1.8707	74.15	−6171	45.1	18.707	55.00	−6171	187.06	35.86	−6172	45.1
230	1.9123	75.15	−5944	45.8	19.123	56.00	−5944	191.22	36.86	−5944	45.8
235	1.9539	76.14	−5713	46.5	19.539	56.99	−5713	195.38	37.85	−5714	46.5
240	1.9954	77.12	−5479	47.2	19.954	57.98	−5479	199.53	38.84	−5479	47.2
245	2.0370	78.11	−5241	47.9	20.370	58.96	−5241	203.69	39.82	−5241	47.9
250	2.0786	79.08	−5000	48.7	20.786	59.94	−5000	207.85	40.79	−5000	48.7
255	2.1202	80.05	−4754	49.4	21.201	60.91	−4754	212.01	41.76	−4755	49.4
260	2.1617	81.02	−4505	50.2	21.617	61.88	−4505	216.16	42.73	−4506	50.2
265	2.2033	81.98	−4253	50.9	22.033	62.84	−4253	220.32	43.69	−4253	50.9
270	2.2449	82.94	−3996	51.7	22.449	63.80	−3996	224.48	44.65	−3996	51.7
275	2.2864	83.90	−3735	52.5	22.864	64.75	−3735	228.64	45.61	−3736	52.5
280	2.3280	84.85	−3471	53.3	23.280	65.71	−3471	232.79	46.56	−3471	53.3
285	2.3696	85.80	−3203	54.0	23.696	66.66	−3203	236.95	47.51	−3203	54.0
290	2.4112	86.75	−2931	54.8	24.111	67.60	−2931	241.11	48.46	−2931	54.8
295	2.4527	87.69	−2655	55.6	24.527	68.55	−2655	245.26	49.40	−2655	55.6
300	2.4943	88.63	−2375	56.4	24.943	69.49	−2375	249.42	50.34	−2375	56.4
305	2.5359	89.57	−2091	57.2	25.359	70.43	−2091	253.58	51.28	−2091	57.2
310	2.5774	90.51	−1803	58.0	25.774	71.36	−1803	257.74	52.22	−1803	58.0

Table 3—continued

$\rho/\text{mol dm}^{-3}$

	1.0 E-6				10.0 E-6				100.0 E-6			
	111.893				**124.095**				**139.775**			
	P_σ 0.93032	S_σ 47.93	U_σ −10498	C_v 32.2	P_σ 10.317	S_σ 32.19	U_σ −10097	C_v 33.5	P_σ 116.19	S_σ 17.11	U_σ −9561	C_v 35.1
$\dfrac{T}{\text{K}}$	$\dfrac{P}{\text{Pa}}$	$\dfrac{S}{\text{J K}^{-1}\text{mol}^{-1}}$	$\dfrac{U}{\text{J mol}^{-1}}$	$\dfrac{C_v}{\text{J K}^{-1}\text{mol}^{-1}}$	$\dfrac{P}{\text{Pa}}$	$\dfrac{S}{\text{J K}^{-1}\text{mol}^{-1}}$	$\dfrac{U}{\text{J mol}^{-1}}$	$\dfrac{C_v}{\text{J K}^{-1}\text{mol}^{-1}}$	$\dfrac{P}{\text{Pa}}$	$\dfrac{S}{\text{J K}^{-1}\text{mol}^{-1}}$	$\dfrac{U}{\text{J mol}^{-1}}$	$\dfrac{C_v}{\text{J K}^{-1}\text{mol}^{-1}}$
315	2.6190	91.44	−1511	58.8	26.190	72.30	−1511	58.8	261.89	53.15	−1512	58.8
320	2.6606	92.37	−1216	59.5	26.606	73.23	−1216	59.5	266.05	54.08	−1216	59.6
325	2.7022	93.30	−916	60.3	27.022	74.16	−916	60.3	270.21	55.01	−916	60.3
330	2.7437	94.23	−612	61.1	27.437	75.08	−612	61.1	274.37	55.94	−612	61.1
335	2.7853	95.15	−304	61.9	27.853	76.01	−304	61.9	278.52	56.86	−305	61.9
340	2.8269	96.08	7	62.7	28.269	76.93	7	62.7	282.68	57.79	7	62.7
345	2.8684	97.00	323	63.5	28.684	77.85	323	63.5	286.84	58.71	323	63.5
350	2.9100	97.92	642	64.3	29.100	78.77	642	64.3	290.99	59.63	642	64.3
355	2.9516	98.84	966	65.1	29.516	79.69	966	65.1	295.15	60.55	966	65.1
360	2.9932	99.75	1293	65.9	29.932	80.61	1293	65.9	299.31	61.46	1293	65.9
365	3.0347	100.67	1625	66.7	30.347	81.52	1625	66.7	303.47	62.38	1625	66.7
370	3.0763	101.58	1960	67.5	30.763	82.43	1960	67.5	307.62	63.29	1960	67.5
380	3.1594	103.40	2643	69.0	31.594	84.25	2643	69.0	315.94	65.11	2642	69.0
390	3.2426	105.21	3341	70.6	32.426	86.07	3341	70.6	324.25	66.92	3341	70.6
400	3.3257	107.02	4054	72.1	33.257	87.87	4054	72.1	332.57	68.73	4054	72.1
410	3.4089	108.82	4783	73.7	34.089	89.67	4783	73.7	340.88	70.53	4783	73.7
420	3.4920	110.61	5527	75.2	34.920	91.47	5527	75.2	349.20	72.32	5527	75.2
430	3.5752	112.40	6287	76.7	35.752	93.25	6287	76.7	357.51	74.11	6286	76.7
440	3.6583	114.18	7061	78.2	36.583	95.03	7061	78.2	365.83	75.89	7061	78.2
450	3.7415	115.95	7850	79.6	37.414	96.81	7850	79.6	374.14	77.66	7850	79.6

475	3.9493	120.35	9886	83.2	39.493	101.21	9886	83.2	394.93	82.06	9886	83.2
500	4.1572	124.71	12011	86.8	41.572	105.57	12011	86.8	415.71	86.42	12011	86.8
525	4.3650	129.03	14223	90.1	43.650	109.88	14223	90.1	436.50	90.74	14223	90.1
550	4.5729	133.30	16518	93.4	45.729	114.15	16518	93.4	457.28	95.01	16518	93.4
575	4.7807	137.52	18893	96.6	47.807	118.38	18893	96.6	478.07	99.23	18893	96.6

Table 3—*continued*

$\rho/\text{mol dm}^{-3}$	1.0 E-3				10.0 E-3				100.0 E-3			
T/K	160.928				191.535				240.127			
	P_σ 1.3363	S_σ 3.05	U_σ −8800	C_v 37.2	P_σ 15.783	S_σ −9.41	U_σ −7636	C_v 40.9	P_σ 188.69	S_σ −19.11	U_σ −5711	C_v 48.4
$\dfrac{T}{\text{K}}$	$\dfrac{P}{\text{kPa}}$	$\dfrac{S}{\text{J K}^{-1}\text{mol}^{-1}}$	$\dfrac{U}{\text{J mol}^{-1}}$	$\dfrac{C_v}{\text{J K}^{-1}\text{mol}^{-1}}$	$\dfrac{P}{\text{kPa}}$	$\dfrac{S}{\text{J K}^{-1}\text{mol}^{-1}}$	$\dfrac{U}{\text{J mol}^{-1}}$	$\dfrac{C_v}{\text{J K}^{-1}\text{mol}^{-1}}$	$\dfrac{P}{\text{kPa}}$	$\dfrac{S}{\text{J K}^{-1}\text{mol}^{-1}}$	$\dfrac{U}{\text{J mol}^{-1}}$	$\dfrac{C_v}{\text{J K}^{-1}\text{mol}^{-1}}$
165	1.3702	3.98	−8647	37.7								
170	1.4118	5.11	−8458	38.2								
175	1.4534	6.23	−8265	38.8								
180	1.4951	7.33	−8070	39.3								
185	1.5367	8.42	−7872	39.9								
190	1.5783	9.49	−7670	40.5								
195	1.6199	10.55	−7466	41.1	16.074	−8.67	−7493	41.3				
200	1.6615	11.60	−7259	41.8	16.494	−7.62	−7285	41.9				
205	1.7031	12.64	−7049	42.4	16.913	−6.57	−7074	42.6				
210	1.7447	13.67	−6835	43.1	17.332	−5.54	−6860	43.2				
215	1.7863	14.69	−6618	43.7	17.752	−4.52	−6642	43.9				
220	1.8279	15.70	−6398	44.4	18.171	−3.50	−6421	44.5				
225	1.8695	16.71	−6174	45.1	18.589	−2.49	−6197	45.2				
230	1.9112	17.71	−5947	45.8	19.008	−1.49	−5969	45.9				
235	1.9528	18.70	−5716	46.5	19.427	−.49	−5738	46.6				

240	1.9943	19.69	−5481	47.2	19.845	.49	−5503	47.3	193.00	−18.13	−5474	49.1
245	2.0359	20.67	−5244	48.0	20.263	1.48	−5264	48.1	197.42	−17.13	−5227	49.7
250	2.0775	21.64	−5002	48.7	20.682	2.46	−5022	48.8	201.82	−16.14	−4976	50.4
255	2.1191	22.62	−4757	49.4	21.100	3.43	−4776	49.5	206.21	−15.15	−4722	51.2
260	2.1607	23.58	−4507	50.2	21.518	4.40	−4527	50.3				
265	2.2023	24.55	−4255	51.0	21.936	5.36	−4274	51.0	210.59	−14.17	−4465	51.9
270	2.2439	25.51	−3998	51.7	22.353	6.32	−4017	51.8	214.96	−13.20	−4204	52.6
275	2.2855	26.46	−3737	52.5	22.771	7.28	−3756	52.6	219.32	−12.22	−3939	53.3
280	2.3271	27.41	−3473	53.3	23.189	8.24	−3491	53.3	223.68	−11.26	−3670	54.1
285	2.3687	28.36	−3205	54.0	23.607	9.19	−3222	54.1	228.02	−10.29	−3398	54.8
290	2.4103	29.31	−2933	54.8	24.024	10.13	−2950	54.9	232.37	−9.33	−3122	55.6
295	2.4519	30.25	−2657	55.6	24.442	11.08	−2674	55.7	236.70	−8.38	−2842	56.3
300	2.4935	31.19	−2377	56.4	24.859	12.02	−2393	56.5	241.03	−7.42	−2559	57.1
305	2.5350	32.13	−2093	57.2	25.276	12.96	−2109	57.2	245.35	−6.47	−2272	57.9
310	2.5766	33.07	−1805	58.0	25.694	13.90	−1821	58.0	249.66	−5.53	−1980	58.6
315	2.6182	34.00	−1513	58.8	26.111	14.83	−1529	58.8	253.98	−4.58	−1685	59.4
320	2.6598	34.94	−1217	59.6	26.528	15.76	−1233	59.6	258.28	−3.64	−1386	60.2
325	2.7014	35.86	−918	60.3	26.945	16.70	−933	60.4	262.58	−2.70	−1084	60.9
330	2.7430	36.79	−614	61.1	27.362	17.62	−629	61.2	266.88	−1.77	−777	61.7
335	2.7846	37.72	−306	61.9	27.779	18.55	−321	62.0	271.17	−.83	−467	62.5

Table 3—continued

ρ/mol dm⁻³	1.0 E-3				10.0 E-3				100.0 E-3			
T/K	160.928				191.535				240.127			
	P_σ 1.3363	S_σ 3.05	U_σ -8800	C_v 37.2	P_σ 15.783	S_σ -9.41	U_σ -7636	C_v 40.9	P_σ 188.69	S_σ -19.11	U_σ -5711	C_v 48.4
$\frac{T}{K}$	$\frac{P}{kPa}$	$\frac{S}{J\,K^{-1}\,mol^{-1}}$	$\frac{U}{J\,mol^{-1}}$	$\frac{C_v}{J\,K^{-1}\,mol^{-1}}$	$\frac{P}{kPa}$	$\frac{S}{J\,K^{-1}\,mol^{-1}}$	$\frac{U}{J\,mol^{-1}}$	$\frac{C_v}{J\,K^{-1}\,mol^{-1}}$	$\frac{P}{kPa}$	$\frac{S}{J\,K^{-1}\,mol^{-1}}$	$\frac{U}{J\,mol^{-1}}$	$\frac{C_v}{J\,K^{-1}\,mol^{-1}}$
340	2.8262	38.64	6	62.7	28.196	19.47	-9	62.8	275.46	.10	-152	63.3
345	2.8677	39.56	321	63.5	28.613	20.40	307	63.6	279.74	1.03	166	64.0
350	2.9093	40.48	641	64.3	29.030	21.32	627	64.4	284.02	1.95	488	64.8
355	2.9509	41.40	964	65.1	29.447	22.23	951	65.1	288.30	2.88	814	65.6
360	2.9925	42.32	1292	65.9	29.864	23.15	1278	65.9	292.57	3.80	1144	66.4
365	3.0341	43.23	1623	66.7	30.281	24.07	1610	66.7	296.84	4.72	1478	67.1
370	3.0757	44.14	1959	67.5	30.698	24.98	1946	67.5	301.10	5.64	1815	67.9
380	3.1588	45.96	2641	69.0	31.531	26.80	2629	69.1	309.63	7.47	2502	69.4
390	3.2420	47.78	3339	70.6	32.365	28.61	3327	70.6	318.14	9.30	3204	71.0
400	3.3251	49.58	4053	72.1	33.198	30.42	4041	72.2	326.64	11.11	3922	72.5
410	3.4083	51.38	4782	73.7	34.031	32.22	4770	73.7	335.14	12.92	4654	74.0
420	3.4915	53.18	5526	75.2	34.864	34.02	5515	75.2	343.62	14.72	5402	75.5
430	3.5746	54.96	6285	76.7	35.697	35.80	6274	76.7	352.10	16.52	6165	77.0
440	3.6578	56.74	7060	78.2	36.530	37.59	7049	78.2	360.56	18.31	6942	78.5
450	3.7409	58.52	7849	79.6	37.363	39.36	7838	79.7	369.02	20.09	7734	79.9

475	3.9488	62.92	9885	83.3	39.445	43.76	9875	83.3	390.15	24.50	9777	83.5
500	4.1567	67.28	12010	86.8	41.527	48.12	12001	86.8	411.24	28.88	11909	87.0
525	4.3646	71.59	14222	90.1	43.608	52.44	14213	90.2	432.30	33.20	14126	90.4
550	4.5725	75.86	16517	93.4	45.689	56.71	16509	93.4	453.34	37.48	16426	93.6
575	4.7804	80.09	18892	96.6	47.770	60.94	18885	96.6	474.35	41.71	18806	96.8

Table 3—continued

ρ/mol dm^{-3}	250.0 E-3				500.0 E-3				750.0 E-3			
T/K	267.709				292.435				308.221			
	P_σ 0.49631	S_σ -21.91	U_σ -4642	C_v 53.7	P_σ 1.0012	S_σ -23.69	U_σ -3753	C_v 59.2	P_σ 1.4755	S_σ -24.71	U_σ -3253	C_v 63.1
$\dfrac{T}{K}$	$\dfrac{P}{MPa}$	$\dfrac{S}{JK^{-1}mol^{-1}}$	$\dfrac{U}{Jmol^{-1}}$	$\dfrac{C_v}{JK^{-1}mol^{-1}}$	$\dfrac{P}{MPa}$	$\dfrac{S}{JK^{-1}mol^{-1}}$	$\dfrac{U}{Jmol^{-1}}$	$\dfrac{C_v}{JK^{-1}mol^{-1}}$	$\dfrac{P}{MPa}$	$\dfrac{S}{JK^{-1}mol^{-1}}$	$\dfrac{U}{Jmol^{-1}}$	$\dfrac{C_v}{JK^{-1}mol^{-1}}$
270	.50170	-21.45	-4519	54.0								
275	.51341	-20.45	-4247	54.7								
280	.52506	-19.46	-3972	55.4								
285	.53666	-18.48	-3693	56.1								
290	.54822	-17.50	-3411	56.8								
295	.55974	-16.52	-3126	57.5	1.0142	-23.17	-3601	59.5				
300	.57121	-15.55	-2836	58.2	1.0394	-22.17	-3302	60.1				
305	.58264	-14.58	-2544	58.9	1.0644	-21.17	-2999	60.8				
310	.59404	-13.62	-2247	59.6	1.0892	-20.17	-2694	61.4	1.4898	-24.35	-3141	63.3
315	.60540	-12.66	-1947	60.4	1.1140	-19.19	-2385	62.1	1.5298	-23.33	-2823	63.8
320	.61673	-11.70	-1644	61.1	1.1385	-18.20	-2073	62.7	1.5696	-22.32	-2502	64.4
325	.62803	-10.75	-1336	61.8	1.1630	-17.23	-1758	63.4	1.6090	-21.32	-2179	65.0
330	.63930	-9.80	-1025	62.6	1.1873	-16.25	-1439	64.1	1.6481	-20.32	-1852	65.6
335	.65054	-8.85	-710	63.3	1.2115	-15.28	-1117	64.8	1.6870	-19.33	-1522	66.3
340	.66175	-7.91	-392	64.1	1.2356	-14.32	-791	65.5	1.7256	-18.34	-1189	66.9
345	.67294	-6.97	-70	64.8	1.2596	-13.36	-462	66.2	1.7640	-17.36	-853	67.6
350	.68411	-6.03	256	65.6	1.2835	-12.40	-130	66.9	1.8021	-16.38	-514	68.2
355	.69525	-5.09	586	66.3	1.3072	-11.45	207	67.6	1.8401	-15.41	-171	68.9
360	.70637	-4.16	920	67.1	1.3310	-10.50	546	68.3	1.8778	-14.44	175	69.5
365	.71746	-3.23	1257	67.8	1.3546	-9.55	890	69.0	1.9154	-13.48	524	70.2

370	.72854	-2.30	1598	68.6	1.3781	-8.61	1236	69.7	1.9528	-12.52	877	70.9
380	.75064	-.45	2291	70.1	1.4250	-6.73	1941	71.2	2.0270	-10.61	1592	72.2
390	.77267	1.39	3000	71.6	1.4715	-4.86	2659	72.6	2.1006	-8.72	2321	73.6
400	.79464	3.22	3723	73.1	1.5178	-3.00	3392	74.0	2.1736	-6.84	3064	75.0
410	.81655	5.04	4461	74.6	1.5639	-1.16	4140	75.4	2.2461	-4.97	3820	76.3
420	.83840	6.86	5214	76.0	1.6098	.68	4901	76.9	2.3181	-3.11	4590	77.7
430	.86020	8.66	5982	77.5	1.6555	2.50	5677	78.3	2.3897	-1.27	5374	79.1
440	.88196	10.46	6764	78.9	1.7009	4.32	6467	79.7	2.4609	.56	6172	80.4
450	.90367	12.25	7560	80.4	1.7462	6.12	7271	81.1	2.5317	2.39	6983	81.8
475	.95777	16.69	9614	83.9	1.8588	10.60	9341	84.5	2.7072	6.90	9070	85.2
500	1.0117	21.08	11755	87.3	1.9706	15.02	11497	87.9	2.8810	11.35	11240	88.5
525	1.0654	25.42	13980	90.7	2.0817	19.39	13736	91.2	3.0532	15.74	13491	91.7
550	1.1189	29.72	16287	93.9	2.1922	23.71	16055	94.4	3.2242	20.08	15823	94.8
575	1.1724	33.96	18674	97.0	2.3021	27.97	18453	97.4	3.3941	24.36	18231	97.8

Table 3—continued

$\rho/\text{mol dm}^{-3}$	1.0				1.5				2.0			
T/K	319.742				335.799				346.411			
σ	P_σ 1.9123	S_σ -25.51	U_σ -2946	C_v 66.1	P_σ 2.6707	S_σ -26.92	U_σ -2657	C_v 70.9	P_σ 3.2825	S_σ -28.28	U_σ -2629	C_v 74.4
$\dfrac{T}{\text{K}}$	$\dfrac{P}{\text{MPa}}$	$\dfrac{S}{\text{J K}^{-1}\text{mol}^{-1}}$	$\dfrac{U}{\text{J mol}^{-1}}$	$\dfrac{C_v}{\text{J K}^{-1}\text{mol}^{-1}}$	$\dfrac{P}{\text{MPa}}$	$\dfrac{S}{\text{J K}^{-1}\text{mol}^{-1}}$	$\dfrac{U}{\text{J mol}^{-1}}$	$\dfrac{C_v}{\text{J K}^{-1}\text{mol}^{-1}}$	$\dfrac{P}{\text{MPa}}$	$\dfrac{S}{\text{J K}^{-1}\text{mol}^{-1}}$	$\dfrac{U}{\text{J mol}^{-1}}$	$\dfrac{C_v}{\text{J K}^{-1}\text{mol}^{-1}}$
320	1.9152	-25.46	-2929	66.2								
325	1.9714	-24.43	-2597	66.7								
330	2.0271	-23.40	-2262	67.2								
335	2.0823	-22.39	-1924	67.8								
340	2.1370	-21.38	-1584	68.4	2.7473	-26.03	-2358	71.2				
345	2.1913	-20.38	-1241	68.9	2.8377	-24.99	-2001	71.7				
350	2.2452	-19.38	-895	69.5	2.9271	-23.96	-1641	72.2	3.3753	-27.51	-2362	74.6
355	2.2988	-18.39	-546	70.1	3.0158	-22.93	-1279	72.6	3.5034	-26.45	-1987	75.0
360	2.3519	-17.41	-193	70.7	3.1036	-21.91	-915	73.2	3.6301	-25.40	-1611	75.4
365	2.4048	-16.43	162	71.4	3.1907	-20.90	-548	73.7	3.7557	-24.36	-1233	75.8
370	2.4573	-15.45	520	72.0	3.2772	-19.89	-178	74.2	3.8801	-23.32	-853	76.3
380	2.5614	-13.52	1247	73.3	3.4481	-17.90	569	75.3	4.1259	-21.27	-85	77.2
390	2.6645	-11.60	1986	74.6	3.6167	-15.93	1328	76.5	4.3678	-19.26	692	78.2
400	2.7665	-9.69	2738	75.9	3.7831	-13.98	2099	77.6	4.6063	-17.26	1479	79.2
410	2.8676	-7.80	3503	77.2	3.9477	$-12.0^{\prime}$	2881	78.8	4.8417	-15.29	2277	80.3

420	2.9679	− 5.93	4282	78.5	4.1105	− 10.13	3675	80.0	5.0744	− 13.34	3085	81.4
430	3.0674	− 4.06	5073	79.8	4.2717	− 8.23	4482	81.3	5.3045	− 11.41	3905	82.6
440	3.1662	− 2.21	5878	81.1	4.4315	− 6.35	5301	82.5	5.5324	− 9.50	4737	83.7
450	3.2644	− .37	6696	82.5	4.5900	− 4.48	6132	83.7	5.7583	− 7.61	5580	84.9
475	3.5074	4.17	8799	85.7	4.9811	.13	8264	86.9	6.3149	− 2.94	7739	87.8
500	3.7473	8.65	10983	89.0	5.3660	4.66	10474	89.9	6.8619	1.64	9971	90.8
525	3.9847	13.07	13247	92.1	5.7458	9.12	12761	93.0	7.4010	6.14	12278	93.7
550	4.2198	17.43	15590	95.2	6.1212	13.52	15124	96.0	7.9334	10.57	14659	96.7
575	4.4532	21.73	18008	98.2	6.4929	17.85	17560	98.9	8.4602	14.93	17111	99.5

Table 3—continued

ρ/mol dm⁻³	2.5				3.0				3.5			
T_σ/K	353.683				358.651				361.924			
P_σ	3.7607				4.1183				4.3694			
S_σ	−29.67				−31.11				−32.60			
U_σ	−2765				−3015				−3348			
C_v	77.1				79.2				80.7			
$\dfrac{T}{K}$	$\dfrac{P}{MPa}$	$\dfrac{S}{J\,K^{-1}mol^{-1}}$	$\dfrac{U}{J\,mol^{-1}}$	$\dfrac{C_v}{J\,K^{-1}mol^{-1}}$	$\dfrac{P}{MPa}$	$\dfrac{S}{J\,K^{-1}mol^{-1}}$	$\dfrac{U}{J\,mol^{-1}}$	$\dfrac{C_v}{J\,K^{-1}mol^{-1}}$	$\dfrac{P}{MPa}$	$\dfrac{S}{J\,K^{-1}mol^{-1}}$	$\dfrac{U}{J\,mol^{-1}}$	$\dfrac{C_v}{J\,K^{-1}mol^{-1}}$
355	3.8054	−29.39	−2664	77.2								
360	3.9741	−28.30	−2277	77.5	4.1754	−30.81	−2908	79.2				
365	4.1412	−27.23	−1889	77.8	4.3858	−29.72	−2512	79.5	4.5256	−31.91	−3100	80.8
370	4.3067	−26.17	−1499	78.1	4.5943	−28.64	−2114	79.7	4.7781	−30.81	−2695	81.0
380	4.6333	−24.08	−714	78.9	5.0062	−26.50	−1313	80.4	5.2777	−28.64	−1883	81.5
390	4.9548	−22.02	80	79.8	5.4119	−24.41	−506	81.1	5.7710	−26.52	−1065	82.1
400	5.2717	−19.99	882	80.7	5.8122	−22.34	309	81.9	6.2586	−24.43	−240	82.8
410	5.5844	−17.98	1693	81.6	6.2076	−20.31	1132	82.7	6.7413	−22.38	592	83.6
420	5.8934	−16.00	2515	82.7	6.5986	−18.31	1964	83.7	7.2196	−20.35	1432	84.5
430	6.1991	−14.05	3346	83.7	6.9858	−16.33	2805	84.6	7.6939	−18.35	2281	85.4
440	6.5017	−12.11	4189	84.8	7.3694	−14.37	3656	85.6	8.1646	−16.38	3140	86.3
450	6.8016	−10.19	5042	85.9	7.7499	−12.43	4518	86.7	8.6321	−14.43	4008	87.3
475	7.5409	−5.48	7223	88.7	8.6889	−7.67	6718	89.4	9.7885	−9.64	6223	89.9
500	8.2675	−.86	9476	91.5	9.6134	−3.02	8987	92.1				
525	8.9838	3.68	11800	94.4								
550	9.6914	8.14	14195	97.2								

Table 3—continued

ρ/mol dm⁻³	4.0				4.5				5.0			
T/K	363.935				365.032				365.501			
	P_σ 4.5302	S_σ −34.11	U_σ −3739	C_v 81.7	P_σ 4.6200	S_σ −35.64	U_σ −4168	C_v 82.3	P_σ 4.6589	S_σ −37.14	U_σ −4616	C_v 82.5
$\dfrac{T}{\text{K}}$	$\dfrac{P}{\text{MPa}}$	$\dfrac{S}{\text{J K}^{-1}\text{mol}^{-1}}$	$\dfrac{U}{\text{J mol}^{-1}}$	$\dfrac{C_v}{\text{J K}^{-1}\text{mol}^{-1}}$	$\dfrac{P}{\text{MPa}}$	$\dfrac{S}{\text{J K}^{-1}\text{mol}^{-1}}$	$\dfrac{U}{\text{J mol}^{-1}}$	$\dfrac{C_v}{\text{J K}^{-1}\text{mol}^{-1}}$	$\dfrac{P}{\text{MPa}}$	$\dfrac{S}{\text{J K}^{-1}\text{mol}^{-1}}$	$\dfrac{U}{\text{J mol}^{-1}}$	$\dfrac{C_v}{\text{J K}^{-1}\text{mol}^{-1}}$
365	4.5936	−33.87	−3652	81.8								
370	4.8903	−32.76	−3243	81.9	4.9589	−34.52	−3759	82.5	5.0066	−36.13	−4244	82.6
380	5.4791	−30.57	−2422	82.3	5.6384	−32.32	−2933	82.8	5.7795	−33.93	−3416	82.9
390	6.0624	−28.42	−1596	82.9	6.3145	−30.16	−2102	83.3	6.5527	−31.77	−2585	83.4
400	6.6409	−26.32	−765	83.5	6.9877	−28.05	−1266	83.9	7.3260	−29.65	−1748	84.0
410	7.2152	−24.25	74	84.2	7.6584	−25.97	−424	84.6	8.0994	−27.56	−904	84.7
420	7.7857	−22.21	920	85.0	8.3268	−23.92	425	85.3	8.8730	−25.51	−54	85.5
430	8.3528	−20.20	1774	85.9	8.9932	−21.90	1283	86.2	9.6465	−23.49	805	86.3
440	8.9169	−18.21	2638	86.8	9.6577	−19.91	2149	87.1	10.420	−21.50	1672	87.2
450	9.4783	−16.25	3511	87.8	10.321	−17.94	3025	88.0	11.194	−19.53	2548	88.1
475	10.871	−11.44	5735	90.3	11.971	−13.12	5255	90.5	13.127	−14.70	4781	90.6

Table 3—continued

ρ/mol dm⁻³

T/K	5.5 365.545				6.0 365.258				6.5 364.608			
	P_σ/MPa 4.6626	S_σ J K⁻¹ mol⁻¹ −38.62	U_σ J mol⁻¹ −5071	C_σ J K⁻¹ mol⁻¹ 82.3	P_σ/MPa 4.6387	S_σ J K⁻¹ mol⁻¹ −40.08	U_σ J mol⁻¹ −5532	C_v J K⁻¹ mol⁻¹ 81.8	P_σ/MPa 4.5851	S_σ J K⁻¹ mol⁻¹ −41.54	U_σ J mol⁻¹ −6007	C_v J K⁻¹ mol⁻¹ 81.1
	P/MPa	S (J K⁻¹ mol⁻¹)	U (J mol⁻¹)	C_v (J K⁻¹ mol⁻¹)	P/MPa	S (J K⁻¹ mol⁻¹)	U (J mol⁻¹)	C_v (J K⁻¹ mol⁻¹)	P/MPa	S (J K⁻¹ mol⁻¹)	U (J mol⁻¹)	C_v (J K⁻¹ mol⁻¹)
365									4.6280	−41.45	−5975	81.1
370	5.0496	−37.62	−4704	82.4	5.1011	−39.02	−5144	82.0	5.1770	−40.35	−5570	81.3
380	5.9223	−35.42	−3878	82.8	6.0845	−36.83	−4322	82.4	6.2885	−38.18	−4755	81.7
390	6.7997	−33.26	−3048	83.3	7.0781	−34.68	−3496	82.9	7.4158	−36.04	−3934	82.3
400	7.6814	−31.15	−2212	83.9	8.0806	−32.58	−2664	83.5	8.5568	−33.95	−3108	83.0
410	8.5667	−29.07	−1370	84.6	9.0909	−30.50	−1825	84.3	9.7099	−31.89	−2274	83.8
420	9.4552	−27.02	−520	85.4	10.108	−28.46	−978	85.1	10.873	−29.86	−1431	84.7
430	10.346	−25.00	337	86.2	11.131	−26.45	−123	86.0	12.046	−27.86	−580	85.6
440	11.240	−23.01	1204	87.1	12.159	−24.47	741	86.9	13.227	−25.88	281	86.6
450	12.136	−21.04	2079	88.0	13.191	−22.50	1615	87.8	14.414	−23.92	1151	87.6
475	14.383	−16.22	4311	90.5	15.789	−17.69	3842	90.4	17.407	−19.12	3373	90.2

Table 3—continued

$\rho/\text{mol dm}^{-3}$	7.0				7.5				8.0			
T/K	363.461				361.633				358.942			
	P_σ 4.4918	S_σ −43.07	U_σ −6514	C_v 80.1	P_σ 4.3465	S_σ −44.72	U_σ −7071	C_v 79.0	P_σ 4.1401	S_σ −46.56	U_σ −7698	C_v 77.7
$\dfrac{T}{K}$	$\dfrac{P}{\text{MPa}}$	$\dfrac{S}{\text{J K}^{-1}\text{mol}^{-1}}$	$\dfrac{U}{\text{J mol}^{-1}}$	$\dfrac{C_v}{\text{J K}^{-1}\text{mol}^{-1}}$	$\dfrac{P}{\text{MPa}}$	$\dfrac{S}{\text{J K}^{-1}\text{mol}^{-1}}$	$\dfrac{U}{\text{J mol}^{-1}}$	$\dfrac{C_v}{\text{J K}^{-1}\text{mol}^{-1}}$	$\dfrac{P}{\text{MPa}}$	$\dfrac{S}{\text{J K}^{-1}\text{mol}^{-1}}$	$\dfrac{U}{\text{J mol}^{-1}}$	$\dfrac{C_v}{\text{J K}^{-1}\text{mol}^{-1}}$
360	4.6822	−42.73	−6390	80.2	4.8218	−43.99	−6805	79.1	4.3108	−46.33	−7616	77.8
365	5.3050	−41.64	−5989	80.4	5.5343	−42.91	−6409	79.4	5.1231	−45.26	−7226	78.1
370	6.5686	−39.49	−5182	81.0	6.9810	−40.78	−5611	80.1	5.9444	−44.19	−6835	78.5
380	7.8534	−37.37	−4369	81.6	8.4533	−38.69	−4806	80.9	7.6114	−42.09	−6046	79.3
390									9.3071	−40.02	−5249	80.1
400	9.1565	−35.30	−3549	82.4	9.9477	−36.63	−3993	81.7	11.027	−37.98	−4443	81.1
410	10.476	−33.25	−2721	83.3	11.461	−34.60	−3171	82.7	12.769	−35.96	−3628	82.1
420	11.808	−31.24	−1884	84.2	12.991	−32.60	−2339	83.7	14.529	−33.97	−2802	83.1
430	13.153	−29.24	−1037	85.2	14.536	−30.62	−1498	84.7	16.304	−32.00	−1965	84.2
440	14.509	−27.27	−180	86.2	16.093	−28.66	−645	85.8	18.092	−30.06	−1117	85.4
450	15.873	−25.33	687	87.2	17.660	−26.72	218	86.8	19.892	−28.12	−258	86.5
475	19.317	−20.54	2901	89.9	21.616	−21.95	2424	89.7	24.430	−23.37	1941	89.4

Table 3—continued

ρ/mol dm⁻³

	8.5				9.0				9.5			
T/K	355.246				350.454				344.523			
	P_σ 3.8702	S_σ −48.63	U_σ −8407	C_v 76.4	P_σ 3.5420	S_σ −50.96	U_σ −9205	C_v 75.0	P_σ 3.1666	S_σ −53.57	U_σ −10093	C_v 73.5
$\dfrac{T}{K}$	$\dfrac{P}{\text{MPa}}$	$\dfrac{S}{\text{J K}^{-1}\text{mol}^{-1}}$	$\dfrac{U}{\text{J mol}^{-1}}$	$\dfrac{C_v}{\text{J K}^{-1}\text{mol}^{-1}}$	$\dfrac{P}{\text{MPa}}$	$\dfrac{S}{\text{J K}^{-1}\text{mol}^{-1}}$	$\dfrac{U}{\text{J mol}^{-1}}$	$\dfrac{C_v}{\text{J K}^{-1}\text{mol}^{-1}}$	$\dfrac{P}{\text{MPa}}$	$\dfrac{S}{\text{J K}^{-1}\text{mol}^{-1}}$	$\dfrac{U}{\text{J mol}^{-1}}$	$\dfrac{C_v}{\text{J K}^{-1}\text{mol}^{-1}}$
345									3.2849	−53.47	−10058	73.5
350									4.5324	−52.41	−9690	74.0
355					4.5209	−49.99	−8863	75.3	5.7912	−51.36	−9319	74.4
360	4.7570	−47.61	−8043	76.7	5.6082	−48.93	−8485	75.8	7.0605	−50.31	−8945	74.9
365	5.6996	−46.55	−7658	77.1	6.7057	−47.88	−8106	76.2	8.3394	−49.27	−8570	75.4
370	6.6516	−45.50	−7272	77.5	7.8127	−46.84	−7723	76.7	9.6269	−48.24	−8191	76.0
380	8.5814	−43.42	−6492	78.4	10.052	−44.79	−6952	77.7	12.225	−46.20	−7426	77.1
390	10.541	−41.37	−5703	79.4	12.320	−42.75	−6169	78.8	14.849	−44.19	−6649	78.2
400	12.526	−39.34	−4904	80.4	14.613	−40.75	−5376	79.9	17.495	−42.19	−5861	79.4
410	14.533	−37.34	−4094	81.5	16.926	−38.76	−4571	81.1	20.158	−40.21	−5060	80.7
420	16.558	−35.37	−3273	82.7	19.256	−36.79	−3755	82.2	22.835	−38.26	−4247	81.9
430	18.599	−33.41	−2441	83.8	21.600	−34.84	−2926	83.5	25.522	−36.31	−3422	83.2
440	20.653	−31.47	−1597	85.0	23.956	−32.91	−2086	84.7	28.218	−34.38	−2583	84.5
450	22.718	−29.54	−741	86.2	26.320	−30.99	−1232	85.9	30.920	−32.47	−1731	85.8
475	27.915	−24.80	1452	89.2	32.258	−26.26	956	89.1	37.685	−27.74	455	89.1

Table 3—continued

10.0 to 11.0 mol dm⁻³

ρ/mol dm⁻³

	10.0				10.5				11.0			
T/K	337.443				329.221				319.874			
	P_σ 2.7593	S_σ −56.48	U_σ −11071	C_v 71.9	P_σ 2.3373	S_σ −59.71	U_σ −12135	C_v 70.3	P_σ 1.9178	S_σ −63.26	U_σ −13279	C_v 68.7
$\dfrac{T}{K}$	$\dfrac{P}{MPa}$	$\dfrac{S}{J\,K^{-1}\,mol^{-1}}$	$\dfrac{U}{J\,mol^{-1}}$	$\dfrac{C_v}{J\,K^{-1}\,mol^{-1}}$	$\dfrac{P}{MPa}$	$\dfrac{S}{J\,K^{-1}\,mol^{-1}}$	$\dfrac{U}{J\,mol^{-1}}$	$\dfrac{C_v}{J\,K^{-1}\,mol^{-1}}$	$\dfrac{P}{MPa}$	$\dfrac{S}{J\,K^{-1}\,mol^{-1}}$	$\dfrac{U}{J\,mol^{-1}}$	$\dfrac{C_v}{J\,K^{-1}\,mol^{-1}}$
320									1.9665	−63.24	−13270	68.7
325									3.9037	−62.17	−12926	69.2
330					2.5970	−59.54	−12080	70.4	5.8517	−61.11	−12578	69.8
335					4.2708	−58.48	−11727	70.9	7.8092	−60.05	−12228	70.3
340	3.4960	−55.94	−10887	72.2	5.9559	−57.43	−11371	71.4	9.7748	−59.01	−11875	70.9
345	4.9454	−54.88	−10525	72.6	7.6510	−56.38	−11013	72.0	11.747	−57.97	−11519	71.5
350	6.4058	−53.83	−10161	73.2	9.3550	−55.34	−10651	72.6	13.726	−56.93	−11160	72.2
355	7.8763	−52.79	−9794	73.7	11.067	−54.31	−10287	73.1	15.710	−55.91	−10797	72.8
360	9.3559	−51.76	−9424	74.2	12.786	−53.28	−9920	73.7	17.698	−54.88	−10432	73.4
365	10.844	−50.73	−9051	74.8	14.511	−52.26	−9550	74.4	19.689	−53.87	−10063	74.1
370	12.339	−49.71	−8676	75.4	16.241	−51.24	−9176	75.0	21.684	−52.85	−9691	74.8
380	15.348	−47.68	−7916	76.6	19.715	−49.23	−8420	76.3	25.677	−50.84	−8936	76.1
390	18.378	−45.68	−7143	77.8	23.202	−47.23	−7650	77.6	29.674	−48.85	−8167	77.5
400	21.424	−43.69	−6359	79.1	26.698	−45.25	−6867	79.0	33.671	−46.86	−7385	79.0
410	24.482	−41.72	−5561	80.4	30.199	−43.28	−6071	80.3	37.663	−44.90	−6588	80.4
420	27.550	−39.77	−4750	81.8	33.702	−41.33	−5261	81.7	41.647	−42.94	−5778	81.8
430	30.623	−37.83	−3926	83.1	37.204	−39.39	−4437	83.1	45.621	−41.00	−4952	83.2
440	33.699	−35.90	−3088	84.4	40.701	−37.46	−3599	84.5	49.582	−39.07	−4113	84.7
450	36.775	−33.99	−2237	85.8	44.193	−35.55	−2747	85.9	53.529	−37.15	−3259	86.1
475	44.457	−29.26	−51	89.1	52.882	−30.81	−557	89.3	63.318	−32.40	−1061	89.7

Table 3—*continued*

$\rho/\text{mol dm}^{-3}$

	11.5				12.0				12.5			
T/K	309.417				297.864				285.229			
	P_σ 1.5170	S_σ −67.16	U_σ −14498	C_v 67.0	P_σ 1.1494	S_σ −71.42	U_σ −15787	C_v 65.3	P_σ 0.82679	S_σ −76.07	U_σ −17140	C_v 63.6
$\dfrac{T}{\text{K}}$	$\dfrac{P}{\text{MPa}}$	$\dfrac{S}{\text{J K}^{-1}\text{mol}^{-1}}$	$\dfrac{U}{\text{J mol}^{-1}}$	$\dfrac{C_v}{\text{J K}^{-1}\text{mol}^{-1}}$	$\dfrac{P}{\text{MPa}}$	$\dfrac{S}{\text{J K}^{-1}\text{mol}^{-1}}$	$\dfrac{U}{\text{J mol}^{-1}}$	$\dfrac{C_v}{\text{J K}^{-1}\text{mol}^{-1}}$	$\dfrac{P}{\text{MPa}}$	$\dfrac{S}{\text{J K}^{-1}\text{mol}^{-1}}$	$\dfrac{U}{\text{J mol}^{-1}}$	$\dfrac{C_v}{\text{J K}^{-1}\text{mol}^{-1}}$
290									3.6979	−75.02	−16836	64.1
295									6.7126	−73.92	−16514	64.7
300					2.2583	−70.95	−15648	65.5	9.7318	−72.82	−16189	65.3
305					4.8606	−69.87	−15319	66.1	12.754	−71.74	−15861	65.9
310	1.7780	−67.03	−14459	67.0	7.4703	−68.79	−14987	66.6	15.777	−70.66	−15530	66.5
315	4.0229	−65.96	−14123	67.6	10.086	−67.72	−14652	67.3	18.799	−69.59	−15196	67.2
320	6.2774	−64.89	−13783	68.2	12.706	−66.65	−14314	67.9	21.820	−68.53	−14858	67.9
325	8.5399	−63.83	−13441	68.7	15.329	−65.60	−13973	68.5	24.839	−67.47	−14517	68.6
330	10.809	−62.77	−13096	69.4	17.954	−64.54	−13629	69.2	27.854	−66.42	−14173	69.3
335	13.084	−61.73	−12747	70.0	20.580	−63.50	−13281	69.9	30.864	−65.37	−13825	70.0
340	15.363	−60.68	−12396	70.6	23.206	−62.46	−12930	70.6	33.869	−64.33	−13473	70.7
345	17.646	−59.65	−12041	71.3	25.830	−61.42	−12576	71.3	36.868	−63.29	−13117	71.5
350	19.931	−58.62	−11683	72.0	28.453	−60.39	−12217	72.0	39.859	−62.26	−12758	72.2
355	22.217	−57.59	−11321	72.6	31.073	−59.37	−11856	72.7	42.844	−61.23	−12395	73.0
360	24.504	−56.57	−10956	73.3	33.690	−58.34	−11490	73.4	45.820	−60.20	−12028	73.7

365	26.792	−55.55	−10588	74.0	36.303	−57.33	−11121	74.2	48.787	−59.18	−11658	74.5
370	29.079	−54.54	−10216	74.8	38.912	−56.31	−10749	74.9	51.746	−58.16	−11283	75.3
380	33.648	−52.53	−9461	76.2	44.113	−54.29	−9992	76.4	57.633	−56.13	−10523	76.8
390	38.209	−50.53	−8692	77.6	49.290	−52.29	−9220	77.9	63.479	−54.12	−9747	78.4
400	42.756	−48.55	−7908	79.1	54.439	−50.30	−8433	79.4	69.280	−52.11	−8955	80.0
410	47.288	−46.58	−7110	80.6	59.559	−48.32	−7631	81.0	75.034	−50.12	−8148	81.5
420	51.800	−44.62	−6297	82.1	64.645	−46.35	−6814	82.5	80.739	−48.14	−7324	83.1
430	56.291	−42.67	−5469	83.5	69.696	−44.39	−5982	84.0	86.392	−46.16	−6486	84.7
440	60.758	−40.73	−4626	85.0	74.710	−42.44	−5134	85.5	91.993	−44.20	−5631	86.2
450	65.199	−38.80	−3768	86.5	79.686	−40.50	−4271	87.1	97.539	−42.24	−4761	87.8
475	76.179	−34.03	−1560	90.2	91.943	−35.69	−2047	90.8	111.16	−37.40	−2519	91.6

369

Table 3—*continued*

ρ/mol dm⁻³	13.0				13.5				14.0			
T/K	271.532				256.812				241.127			
	P_σ 0.55809	S_σ −81.15	U_σ −18851	C_v 61.9	P_σ 0.34800	S_σ −86.69	U_σ −20013	C_v 60.3	P_σ 0.19626	S_σ −92.73	U_σ −21516	C_v 58.8
$\frac{T}{K}$	$\frac{P}{MPa}$	$\frac{S}{J\,K^{-1}\,mol^{-1}}$	$\frac{U}{J\,mol^{-1}}$	$\frac{C_v}{J\,K^{-1}\,mol^{-1}}$	$\frac{P}{MPa}$	$\frac{S}{J\,K^{-1}\,mol^{-1}}$	$\frac{U}{J\,mol^{-1}}$	$\frac{C_v}{J\,K^{-1}\,mol^{-1}}$	$\frac{P}{MPa}$	$\frac{S}{J\,K^{-1}\,mol^{-1}}$	$\frac{U}{J\,mol^{-1}}$	$\frac{C_v}{J\,K^{-1}\,mol^{-1}}$
245									3.8078	−91.79	−21288	59.1
250									8.4680	−90.59	−20991	59.6
255									13.122	−89.41	−20692	60.2
260					2.9176	−85.94	−19820	60.6	17.768	−88.23	−20390	60.7
265					6.9488	−84.78	−19516	61.1	22.403	−87.07	−20084	61.3
270					10.979	−83.64	−19209	61.7	27.025	−85.92	−19776	61.9
275	2.9731	−80.36	−18336	62.3	15.005	−82.50	−18899	62.3	31.632	−84.78	−19465	62.6
280	6.4588	−79.24	−18023	62.8	19.026	−81.37	−18586	62.9	36.222	−83.64	−19150	63.3
285	9.9467	−78.12	−17708	63.4	23.039	−80.25	−18270	63.5	40.794	−82.52	−18832	64.0
290	13.435	−77.01	−17389	64.0	27.043	−79.14	−17951	64.2	45.348	−81.40	−18511	64.7
295	16.921	−75.91	−17068	64.6	31.037	−78.04	−17628	64.9	49.881	−80.29	−18186	65.4
300	20.404	−74.82	−16743	65.3	35.019	−76.94	−17302	65.6	54.393	−79.18	−17857	66.1
305	23.882	−73.74	−16415	66.0	38.987	−75.85	−16972	66.3	58.883	−78.08	−17524	66.9
310	27.354	−72.66	−16083	66.7	42.942	−74.77	−16638	67.1	63.351	−76.99	−17188	67.7
315	30.819	−71.58	−15748	67.4	46.882	−73.69	−16301	67.8	67.796	−75.90	−16847	68.5

13.0 to 14.0 mol dm⁻³

13.0 to 14.0 $mol\ dm^{-3}$

320	34.275	−70.52	−15409	68.1	50.806	−72.61	−15960	68.6	72.217	−74.81	−16503	69.2
325	37.723	−69.46	−15067	68.8	54.714	−71.55	−15616	69.3	76.614	−73.73	−16155	70.0
330	41.160	−68.40	−14721	69.6	58.605	−70.48	−15267	70.1	80.986	−72.66	−15803	70.9
335	44.587	−67.35	−14371	70.3	62.478	−69.42	−14914	70.9	85.333	−71.59	−15446	71.7
340	48.002	−66.30	−14018	71.1	66.333	−68.36	−14558	71.7	89.655	−70.52	−15086	72.5
345	51.405	−65.26	−13660	71.9	70.170	−67.31	−14197	72.5	93.951	−69.45	−14721	73.3
350	54.796	−64.22	−13299	72.7	73.987	−66.26	−13833	73.3	98.221	−68.39	−14353	74.1
355	58.174	−63.18	−12934	73.4	77.785	−65.22	−13464	74.1	102.46	−67.34	−13980	75.0
360	61.538	−62.15	−12564	74.2	81.564	−64.17	−13092	74.9	106.68	−66.28	−13603	75.8
365	64.888	−61.12	−12191	75.0	85.322	−63.14	−12715	75.8	110.87	−65.23	−13222	76.7
370	68.223	−60.09	−11814	75.8	89.060	−62.10	−12334	76.6	115.04	−64.18	−12836	77.5
380	74.849	−58.05	−11048	77.4	96.473	−60.04	−11560	78.2	123.28	−62.09	−12053	79.2
390	81.413	−56.02	−10265	79.0	103.80	−57.98	−10770	79.9	131.42	−60.01	−11252	80.9
400	87.913	−54.00	−9467	80.6	111.04	−55.94	−9963	81.5	139.44	−57.94	−10435	82.6
410	94.346	−51.98	−8652	82.3	118.20	−53.91	−9139	83.2	147.35	−55.88	−9600	84.3
420	100.71	−49.98	−7822	83.9	125.26	−51.88	−8299	84.8	155.14	−53.83	−8749	86.0
430	107.01	−47.99	−6975	85.5	132.23	−49.87	−7443	86.5	162.83	−51.79	−7881	87.7
440	113.23	−46.01	−6112	87.1	139.11	−47.86	−6570	88.1	170.39	−49.75	−6996	89.3
450	119.38	−44.03	−5234	88.7	145.90	−45.86	−5681	89.7	177.84	−47.73	−6094	91.0
475	134.43	−39.13	−2968	92.6	162.45	−40.90	−3387	93.8	195.94	−42.70	−3768	95.1

Table 3—continued

ρ/mol dm⁻³	14.5				15.0				15.5			
T/K	224.565				207.247				189.328			
	P_σ 0.09727	S_σ −99.34	U_σ −23054	C_σ 57.5	P_σ 0.04071	S_σ −106.57	U_σ −24615	C_σ 56.4	P_σ 0.01362	S_σ −114.52	U_σ −26190	C_σ 55.6
$\dfrac{T}{K}$	$\dfrac{P}{MPa}$	$\dfrac{S}{J\,K^{-1}\,mol^{-1}}$	$\dfrac{U}{J\,mol^{-1}}$	$\dfrac{C_v}{J\,K^{-1}\,mol^{-1}}$	$\dfrac{P}{MPa}$	$\dfrac{S}{J\,K^{-1}\,mol^{-1}}$	$\dfrac{U}{J\,mol^{-1}}$	$\dfrac{C_v}{J\,K^{-1}\,mol^{-1}}$	$\dfrac{P}{MPa}$	$\dfrac{S}{J\,K^{-1}\,mol^{-1}}$	$\dfrac{U}{J\,mol^{-1}}$	$\dfrac{C_v}{J\,K^{-1}\,mol^{-1}}$
190									.98207	−114.32	−26153	55.6
195									8.1805	−112.87	−25874	55.9
200									15.358	−111.45	−25594	56.3
205									22.508	−110.06	−25311	56.6
210					3.4724	−105.83	−24460	56.6	29.625	−108.69	−25027	57.0
215					9.6966	−104.49	−24176	57.0	36.704	−107.34	−24741	57.5
220					15.905	−103.18	−23890	57.4	43.743	−106.02	−24452	58.0
225	.56631	−99.23	−23029	57.5	22.092	−101.88	−23602	57.8	50.738	−104.71	−24161	58.5
230	5.9566	−97.96	−22740	57.9	28.255	−100.61	−23312	58.3	57.688	−103.42	−23867	59.0
235	11.339	−96.71	−22450	58.4	34.390	−99.35	−23019	58.8	(64.591)	(−102.14)	(−23571)	(59.6)
240	16.709	−95.47	−22156	58.9	40.494	−98.10	−22723	59.4	(71.447)	(−100.88)	(−23271)	(60.2)
245	22.064	−94.25	−21861	59.4	46.566	−96.87	−22425	60.0	(78.254)	(−99.63)	(−22968)	(60.9)
250	27.401	−93.05	−21563	60.0	52.605	−95.65	−22124	60.6	(85.012)	(−98.40)	(−22663)	(61.5)
255	32.717	−91.86	−21261	60.5	58.607	−94.45	−21819	61.2	(91.720)	(−97.17)	(−22353)	(62.2)
260	38.011	−90.67	−20957	61.2	(64.574)	(−93.25)	(−21511)	(61.9)	(98.378)	(−95.96)	(−22041)	(62.9)
265	43.280	−89.50	−20650	61.8	(70.502)	(−92.07)	(−21200)	(62.6)	(104.99)	(−94.75)	(−21724)	(63.6)
270	48.524	−88.34	−20339	62.5	(76.393)	(−90.89)	(−20885)	(63.3)	(111.54)	(−93.56)	(−21405)	(64.3)
275	53.741	−87.19	−20025	63.2	(82.245)	(−89.72)	(−20567)	(64.0)	(118.05)	(−92.37)	(−21081)	(65.1)
280	58.930	−86.04	−19707	63.9	(88.058)	(−88.56)	(−20245)	(64.8)	(124.51)	(−91.19)	(−20754)	(65.9)
285	(64.091)	(−84.91)	(−19385)	(64.6)	(93.832)	(−87.41)	(−19919)	(65.5)	(130.92)	(−90.02)	(−20422)	(66.6)

290	(69.222)	(−83.77)	(−19060)	(65.4)	(99.565)	(−86.26)	(−19589)	(66.3)	(137.27)	(−88.85)	(−20087)	(67.4)
295	(74.323)	(−82.65)	(−18732)	(66.1)	(105.26)	(−85.12)	(−19256)	(67.1)	(143.58)	(−87.69)	(−19748)	(68.3)
300	79.393	−81.53	−18399	66.9	110.91	−83.99	−18918	67.9	149.84	−86.54	−19405	69.1
305	84.433	−80.42	−18062	67.7	116.52	−82.86	−18577	68.7	156.05	−85.39	−19057	69.9
310	89.441	−79.31	−17722	68.5	122.10	−81.73	−18231	69.5	162.20	−84.24	−18705	70.8
315	94.417	−78.21	−17377	69.3	127.63	−80.61	−17881	70.4	168.31	−83.10	−18349	71.6
320	99.360	−77.11	−17029	70.1	133.12	−79.50	−17527	71.2	174.37	−81.97	−17989	72.5
325	104.27	−76.02	−16676	71.0	138.56	−78.39	−17169	72.1	180.37	−80.84	−17625	73.3
330	109.15	−74.93	−16319	71.8	143.97	−77.28	−16807	72.9	186.33	−79.71	−17256	74.2
335	114.00	−73.84	−15958	72.6	149.34	−76.18	−16440	73.8	192.24	−78.59	−16883	75.1
340	118.81	−72.76	−15593	73.5	154.66	−75.08	−16069	74.6	198.09	−77.47	−16505	76.0
345	123.59	−71.68	−15223	74.3	159.95	−73.98	−15693	75.5	203.90	−76.36	−16123	76.9
350	128.33	−70.60	−14849	75.2	165.19	−72.89	−15314	76.4	209.66	−75.24	−15736	77.8
355	133.05	−69.53	−14471	76.0	170.39	−71.80	−14929	77.3	215.37	−74.13	−15345	78.7
360	137.72	−68.46	−14089	76.9	175.55	−70.71	−14541	78.1	221.02	−73.03	−14950	79.6
365	142.37	−67.40	−13702	77.8	180.66	−69.63	−14148	79.0	226.63	−71.92	−14549	80.5
370	146.98	−66.33	−13311	78.6	185.74	−68.55	−13751	79.9	232.19	−70.82	−14145	81.4
380	156.09	−64.21	−12516	80.4	195.76	−66.39	−12942	81.7	243.15	−68.63	−13322	83.2
390	165.07	−62.10	−11704	82.1	205.62	−64.25	−12117	83.5	253.91	−66.44	−12481	85.0
400	173.91	−60.00	−10874	83.8	215.30	−62.11	−11273	85.3	264.47	−64.27	−11621	86.8
410	182.61	−57.91	−10027	85.6	224.82	−59.99	−10412	87.0	274.83	−62.10	−10744	88.7
420	191.17	−55.83	−9163	87.3	234.16	−57.87	−9532	88.8	284.99	−59.94	−9848	90.5
430	199.58	−53.75	−8282	89.0	243.34	−55.76	−8636	90.6	294.94	−57.79	−8934	92.3
440	207.86	−51.69	−7383	90.7	252.34	−53.65	−7721	92.3	304.69	−55.65	−8002	94.1
450	215.99	−49.63	−6467	92.4	261.17	−51.56	−6789	94.1	314.24	−53.51	−7052	95.9
475	235.68	−44.52	−4103	96.7	282.49	−46.36	−4383	98.4	337.20	−48.21	−4598	100.4

Table 3—continued

$\rho/\text{mol dm}^{-3}$		14.5				15.0				15.5		
T/K		224.565				207.247				189.328		
	P_σ 0.09727	S_σ −99.34	U_σ −23054	C_v 57.5	P_σ 0.04071	S_σ −106.57	U_σ −24615	C_v 56.4	P_σ 0.01362	S_σ −114.52	U_σ −26190	C_v 55.6
$\dfrac{T}{\text{K}}$	$\dfrac{P}{\text{MPa}}$	$\dfrac{S}{\text{J K}^{-1}\text{mol}^{-1}}$	$\dfrac{U}{\text{J mol}^{-1}}$	$\dfrac{C_v}{\text{J K}^{-1}\text{mol}^{-1}}$	$\dfrac{P}{\text{MPa}}$	$\dfrac{S}{\text{J K}^{-1}\text{mol}^{-1}}$	$\dfrac{U}{\text{J mol}^{-1}}$	$\dfrac{C_v}{\text{J K}^{-1}\text{mol}^{-1}}$	$\dfrac{P}{\text{MPa}}$	$\dfrac{S}{\text{J K}^{-1}\text{mol}^{-1}}$	$\dfrac{U}{\text{J mol}^{-1}}$	$\dfrac{C_v}{\text{J K}^{-1}\text{mol}^{-1}}$
135									2.2401	−143.05	−30778	51.6
140									13.475	−141.16	−30518	52.5
145									24.640	−139.30	−30253	53.4
150									35.712	−137.48	−29984	54.2
155					4.8715	−132.02	−29187	54.0	46.678	−135.69	−29711	54.9
160					14.505	−130.30	−28916	54.5	57.529	−133.94	−29436	55.5
165					24.089	−128.62	−28642	54.9	(68.261)	(−132.22)	(−29157)	(56.0)
170					33.609	−126.97	−28366	55.4	(78.873)	(−130.54)	(−28875)	(56.6)
175	6.6698	−122.00	−27546	55.1	43.057	−125.36	−28088	55.8	(89.365)	(−128.89)	(−28591)	(57.1)
180	14.975	−120.44	−27270	55.4	52.427	−123.78	−27808	56.2	(99.739)	(−127.28)	(−28305)	(57.5)
185	23.244	−118.92	−26992	55.8	(61.715)	(−122.23)	(−27526)	(56.7)	(110.00)	(−125.70)	(−28016)	(58.0)
190	31.467	−117.43	−26712	56.1	(70.920)	(−120.71)	(−27241)	(57.1)	(120.14)	(−124.15)	(−27725)	(58.4)
195	39.639	−115.96	−26430	56.5	(80.040)	(−119.22)	(−26955)	(57.6)	(130.18)	(−122.62)	(−27432)	(58.9)
200	47.755	−114.53	−26146	57.0	(89.075)	(−117.76)	(−26665)	(58.0)	(140.11)	(−121.12)	(−27136)	(59.4)
205	55.813	−113.12	−25861	57.4	(98.026)	(−116.32)	(−26374)	(58.5)	(149.94)	(−119.65)	(−26838)	(59.9)

210	(63.810)	(−111.73)	(−25572)	(57.9)	(106.89)	(−114.91)	(−26080)	(59.0)	(159.66)	(−118.20)	(−26537)	(60.4)
215	(71.745)	(−110.36)	(−25282)	(58.4)	(115.68)	(−113.51)	(−25784)	(59.6)	(169.29)	(−116.77)	(−26234)	(60.9)
220	(79.617)	(−109.01)	(−24988)	(58.9)	(124.38)	(−112.13)	(−25484)	(60.1)	(178.82)	(−115.37)	(−25928)	(61.5)
225	(87.426)	(−107.68)	(−24692)	(59.5)	(133.00)	(−110.78)	(−25182)	(60.7)	(188.26)	(−113.98)	(−25619)	(62.1)
230	(95.170)	(−106.37)	(−24394)	(60.1)	(141.55)	(−109.44)	(−24877)	(61.3)	(197.61)	(−112.61)	(−25307)	(62.7)
235	(102.85)	(−105.07)	(−24092)	(60.7)	(150.01)	(−108.11)	(−24569)	(61.9)	(206.87)	(−111.25)	(−24992)	(63.3)
240	(110.47)	(−103.78)	(−23787)	(61.3)	(158.40)	(−106.80)	(−24258)	(62.6)	(216.04)	(−109.91)	(−24674)	(64.0)
245	(118.02)	(−102.51)	(−23479)	(62.0)	(166.72)	(−105.50)	(−23944)	(63.2)	(225.13)	(−108.59)	(−24352)	(64.6)
250	(125.52)	(−101.25)	(−23167)	(62.6)	(174.96)	(−104.22)	(−23626)	(63.9)	(234.13)	(−107.28)	(−24027)	(65.3)
255	(132.94)	(−100.01)	(−22852)	(63.3)	(183.12)	(−102.95)	(−23304)	(64.6)	(243.05)	(−105.98)	(−23699)	(66.1)
260	(140.31)	(−98.77)	(−22534)	(64.1)	(191.21)	(−101.68)	(−22979)	(65.4)	(251.89)	(−104.69)	(−23367)	(66.8)
265	(147.62)	(−97.54)	(−22212)	(64.8)	(199.23)	(−100.43)	(−22650)	(66.1)	(260.64)	(−103.41)	(−23031)	(67.6)
270	(154.86)	(−96.33)	(−21886)	(65.6)	(207.18)	(−99.19)	(−22318)	(66.9)	(269.32)	(−102.14)	(−22691)	(68.4)
275	(162.04)	(−95.12)	(−21556)	(66.3)	(215.06)	(−97.95)	(−21981)	(67.7)	(277.91)	(−100.87)	(−22347)	(69.2)
280	(169.16)	(−93.91)	(−21222)	(67.1)	(222.86)	(−96.73)	(−21641)	(68.5)	(286.43)	(−99.62)	(−21999)	(70.0)
285	(176.22)	(−92.72)	(−20885)	(67.9)	(230.60)	(−95.51)	(−21296)	(69.3)	(294.86)	(−98.37)	(−21647)	(70.8)
290	(183.23)	(−91.53)	(−20543)	(68.7)	(238.26)	(−94.29)	(−20948)	(70.2)	(303.22)	(−97.13)	(−21291)	(71.7)
295	(190.17)	(−90.35)	(−20198)	(69.6)	(245.86)	(−93.09)	(−20595)	(71.0)	(311.50)	(−95.90)	(−20931)	(72.5)
300	197.05	−89.17	−19848	70.4	253.39	−91.89	−20238	71.9	319.70	−94.68	−20566	73.4
305	203.87	−88.00	−19493	71.3	260.85	−90.69	−19876	72.7	327.83	−93.45	−20196	74.3

Table 3—*continued*

ρ/mol dm⁻³		16.0					16.5					17.0			
T/K		170.999					152.479					134.006			
	P_σ 0.00337	S_σ −123.27	U_σ −27766		C_v 54.9	P_σ 0.00055	S_σ −132.91	U_σ −29323		C_v 53.8	P_σ 0.51E-4	S_σ −143.43	U_σ −30829		C_v 51.4
$\frac{T}{K}$	$\frac{P}{MPa}$	$\frac{S}{J\,K^{-1}mol^{-1}}$	$\frac{U}{J\,mol^{-1}}$		$\frac{C_v}{J\,K^{-1}mol^{-1}}$	$\frac{P}{MPa}$	$\frac{S}{J\,K^{-1}mol^{-1}}$	$\frac{U}{J\,mol^{-1}}$		$\frac{C_v}{J\,K^{-1}mol^{-1}}$	$\frac{P}{MPa}$	$\frac{S}{J\,K^{-1}mol^{-1}}$	$\frac{U}{J\,mol^{-1}}$		$\frac{C_v}{J\,K^{-1}mol^{-1}}$
310	210.63	−86.84	−19135		72.1	268.24	−89.50	−19510		73.6	335.87	−92.24	−19822		75.2
315	217.34	−85.67	−18772		73.0	275.57	−88.32	−19140		74.5	343.84	−91.03	−19444		76.1
320	223.98	−84.52	−18405		73.9	282.82	−87.14	−18765		75.4	351.74	−89.82	−19061		77.1
325	230.57	−83.37	−18033		74.8	290.01	−85.96	−18386		76.3	359.55	−88.62	−18673		78.0
330	237.09	−82.22	−17657		75.7	297.13	−84.79	−18002		77.2	367.29	−87.42	−18281		78.9
335	243.56	−81.07	−17277		76.6	304.18	−83.62	−17613		78.2	374.95	−86.23	−17884		79.9
340	249.97	−79.93	−16892		77.5	311.16	−82.45	−17220		79.1	382.53	−85.04	−17482		80.9
345	256.32	−78.79	−16502		78.4	318.08	−81.29	−16822		80.0	390.04	−83.85	−17075		81.8
350	262.61	−77.66	−16108		79.3	324.93	−80.13	−16420		81.0	397.46	−82.66	−16664		82.8
355	268.85	−76.53	−15709		80.2	331.71	−78.98	−16013		81.9	404.81	−81.48	−16247		83.8
360	275.02	−75.40	−15306		81.1	338.42	−77.83	−15601		82.9	412.09	−80.31	−15826		84.7
365	281.14	−74.27	−14898		82.1	345.06	−76.68	−15184		83.8	419.28	−79.13	−15400		85.7
370	287.20	−73.15	−14485		83.0	351.64	−75.53	−14762		84.8	426.40	−77.96	−14969		86.7
380	299.13	−70.91	−13646		84.9	364.58	−73.24	−13905		86.7	440.39	−75.62	−14092		88.7
390	310.83	−68.68	−12788		86.7	377.26	−70.97	−13028		88.6	454.08	−73.29	−13195		90.7

400	322.30	−66.46	−11911	88.6	389.65	−68.70	−12133	90.6	467.44	−70.97	−12278	92.7
410	333.53	−64.25	−11015	90.5	401.78	−66.44	−11217	92.5	480.49	−68.65	−11342	94.7
420	344.52	−62.05	−10101	92.4	413.62	−64.19	−10283	94.4	493.22	−66.35	−10385	96.6
430	355.27	−59.86	−9168	94.2	425.19	−61.94	−9329	96.3	505.63	−64.05	−9409	98.6
440	365.78	−57.67	−8217	96.1	436.48	−59.71	−8356	98.2	517.71	−61.76	−8413	100.6
450	376.05	−55.49	−7247	97.9	447.49	−57.48	−7364	100.2	529.47	−59.48	−7397	102.6
475	400.67	−50.07	−4741	102.5	473.78	−51.94	−4801	104.9	557.44	−53.80	−4771	107.5

Table 3—continued

ρ/mol dm^{-3}	16.0				16.5				17.0			
T/K	170.999				152.479				134.006			
	P_σ 0.00337	S_σ -123.27	U_σ -27766	C_v 54.9	P_σ 0.00055	S_σ -132.91	U_σ -29323	C_v 53.8	P_σ 0.51 E-4	S_σ -143.43	U_σ -30829	C_v 51.4
$\dfrac{T}{\text{K}}$	$\dfrac{P}{\text{MPa}}$	$\dfrac{S}{\text{J K}^{-1}\text{mol}^{-1}}$	$\dfrac{U}{\text{J mol}^{-1}}$	$\dfrac{C_v}{\text{J K}^{-1}\text{mol}^{-1}}$	$\dfrac{P}{\text{MPa}}$	$\dfrac{S}{\text{J K}^{-1}\text{mol}^{-1}}$	$\dfrac{U}{\text{J mol}^{-1}}$	$\dfrac{C_v}{\text{J K}^{-1}\text{mol}^{-1}}$	$\dfrac{P}{\text{MPa}}$	$\dfrac{S}{\text{J K}^{-1}\text{mol}^{-1}}$	$\dfrac{U}{\text{J mol}^{-1}}$	$\dfrac{C_v}{\text{J K}^{-1}\text{mol}^{-1}}$
95									57.623	-174.05	-34189	
100					7.0107	-166.29	-33479		(75.230)	(-171.29)	(-33921)	
105					22.630	-163.94	-33238		(92.369)	(-168.84)	(-33669)	
110					38.062	-161.73	-33001	47.5	(109.04)	(-166.53)	(-33421)	(50.0)
115					53.235	-159.60	-32761	48.6	(125.25)	(-164.28)	(-33168)	(51.2)
120	11.174	-153.05	-32041	48.6	(68.119)	(-157.50)	(-32514)	(50.2)	(141.05)	(-162.07)	(-32909)	(52.7)
125	24.305	-151.04	-31795	50.0	(82.704)	(-155.42)	(-32260)	(51.8)	(156.46)	(-159.89)	(-32641)	(54.2)
130	37.297	-149.06	-31542	51.3	(96.995)	(-153.36)	(-31997)	(53.2)	(171.53)	(-157.73)	(-32367)	(55.6)
135	50.122	-147.09	-31282	52.6	(111.00)	(-151.33)	(-31727)	(54.6)	(186.28)	(-155.61)	(-32085)	(56.9)
140	(62.769)	(-145.16)	(-31016)	(53.8)	(124.75)	(-149.32)	(-31452)	(55.7)	(200.75)	(-153.52)	(-31798)	(57.9)
145	(75.231)	(-143.25)	(-30744)	(54.8)	(138.24)	(-147.35)	(-31170)	(56.7)	(214.97)	(-151.48)	(-31507)	(58.8)
150	(87.511)	(-141.38)	(-30468)	(55.6)	(151.51)	(-145.41)	(-30885)	(57.5)	(228.95)	(-149.47)	(-31211)	(59.5)
155	(99.612)	(-139.54)	(-30188)	(56.4)	(164.56)	(-143.51)	(-30595)	(58.2)	(242.73)	(-147.51)	(-30912)	(60.1)
160	(111.54)	(-137.74)	(-29905)	(57.0)	(177.41)	(-141.65)	(-30303)	(58.8)	(256.32)	(-145.60)	(-30610)	(60.6)
165	(123.30)	(-135.98)	(-29618)	(57.6)	(190.07)	(-139.83)	(-30007)	(59.4)	(269.73)	(-143.73)	(-30306)	(61.0)

170	(134.91)	(−134.25)	(−29329)	(58.1)	(202.56)	(−138.05)	(−29709)	(59.8)	(282.98)	(−141.90)	(−30000)	(61.4)
175	(146.36)	(−132.56)	(−29037)	(58.6)	(214.90)	(−136.31)	(−29409)	(60.3)	(296.08)	(−140.11)	(−29692)	(61.8)
180	(157.68)	(−130.90)	(−28743)	(59.1)	(227.07)	(−134.61)	(−29107)	(60.7)	(309.05)	(−138.37)	(−29382)	(62.2)
185	(168.85)	(−129.28)	(−28446)	(59.5)	(239.11)	(−132.94)	(−28802)	(61.1)	(321.88)	(−136.66)	(−29071)	(62.5)
190	(179.90)	(−127.69)	(−28148)	(60.0)	(251.01)	(−131.31)	(−28496)	(61.5)	(334.59)	(−134.99)	(−28757)	(62.9)
195	(190.82)	(−126.12)	(−27847)	(60.4)	(262.79)	(−129.71)	(−28187)	(61.9)	(347.17)	(−133.35)	(−28441)	(63.3)
200	(201.62)	(−124.59)	(−27544)	(60.9)	(274.44)	(−128.13)	(−27877)	(62.4)	(359.65)	(−131.74)	(−28124)	(63.7)
205	(212.31)	(−123.08)	(−27238)	(61.3)	(285.97)	(−126.59)	(−27564)	(62.8)	(372.01)	(−130.16)	(−27804)	(64.2)
210	(222.88)	(−121.60)	(−26930)	(61.9)	(297.39)	(−125.07)	(−27248)	(63.3)	(384.26)	(−128.61)	(−27482)	(64.7)
215	(233.35)	(−120.13)	(−26619)	(62.4)	(308.70)	(−123.57)	(−26931)	(63.8)	(396.41)	(−127.08)	(−27157)	(65.2)
220	(243.71)	(−118.69)	(−26306)	(62.9)	(319.90)	(−122.10)	(−26610)	(64.4)	(408.46)	(−125.57)	(−26830)	(65.7)
225	(253.98)	(−117.27)	(−25990)	(63.5)	(331.00)	(−120.65)	(−26287)	(64.9)	(420.40)	(−124.09)	(−26500)	(66.3)
230	(264.14)	(−115.87)	(−25671)	(64.1)	(341.99)	(−119.21)	(−25961)	(65.6)	(432.25)	(−122.62)	(−26167)	(66.9)
235	(274.21)	(−114.49)	(−25349)	(64.7)	(352.89)	(−117.80)	(−25631)	(66.2)	(443.99)	(−121.18)	(−25830)	(67.6)
240	(284.18)	(−113.12)	(−25024)	(65.4)	(363.69)	(−116.39)	(−25299)	(66.9)	(455.63)	(−119.75)	(−25490)	(68.3)
245	(294.06)	(−111.76)	(−24695)	(66.1)	(374.38)	(−115.01)	(−24963)	(67.6)	(467.18)	(−118.33)	(−25147)	(69.0)
250	(303.85)	(−110.42)	(−24363)	(66.8)	(384.99)	(−113.64)	(−24623)	(68.3)	(478.63)	(−116.93)	(−24800)	(69.8)
255	(313.55)	(−109.09)	(−24027)	(67.5)	(395.49)	(−112.28)	(−24280)	(69.0)	(489.97)	(−115.54)	(−24450)	(70.5)
260	(323.15)	(−107.77)	(−23687)	(68.3)	(405.90)	(−110.93)	(−23933)	(69.8)	(501.22)	(−114.16)	(−24095)	(71.3)
265	(332.67)	(−106.46)	(−23344)	(69.1)	(416.22)	(−109.59)	(−23581)	(70.6)	(512.37)	(−112.80)	(−23736)	(72.2)
270	(342.10)	(−105.16)	(−22996)	(69.9)	(426.44)	(−108.26)	(−23226)	(71.4)	(523.42)	(−111.44)	(−23373)	(73.0)
275	(351.44)	(−103.87)	(−22645)	(70.7)	(436.56)	(−106.95)	(−22867)	(72.3)	(534.36)	(−110.09)	(−23006)	(73.9)
280	(360.70)	(−102.59)	(−22289)	(71.5)	(446.59)	(−105.63)	(−22503)	(73.2)	(545.21)	(−108.75)	(−22634)	(74.8)
285	(369.87)	(−101.32)	(−21930)	(72.4)	(456.53)	(−104.33)	(−22135)	(74.0)	(555.95)	(−107.42)	(−22258)	(75.7)
290	(378.95)	(−100.05)	(−21565)	(73.3)	(466.37)	(−103.04)	(−21763)	(75.0)	(566.59)	(−106.10)	(−21877)	(76.7)

Table 3—*continued*

ρ/mol dm⁻³	17.5				18.0				18.5			
T/K	115.773				97.765				97.765			
	P_σ 0.21 E-5	S_σ −154.78	U_σ −32245	C_σ 47.6	P_σ 0.25 E-7	S_σ −167.43	U_σ −33592		P_σ (577.13)	S_σ −104.78	U_σ −21491	
$\dfrac{T}{\text{K}}$	$\dfrac{P}{\text{MPa}}$	$\dfrac{S}{\text{J K}^{-1}\text{mol}^{-1}}$	$\dfrac{U}{\text{J mol}^{-1}}$	$\dfrac{C_v}{\text{J K}^{-1}\text{mol}^{-1}}$	$\dfrac{P}{\text{MPa}}$	$\dfrac{S}{\text{J K}^{-1}\text{mol}^{-1}}$	$\dfrac{U}{\text{J mol}^{-1}}$	$\dfrac{C_v}{\text{J K}^{-1}\text{mol}^{-1}}$	$\dfrac{P}{\text{MPa}}$	$\dfrac{S}{\text{J K}^{-1}\text{mol}^{-1}}$	$\dfrac{U}{\text{J mol}^{-1}}$	$\dfrac{C_v}{\text{J K}^{-1}\text{mol}^{-1}}$
295	(387.94)	(−98.79)	(−21197)	(74.2)	(476.12)	(−101.75)	(−21386)	(75.9)	(577.13)	(−104.78)	(−21491)	(77.6)
300	396.85	−97.54	−20824	75.1	485.77	−100.46	−21004	76.8	587.57	−103.46	−21100	78.6
305	405.67	−96.29	−20446	76.0	495.32	−99.19	−20618	77.8	597.90	−102.16	−20705	79.6
310	414.40	−95.04	−20064	76.9	504.78	−97.91	−20226	78.7	608.12	−100.85	−20304	80.6
315	423.05	−93.80	−19677	77.9	514.14	−96.65	−19830	79.7	618.24	−99.55	−19899	81.6
320	431.61	−92.57	−19285	78.8	523.41	−95.38	−19429	80.7	628.25	−98.26	−19488	82.7
325	440.08	−91.34	−18888	79.8	532.58	−94.13	−19023	81.7	638.15	−96.97	−19072	83.7
330	448.47	−90.12	−18487	80.8	541.65	−92.87	−18612	82.7	647.95	−95.69	−18651	84.8
335	456.77	−88.89	−18081	81.8	550.62	−91.62	−18196	83.7	657.63	−94.40	−18224	85.8
340	464.98	−87.68	−17669	82.7	559.50	−90.37	−17775	84.8	667.21	−93.12	−17793	86.9
345	473.11	−86.46	−17253	83.7	568.28	−89.13	−17349	85.8	676.67	−91.85	−17355	88.0
350	481.14	−85.25	−16832	84.7	576.95	−87.88	−16917	86.8	686.02	−90.57	−16913	89.0
355	489.09	−84.04	−16406	85.7	585.53	−86.65	−16480	87.9	695.26	−89.30	−16465	90.1
360	496.95	−82.83	−15974	86.8	594.01	−85.41	−16038	88.9	704.39	−88.03	−16011	91.2
365	504.72	−81.63	−15538	87.8	602.38	−84.18	−15591	90.0	713.41	−86.77	−15553	92.3
370	512.40	−80.43	−15097	88.8	610.66	−82.94	−15139	91.0	722.31	−85.50	−15088	93.4
380	527.50	−78.03	−14199	90.8	626.91	−80.49	−14218	93.2	739.76	−82.98	−14143	95.6
390	542.23	−75.65	−13280	92.9	642.74	−78.04	−13276	95.3	756.75	−80.47	−13175	97.9
400	556.61	−73.27	−12341	95.0	658.17	−75.60	−12312	97.4	773.27	−77.96	−12185	100.1
410	570.61	−70.90	−11381	97.0	673.17	−73.17	−11327	99.6	789.32	−75.47	−11173	102.3

420	584.25	−68.54	−10400	99.1	687.76	−70.74	−10320	101.7	804.88	−72.97	−10139	104.6
430	597.53	−66.18	−9399	101.1	701.92	−68.33	−9293	103.9	819.97	−70.49	−9082	106.8
440	610.43	−63.83	−8378	103.2	715.66	−65.91	−8243	106.0	834.56	−68.01	−8003	109.0
450	622.95	−61.49	−7335	105.2	728.97	−63.51	−7173	108.1	848.67	−65.53	−6901	111.2
475	652.62	−55.66	−4641	110.3	760.34	−57.52	−4404	113.4	881.76	−59.37	−4051	116.7

Table 3—continued

	17.5				18.0				18.5			
p/mol dm⁻³	115.773				97.765							
σ	P_σ 0.21 E-5	S_σ −154.78	U_σ −32245	C_v 47.6	P_σ 0.25 E-7	S_σ −167.43	U_σ −33592	C_v	P_σ	S_σ	U_σ	C_v
$\dfrac{T}{K}$	$\dfrac{P}{MPa}$	$\dfrac{S}{J\,K^{-1}\,mol^{-1}}$	$\dfrac{U}{J\,mol^{-1}}$	$\dfrac{C_v}{J\,K^{-1}\,mol^{-1}}$	$\dfrac{P}{MPa}$	$\dfrac{S}{J\,K^{-1}\,mol^{-1}}$	$\dfrac{U}{J\,mol^{-1}}$	$\dfrac{C_v}{J\,K^{-1}\,mol^{-1}}$	$\dfrac{P}{MPa}$	$\dfrac{S}{J\,K^{-1}\,mol^{-1}}$	$\dfrac{U}{J\,mol^{-1}}$	$\dfrac{C_v}{J\,K^{-1}\,mol^{-1}}$
100	(167.15)	(−176.27)	(−34250)									
105	(184.19)	(−173.67)	(−33984)									
110	(200.75)	(−171.22)	(−33720)	(52.8)	(318.53)	(−175.45)	(−33839)	(54.6)				
115	(216.90)	(−168.85)	(−33454)	(53.9)	(333.09)	(−173.00)	(−33564)	(55.4)				
120	(232.67)	(−166.53)	(−33181)	(55.2)	(347.58)	(−170.62)	(−33284)	(56.6)	(493.30)	(−174.01)	(−33156)	(55.5)
125	(248.12)	(−164.25)	(−32902)	(56.5)	(361.99)	(−168.29)	(−32998)	(57.8)	(505.10)	(−171.71)	(−32875)	(56.8)
130	(263.30)	(−162.01)	(−32616)	(57.8)	(376.36)	(−166.00)	(−32707)	(58.8)	(517.29)	(−169.46)	(−32588)	(58.0)
135	(278.23)	(−159.81)	(−32325)	(58.8)	(390.69)	(−163.76)	(−32410)	(59.8)	(529.83)	(−167.26)	(−32296)	(59.0)
140	(292.95)	(−157.66)	(−32028)	(59.7)	(404.98)	(−161.57)	(−32109)	(60.6)	(542.67)	(−165.10)	(−31998)	(59.8)
145	(307.48)	(−155.55)	(−31728)	(60.4)	(419.25)	(−159.44)	(−31804)	(61.3)	(555.78)	(−162.98)	(−31697)	(60.6)
150	(321.85)	(−153.49)	(−31424)	(61.0)	(433.49)	(−157.35)	(−31497)	(61.8)	(569.12)	(−160.92)	(−31393)	(61.2)
155	(336.06)	(−151.48)	(−31118)	(61.6)	(447.71)	(−155.31)	(−31186)	(62.3)	(582.65)	(−158.90)	(−31085)	(61.8)
160	(350.14)	(−149.52)	(−30809)	(62.0)	(461.91)	(−153.33)	(−30874)	(62.7)	(596.36)	(−156.93)	(−30775)	(62.2)
165	(364.09)	(−147.60)	(−30498)	(62.4)	(476.08)	(−151.40)	(−30560)	(63.0)	(610.22)	(−155.01)	(−30463)	(62.7)
170	(377.93)	(−145.74)	(−30185)	(62.7)	(490.22)	(−149.51)	(−30244)	(63.4)	(624.20)	(−153.13)	(−30149)	(63.1)

175	(391.66)	(−143.91)	(−29871)	(63.0)	(504.33)	(−147.67)	(−29926)	(63.7)	(638.28)	(−151.30)	(−29832)	(63.5)
180	(405.28)	(−142.13)	(−29555)	(63.4)	(518.41)	(−145.87)	(−29607)	(64.1)	(652.45)	(−149.50)	(−29513)	(64.0)
185	(418.81)	(−140.39)	(−29237)	(63.7)	(532.45)	(−144.11)	(−29285)	(64.4)	(666.69)	(−147.74)	(−29192)	(64.4)
190	(432.24)	(−138.69)	(−28918)	(64.1)	(546.45)	(−142.38)	(−28962)	(64.8)	(680.97)	(−146.02)	(−28869)	(64.9)
195	(445.57)	(−137.02)	(−28596)	(64.5)	(560.41)	(−140.70)	(−28637)	(65.2)	(695.29)	(−144.33)	(−28543)	(65.4)
200	(458.82)	(−135.38)	(−28273)	(64.9)	(574.31)	(−139.04)	(−28310)	(65.7)	(709.64)	(−142.66)	(−28215)	(66.0)
205	(471.97)	(−133.77)	(−27947)	(65.4)	(588.16)	(−137.41)	(−27980)	(66.2)	(723.99)	(−141.03)	(−27883)	(66.6)
210	(485.03)	(−132.19)	(−27619)	(65.9)	(601.95)	(−135.81)	(−27648)	(66.7)	(738.33)	(−139.41)	(−27549)	(67.2)
215	(498.00)	(−130.64)	(−27289)	(66.4)	(615.68)	(−134.23)	(−27312)	(67.3)	(752.67)	(−137.83)	(−27211)	(67.9)
220	(510.88)	(−129.10)	(−26955)	(67.0)	(629.34)	(−132.68)	(−26974)	(67.9)	(766.97)	(−136.26)	(−26870)	(68.6)
225	(523.67)	(−127.59)	(−26619)	(67.6)	(642.93)	(−131.14)	(−26633)	(68.6)	(781.25)	(−134.71)	(−26526)	(69.3)
230	(536.37)	(−126.10)	(−26279)	(68.2)	(656.45)	(−129.63)	(−26288)	(69.3)	(795.47)	(−133.18)	(−26177)	(70.1)
235	(548.97)	(−124.63)	(−25937)	(68.9)	(669.88)	(−128.13)	(−25940)	(70.0)	(809.65)	(−131.66)	(−25825)	(70.9)
240	(561.47)	(−123.17)	(−25590)	(69.6)	(683.23)	(−126.64)	(−25588)	(70.8)	(823.77)	(−130.16)	(−25468)	(71.8)
245	(573.88)	(−121.72)	(−25240)	(70.4)	(696.50)	(−125.18)	(−25231)	(71.6)	(837.82)	(−128.67)	(−25107)	(72.7)
250	(586.20)	(−120.29)	(−24886)	(71.2)	(709.67)	(−123.72)	(−24871)	(72.5)	(851.79)	(−127.19)	(−24741)	(73.6)
255	(598.41)	(−118.88)	(−24529)	(72.0)	(722.75)	(−122.28)	(−24507)	(73.4)	(865.69)	(−125.72)	(−24370)	(74.6)
260	(610.52)	(−117.47)	(−24167)	(72.8)	(735.73)	(−120.84)	(−24138)	(74.3)	(879.50)	(−124.27)	(−23995)	(75.6)
265	(622.53)	(−116.08)	(−23800)	(73.7)	(748.61)	(−119.42)	(−23764)	(75.2)	(893.21)	(−122.82)	(−23615)	(76.6)
270	(634.44)	(−114.69)	(−23429)	(74.6)	(761.39)	(−118.01)	(−23386)	(76.2)	(906.83)	(−121.38)	(−23229)	(77.6)

Table 3—continued

ρ/mol dm⁻³	19.0				19.5				20.0			
$\dfrac{T}{\text{K}}$	$\dfrac{P_\sigma}{\text{MPa}}$	$\dfrac{S_\sigma}{\text{J K}^{-1}\text{mol}^{-1}}$	$\dfrac{U_\sigma}{\text{J mol}^{-1}}$	$\dfrac{C_v}{\text{J K}^{-1}\text{mol}^{-1}}$	$\dfrac{P_\sigma}{\text{MPa}}$	$\dfrac{S_\sigma}{\text{J K}^{-1}\text{mol}^{-1}}$	$\dfrac{U_\sigma}{\text{J mol}^{-1}}$	$\dfrac{C_v}{\text{J K}^{-1}\text{mol}^{-1}}$	$\dfrac{P_\sigma}{\text{MPa}}$	$\dfrac{S_\sigma}{\text{J K}^{-1}\text{mol}^{-1}}$	$\dfrac{U_\sigma}{\text{J mol}^{-1}}$	$\dfrac{C_v}{\text{J K}^{-1}\text{mol}^{-1}}$
275	(646.24)	(−113.31)	(−23054)	(75.5)	(774.06)	(−116.60)	(−23002)	(77.1)	(920.35)	(−119.94)	(−22839)	(78.7)
280	(657.94)	(−111.94)	(−22674)	(76.5)	(786.62)	(−115.20)	(−22614)	(78.2)	(933.75)	(−118.52)	(−22443)	(79.8)
285	(669.52)	(−110.58)	(−22289)	(77.5)	(799.08)	(−113.81)	(−22221)	(79.2)	(947.05)	(−117.09)	(−22041)	(80.9)
290	(681.00)	(−109.22)	(−21899)	(78.4)	(811.41)	(−112.42)	(−21822)	(80.2)	(960.23)	(−115.68)	(−21634)	(82.0)
295	(692.37)	(−107.88)	(−21505)	(79.5)	(823.63)	(−111.04)	(−21418)	(81.3)	(973.29)	(−114.27)	(−21221)	(83.1)
300	703.63	−106.53	−21105	80.5	835.73	−109.67	−21009	82.4	986.23	−112.86	−20802	84.3
305	714.77	−105.19	−20700	81.5	847.71	−108.29	−20595	83.5	999.04	−111.46	−20378	85.5
310	725.79	−103.86	−20290	82.6	859.56	−106.93	−20174	84.6				
315	736.70	−102.53	−19874	83.6	871.29	−105.57	−19749	85.7				
320	747.50	−101.20	−19453	84.7	882.89	−104.21	−19317	86.9				
325	758.17	−99.88	−19027	85.8	894.36	−102.85	−18880	88.0				
330	768.72	−98.56	−18595	86.9	905.70	−101.50	−18437	89.2				
335	779.16	−97.25	−18158	88.0	916.90	−100.15	−17988	90.4				
340	789.47	−95.93	−17715	89.2	927.97	−98.80	−17533	91.5				
345	799.65	−94.62	−17266	90.3	938.91	−97.46	−17072	92.7				
350	809.71	−93.32	−16812	91.4	949.70	−96.11	−16606	93.9				
355	819.65	−92.01	−16352	92.6	960.36	−94.77	−16133	95.1				
360	829.46	−90.71	−15886	93.7	970.87	−93.43	−15655	96.3				
365	839.14	−89.41	−15415	94.9	981.24	−92.10	−15170	97.5				
370	848.69	−88.11	−14938	96.0	991.47	−90.76	−14679	98.7				

380	867.41	−85.52	−13966	98.3
390	885.60	−82.94	−12971	100.6
400	903.26	−80.36	−11953	103.0
410	920.38	−77.79	−10912	105.3
420	936.96	−75.22	−9847	107.6
430	952.98	−72.66	−8759	110.0
440	968.46	−70.11	−7647	112.3
450	983.37	−67.56	−6513	114.6

Table 3—continued

ρ/mol dm⁻³	20.5				21.0							
T/K	$\dfrac{P_\sigma}{\text{MPa}}$	$\dfrac{S_\sigma}{\text{J K}^{-1}\text{mol}^{-1}}$	$\dfrac{U_\sigma}{\text{J mol}^{-1}}$	$\dfrac{C_v}{\text{J K}^{-1}\text{mol}^{-1}}$	$\dfrac{P_\sigma}{\text{MPa}}$	$\dfrac{S_\sigma}{\text{J K}^{-1}\text{mol}^{-1}}$	$\dfrac{U_\sigma}{\text{J mol}^{-1}}$	$\dfrac{C_v}{\text{J K}^{-1}\text{mol}^{-1}}$	$\dfrac{P_\sigma}{\text{MPa}}$	$\dfrac{S_\sigma}{\text{J K}^{-1}\text{mol}^{-1}}$	$\dfrac{U_\sigma}{\text{J mol}^{-1}}$	$\dfrac{C_v}{\text{J K}^{-1}\text{mol}^{-1}}$
130	(696.24)	(−172.06)	(−32190)	(53.9)								
135	(705.17)	(−169.99)	(−31917)	(55.3)								
140	(714.95)	(−167.96)	(−31638)	(56.5)								
145	(725.49)	(−165.96)	(−31352)	(57.6)	(940.63)	(−168.02)	(−30690)	(51.2)				
150	(736.67)	(−163.99)	(−31062)	(58.5)	(947.70)	(−166.26)	(−30430)	(52.8)				
155	(748.42)	(−162.06)	(−30767)	(59.3)	(955.91)	(−164.51)	(−30163)	(54.2)				
160	(760.66)	(−160.16)	(−30468)	(60.1)	(965.12)	(−162.77)	(−29889)	(55.4)				
165	(773.32)	(−158.30)	(−30166)	(60.8)	(975.19)	(−161.04)	(−29609)	(56.6)				
170	(786.36)	(−156.48)	(−29861)	(61.4)	(986.02)	(−159.34)	(−29323)	(57.7)				
175	(799.72)	(−154.69)	(−29552)	(62.1)	(997.50)	(−157.65)	(−29032)	(58.7)				
180	(813.35)	(−152.93)	(−29240)	(62.7)								
185	(827.21)	(−151.20)	(−28925)	(63.4)								
190	(841.28)	(−149.50)	(−28606)	(64.0)								
195	(855.51)	(−147.83)	(−28284)	(64.7)								
200	(869.87)	(−146.18)	(−27959)	(65.4)								

205	(884.35)	(−144.56)	(−27630)	(66.2)
210	(898.92)	(−142.95)	(−27297)	(67.0)
215	(913.55)	(−141.37)	(−26960)	(67.8)
220	(928.23)	(−139.80)	(−26619)	(68.6)
225	(942.94)	(−138.25)	(−26274)	(69.5)
230	(957.65)	(−136.71)	(−25924)	(70.4)
235	(972.37)	(−135.18)	(−25569)	(71.4)
240	(987.06)	(−133.67)	(−25210)	(72.4)

Tables 4 and 5

THE VARIATION OF
MOLAR VOLUME,
MOLAR ENTHALPY
AND
MOLAR ENTROPY
WITH
TEMPERATURE AND PRESSURE
ALONG THE
SATURATION CURVE
FROM THE
TRIPLE POINT
TO THE CRITICAL POINT

Notes:

1. Expressions such as $1.0\,E-9$ are to be read as 1.0×10^{-9}.
2. Interpolation between 360 K and 365.57 K (T_c) or between 4.6 MPa and 4.6646 MPa (P_c) can only be approximate.
3. In the units used in this table,

$$\frac{P}{\text{MPa}} \times \frac{V}{\text{cm}^3\,\text{mol}^{-1}} = \frac{PV}{\text{J}\,\text{mol}^{-1}}.$$

4. In this table the molar entropy is given a value of zero at 298.15 K and 1 atm (0.101 325 MPa) in the ideal gas state.

$$S^{\text{id}}(298.15\,\text{K},\,1\,\text{atm}) - S^{\text{ic}}(0\,\text{K}) = 266.62\,\text{J}\,\text{K}^{-1}\,\text{mol}^{-1}.$$

5. In this table the molar enthalpy is given a value of zero at 298.15 K in the ideal gas state.

$$H^{\text{id}}(298\,15\,\text{K}) - H^{\text{id}}(0\,\text{K}) = 13\,546\,\text{J}\,\text{mol}^{-1}.$$

$$H^{\text{id}}(298.15\,\text{K}) - H^{\text{ic}}(0\,\text{K}) = 40\,660\,\text{J}\,\text{mol}^{-1}.$$

Table 4. SATURATION CURVE: TEMPERATURE

$\dfrac{T_\sigma}{K}$	$\dfrac{P_\sigma}{MPa}$	Molar volume cm³ mol⁻¹			Molar enthalpy J mol⁻¹			Molar entropy J K⁻¹ mol⁻¹		
		V_l	ΔV	V_g	H_l	ΔH	H_g	S_l	ΔS	S_g
87.890	.95402E − 09	54.732	.76596E + 12	.76596E + 12	− 34458	23953	− 10505	− 176.79	272.54	95.75
90	.20530E − 08	54.904	.36448E + 12	.36448E + 12	− 34250	23825	− 10425	− 174.45	264.72	90.27
95	.10871E − 07	55.320	.72655E + 11	.72655E + 11	− 33813	23578	− 10235	− 169.72	248.19	78.47
100	.48055E − 07	55.748	.17302E + 11	.17302E + 11	− 33420	23379	− 10041	− 165.69	233.80	68.10
105	.18242E − 06	56.183	.47857E + 10	.47857E + 10	− 33046	23202	− 9844	− 162.04	220.97	58.93
110	.60789E − 06	56.625	.15045E + 10	.15045E + 10	− 32676	23032	− 9644	− 158.60	209.38	50.78
115	.18097E − 05	57.073	.52834E + 09	.52834E + 09	− 32303	22862	− 9441	− 155.28	198.80	43.51
120	.48828E − 05	57.526	.20433E + 09	.20433E + 09	− 31924	22688	− 9236	− 152.06	189.07	37.01
125	.12084E − 04	57.985	.86005E + 08	.86005E + 08	− 31539	22511	− 9028	− 148.92	180.09	31.18
130	.27707E − 04	58.448	.39008E + 08	.39008E + 08	− 31148	22331	− 8817	− 145.84	171.77	25.93
135	.59371E − 04	58.917	.18904E + 08	.18904E + 08	− 30750	22146	− 8604	− 142.84	164.04	21.20
140	.11977E − 03	59.392	.97165E + 07	.97166E + 07	− 30347	21958	− 8389	− 139.91	156.84	16.93
145	.22895E − 03	59.873	.52641E + 07	.52641E + 07	− 29940	21769	− 8171	− 137.05	150.13	13.07
150	.41697E − 03	60.361	.29894E + 07	.29895E + 07	− 29528	21577	− 7951	− 134.26	143.85	9.58
155	.72705E − 03	60.857	.17711E + 07	.17711E + 07	− 29113	21384	− 7729	− 131.54	137.96	6.42
160	.12189E − 02	61.360	.10900E + 07	.10900E + 07	− 28695	21190	− 7505	− 128.89	132.44	3.55
165	.19723E − 02	61.873	.69426E + 06	.69432E + 06	− 28274	20995	− 7279	− 126.30	127.24	.94
170	.30903E − 02	62.395	.45617E + 06	.45623E + 06	− 27851	20799	− 7052	− 123.77	122.35	− 1.42
175	.47023E − 02	62.927	.30828E + 06	.30834E + 06	− 27425	20602	− 6823	− 121.30	117.72	− 3.58
180	.69671E − 02	63.470	.21372E + 06	.21378E + 06	− 26996	20404	− 6592	− 118.89	113.35	− 5.53
185	.10075E − 01	64.025	.15164E + 06	.15170E + 06	− 26565	20204	− 6361	− 116.52	109.21	− 7.31
190	.14249E − 01	64.593	.10989E + 06	.10995E + 06	− 26131	20002	− 6129	− 114.21	105.27	− 8.94
195	.19747E − 01	65.175	81172	81237	− 25694	19798	− 5896	− 111.95	101.53	− 10.42
200	.26860E − 01	65.772	61019	61084	− 25255	19592	− 5663	− 109.72	97.96	− 11.76
205	.35915E − 01	66.386	46607	46673	− 24812	19382	− 5430	− 107.54	94.55	− 12.99
210	.47272E − 01	67.016	36121	36188	− 24367	19169	− 5198	− 105.40	91.28	− 14.12
215	.61322E − 01	67.665	28369	28437	− 23917	18951	− 4966	− 103.29	88.15	− 15.14

220		68.335	22554	22622	−23464	18729	−4733	−101.21	83.15	−16.68
225	.78488E − 01	69.027	18131	18200	−23007	18501	−4506	−99.16	82.23	−16.93
225.460	.99223E − 01	69.091	17779	17848	−22965	18480	−4485	−98.97	81.97	−17.01
	.101325									
230	.12401	69.742	14724	14793	−22546	18267	−4279	−97.14	79.42	−17.72
235	.15334	70.483	12068	12139	−22080	18027	−4053	−95.14	76.71	−18.43
240	.18775	71.251	9975.9	10047	−21609	17779	−3830	−93.17	74.08	−19.09
245	.22778	72.050	8310.2	8382.3	−21133	17523	−3610	−91.22	71.52	−19.70
250	.27401	72.882	6971.6	7044.5	−20652	17259	−3393	−89.29	69.04	−20.25
255	.32701	73.750	5886.1	5959.8	−20165	16985	−3180	−87.38	66.61	−20.77
260	.38737	74.657	4998.4	5073.0	−19672	16702	−2970	−85.48	64.24	−21.24
265	.45573	75.608	4266.7	4342.3	−19172	16406	−2766	−83.59	61.91	−21.68
270	.53269	76.607	3659.2	3735.8	−18665	16099	−2566	−81.72	59.63	−22.10
275	.61889	77.658	3151.1	3228.8	−18151	15779	−2372	−79.86	57.38	−22.48
280	.71499	78.769	2723.4	2802.2	−17629	15444	−2185	−78.01	55.16	−22.85
285	.82165	79.945	2361.0	2441.0	−17099	15095	−2004	−76.16	52.96	−23.20
290	.93954	81.194	2052.1	2133.3	−16560	14729	−1831	−74.32	50.79	−23.53
295	1.0694	82.528	1787.3	1869.8	−16011	14343	−1668	−72.48	48.62	−23.86
300	1.2118	83.956	1558.8	1642.8	−15451	13937	−1514	−70.64	46.46	−24.18
305	1.3676	85.493	1360.7	1446.2	−14881	13510	−1371	−68.79	44.29	−24.50
310	1.5376	87.158	1187.9	1275.1	−14298	13057	−1241	−66.94	42.12	−24.83
315	1.7225	88.971	1036.4	1125.3	−13701	12575	−1126	−65.09	39.92	−25.17
320	1.9231	90.962	902.64	993.60	−13089	12060	−1029	−63.22	37.69	−25.53
325	2.1403	93.169	783.92	877.09	−12459	11507	−952	−61.33	35.41	−25.92
330	2.3751	95.644	677.79	773.43	−11809	10910	−899	−59.41	33.06	−26.35
335	2.6284	98.464	582.14	680.61	−11135	10258	−877	−57.46	30.62	−26.83
340	2.9015	101.74	495.09	596.84	−10431	9539	−892	−55.45	28.06	−27.40
345	3.1956	105.67	414.77	520.45	−9687	8729	−958	−53.37	25.30	−28.07
350	3.5121	110.60	339.12	449.72	−8887	7793	−1094	−51.17	22.27	−28.90
355	3.8529	117.25	265.21	382.46	−7999	6659	−1340	−48.75	18.76	−30.00
360	4.2202	127.82	186.64	314.46	−6929	5129	−1800	−45.89	14.25	−31.64
365	4.6174	160.31	62.952	223.26	−5015	1896	−3119	−40.77	5.20	−35.57
365.570	4.6646	188.37	0	188.374	−4017	0	−4017	−38.06	0	−38.06

Table 5. SATURATION CURVE: PRESSURE

$\frac{P_\sigma}{\text{MPa}}$	$\frac{T_\sigma}{\text{K}}$	Molar volume $\text{cm}^3\,\text{mol}^{-1}$			Molar enthalpy J mol^{-1}			Molar entropy $\text{J K}^{-1}\,\text{mol}^{-1}$		
		V_l	ΔV	V_g	H_l	ΔH	H_g	S_l	ΔS	S_g
.95402E − 09	87.890	54.732	.76596E + 12	.76596E + 12	− 34458	23953	− 10505	− 176.79	272.54	95.75
.10E − 08	88.016	54.742	.73180E + 12	.73180E + 12	− 34445	23945	− 10500	− 176.65	272.06	95.41
.10E − 07	94.735	55.298	.78766E + 11	.78766E + 11	− 33834	23589	− 10245	− 169.95	249.01	79.06
.10E − 06	102.682	55.980	.85373E + 10	.85373E + 10	− 33218	23283	− 9935	− 163.70	226.75	63.05
.50E − 06	109.154	56.550	.18151E + 10	.18151E + 10	− 32738	23060	− 9678	− 159.17	211.27	52.10
.10E − 05	112.222	56.824	.93305E + 09	.93305E + 09	− 32510	22956	− 9554	− 157.11	204.56	47.45
.50E − 05	120.125	57.538	.19975E + 09	.19975E + 09	− 31915	22684	− 9231	− 151.98	188.84	36.86
.10E − 04	123.918	57.885	.10303E + 09	.10303E + 09	− 31623	22550	− 9073	− 149.59	181.98	32.39
.25E − 04	129.357	58.388	.43018E + 08	.43019E + 08	− 31198	22354	− 8844	− 146.24	172.81	26.57
.50E − 04	133.836	58.808	.22253E + 08	.22253E + 08	− 30843	22189	− 8654	− 143.54	165.79	22.26
.75E − 04	136.620	59.070	.15144E + 08	.15144E + 08	− 30620	22085	− 8535	− 141.89	161.65	19.77
.10E − 03	138.675	59.266	.11528E + 08	.11528E + 08	− 30454	22008	− 8446	− 140.68	158.70	18.02
.25E − 03	145.710	59.942	.48443E + 07	.48443E + 07	− 29881	21741	− 8140	− 136.65	149.21	12.56
.50E − 03	151.593	60.518	.25192E + 07	.25193E + 07	− 29396	21515	− 7881	− 133.39	141.93	8.54
.75E − 03	155.291	60.886	.17201E + 07	.17201E + 07	− 29089	21373	− 7716	− 131.39	137.63	6.24
.10E − 02	158.041	61.162	.13126E + 07	.13127E + 07	− 28859	21266	− 7593	− 129.92	134.56	4.64
.25E − 02	167.597	62.143	.55613E + 06	.55619E + 06	− 28054	20893	− 7161	− 124.98	124.66	− .31
.50E − 02	175.760	63.008	.29113E + 06	.29119E + 06	− 27360	20572	− 6788	− 120.93	117.04	− 3.88
.75E − 02	180.974	63.577	.19955E + 06	.19961E + 06	− 26912	20365	− 6547	− 118.42	112.53	− 5.89
.10E − 01	184.896	64.013	.15270E + 06	.15276E + 06	− 26574	20208	− 6366	− 116.57	109.29	− 7.28
.25E − 01	198.807	65.629	65218	65284	− 25360	19641	− 5719	− 110.25	98.80	− 11.45
.50E − 01	211.056	67.152	34289	34356	− 24272	19123	− 5149	− 104.95	90.61	− 14.34
.75E − 01	219.059	68.207	23528	23596	− 23550	18771	− 4779	− 101.60	85.69	− 15.91
.1	225.171	69.051	17999	18068	− 22992	18494	− 4498	− 99.09	82.13	− 16.96
.101325	225.460	69.091	17780	17849	− 22965	18480	− 4485	− 98.97	81.97	− 17.01

.2	241.610	71.505	9398.3	9469.8	−21457	17698	−3759	−92.54	73.25	−19.29
.3	252.536	73.317	6393.3	6466.6	−20406	17122	−3284	−88.32	67.80	−20.52
.4	260.971	74.838	4845.2	4920.1	−19575	16645	−2930	−85.11	63.78	−21.33
.5	267.947	76.191	3895.4	3971.6	−18874	16227	−2647	−82.49	60.56	−21.93
.6	273.951	77.433	3250.5	3327.9	−18260	15848	−2412	−80.25	57.85	−22.40
.7	279.254	78.599	2782.7	2861.3	−17707	15495	−2212	−78.28	55.49	−22.79
.8	284.025	79.710	2427.1	2506.9	−17203	15164	−2039	−76.52	53.39	−23.13
.9	288.377	80.780	2147.2	2228.0	−16736	14849	−1887	−74.91	51.49	−23.42
1.0	292.387	81.820	1920.7	2002.5	−16299	14547	−1752	−73.44	49.75	−23.69
1.2	299.602	83.838	1575.8	1659.7	−15496	13971	−1525	−70.78	46.63	−24.15
1.4	305.985	85.811	1324.8	1410.6	−14767	13423	−1344	−68.43	43.87	−24.56
1.6	311.735	87.768	1133.1	1220.9	−14092	12892	−1200	−66.30	41.36	−24.94
1.8	316.981	89.736	981.42	1071.2	−13460	12374	−1086	−64.35	39.04	−25.31
2.0	321.816	91.736	857.92	949.65	−12862	11864	−998	−62.53	36.87	−25.67
2.2	326.309	93.788	755.02	848.81	−12291	11356	−935	−60.83	34.80	−26.03
2.4	330.509	95.914	667.60	763.52	−11742	10847	−895	−59.21	32.82	−26.39
2.6	334.457	98.139	592.08	690.22	−11210	10333	−877	−57.67	30.89	−26.78
2.8	338.185	100.49	525.82	626.31	−10691	9810	−881	−56.19	29.01	−27.18
3.0	341.716	103.01	466.86	569.87	−10181	9273	−908	−54.75	27.14	−27.61
3.2	345.073	105.74	413.64	519.38	−9676	8717	−959	−53.34	25.26	−28.08
3.4	348.271	108.75	364.90	473.65	−9172	8135	−1037	−51.94	23.36	−28.59
3.6	351.326	112.14	319.51	431.66	−8663	7517	−1146	−50.55	21.39	−29.16
3.8	354.248	116.09	276.38	392.47	−8140	6847	−1293	−49.14	19.33	−29.81
4.0	357.049	120.86	234.23	355.09	−7592	6100	−1492	−47.66	17.09	−30.58
4.2	359.734	127.07	191.16	318.23	−6994	5228	−1766	−46.06	14.53	−31.53
4.4	362.312	136.28	143.19	279.47	−6284	4108	−2176	−44.17	11.34	−32.83
4.6	364.789	156.52	73.121	229.64	−5180	2196	−2984	−41.21	6.02	−35.19
4.6646	365.570	188.37	0	188.37	−4017	0	−4017	−38.06	0	−38.06

Tables 6 and 7

THE VARIATION OF
MOLAR VOLUME,
MOLAR ENTHALPY
AND
MOLAR ENTROPY
WITH
TEMPERATURE AND PRESSURE
ALONG THE MELTING CURVE
FROM THE TRIPLE POINT
$(87.89\ \text{K},\ 0.954\,02 \times 10^{-9}\ \text{MPa})$
TO 145 K OR 1000 MPa

Notes:
1. Expressions such as $1.0\,\text{E} - 9$ are to be read as 1.0×10^{-9}.
2. In the units used in this table,

$$\frac{P}{\text{MPa}} \times \frac{V}{\text{cm}^3\,\text{mol}^{-1}} = \frac{PV}{\text{J}\,\text{mol}^{-1}}.$$

3. In this table the molar entropy is given a value of zero at 298.15 K and 1 atm (0.101 325 MPa) in the ideal gas state.

$$S^{\text{id}}(298.15\ \text{K},\ 1\ \text{atm}) - S^{\text{ic}}(0\ \text{K}) = 266.62\ \text{J}\,\text{K}^{-1}\,\text{mol}^{-1}.$$

4. In this table the molar enthalpy is given a value of zero at 298.15 K in the ideal gas state.

$$H^{\text{id}}(298.15\ \text{K}) - H^{\text{id}}(0\ \text{K}) = 13\,546\ \text{J}\,\text{mol}^{-1}.$$

$$H^{\text{id}}(298.15\ \text{K}) - H^{\text{ic}}(0\ \text{K}) = 40\,660\ \text{J}\,\text{mol}^{-1}.$$

Table 6. MELTING CURVE: TEMPERATURE

$\dfrac{T_m}{K}$	$\dfrac{P_m}{MPa}$	$\dfrac{\text{Molar volume}}{cm^3\,mol^{-1}}$	$\dfrac{\text{Molar enthalpy}}{J\,mol^{-1}}$	$\dfrac{\text{Molar entropy}}{J\,K^{-2}\,mol^{-1}}$
87.890	.95402E − 09	54.732	− 34461	− 350.59
88	1.7075	54.700	− 34366	− 176.80
89	19.962	54.374	− 33395	− 177.09
90	36.959	54.105	− 32488	− 177.25
91	52.997	53.874	− 31630	− 177.34
92	68.425	53.668	− 30804	− 177.38
93	83.461	53.480	− 29999	− 177.39
94	98.246	53.305	− 29209	− 177.37
95	112.88	53.139	− 28427	− 177.35
98	149.26	52.753	− 26491	− 177.24
100	185.90	52.393	− 24550	− 177.09
105	261.66	51.710	− 20565	− 176.68
110	342.43	51.053	− 16353	− 176.10
115	429.61	50.407	− 11850	− 175.38
120	524.26	49.771	− 7008	− 174.52
125	627.31	49.146	− 1791	− 173.53
130	717.72	48.656	2793	− 172.26
135	807.24	48.210	7304	− 170.93
140	905.96	47.753	12218	− 169.64
145	1014.7	47.289	17563	− 168.39

Table 7. MELTING CURVE: PRESSURE

$\dfrac{P_m}{\text{MPa}}$	$\dfrac{T_m}{\text{K}}$	Molar volume $\text{cm}^3\,\text{mol}^{-1}$	Molar enthalpy $\text{J}\,\text{mol}^{-1}$	Molar entropy $\text{J}\,\text{K}^{-1}\,\text{mol}^{-1}$
.95402E − 09	87.890	54.732	− 34461	− 350.59
.1	87.900	54.730	− 34452	− 176.79
1	87.959	54.714	− 34403	− 176.79
10	88.448	54.547	− 33925	− 176.94
20	89.002	54.373	− 33393	− 177.09
30	89.582	54.211	− 32860	− 177.19
40	90.186	54.059	− 32325	− 177.27
50	90.810	53.915	− 31790	− 177.33
60	91.450	53.779	− 31255	− 177.36
70	92.104	53.648	− 30720	− 177.38
80	92.768	53.523	− 30184	− 177.39
90	93.441	53.402	− 29650	− 177.38
100	94.119	53.285	− 29115	− 177.37
200	100.95	52.260	− 23806	− 177.02
300	107.42	51.390	− 18561	− 176.42
400	113.35	50.620	− 13375	− 175.64
500	118.76	49.928	− 8244	− 174.75
600	123.72	49.306	− 3168	− 173.80
700	128.28	48.742	1856	− 172.83
800	134.61	48.245	6941	− 171.03
900	139.71	47.779	11923	−169.72
1000	144.35	47.349	16844	−168.55

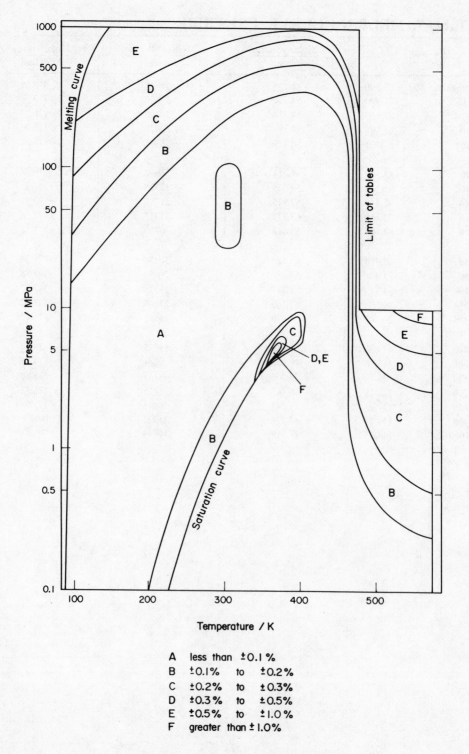

Figure 18. Tolerance diagram for density

A   less than ±0.1%
B   ±0.1%   to   ±0.2%
C   ±0.2%   to   ±0.3%
D   ±0.3%   to   ±0.5%
E   ±0.5%   to   ±1.0%
F   greater than ±1.0%

398

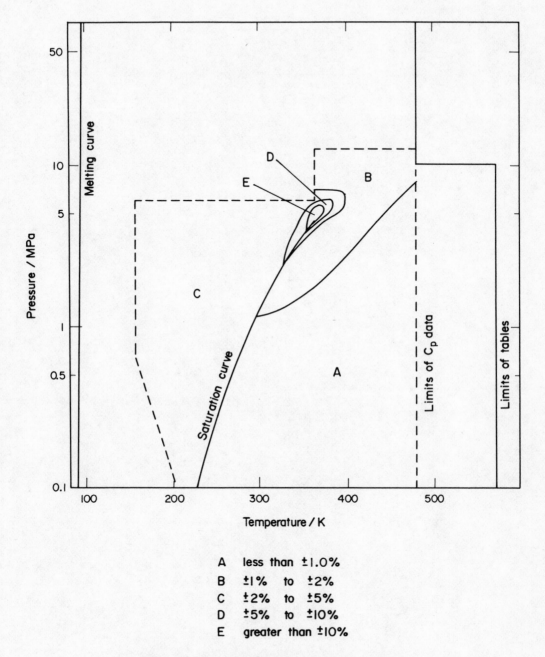

A   less than ±1.0%
B   ±1%   to   ±2%
C   ±2%   to   ±5%
D   ±5%   to   ±10%
E   greater than ±10%

Figure 19. Tolerance diagram for isobaric heat capacity

OTHER TITLES IN THE CHEMICAL DATA SERIES